About Island Press

Island Press is the only nonprofit organization in the United States whose principal purpose is the publication of books on environmental issues and natural resource management. We provide solutions-oriented information to professionals, public officials, business and community leaders, and concerned citizens who are shaping responses to environmental problems.

In 2003, Island Press celebrates its nineteenth anniversary as the leading provider of timely and practical books that take a multidisciplinary approach to critical environmental concerns. Our growing list of titles reflects our commitment to bringing the best of an expanding body of literature to the environmental community throughout North America and the world.

Support for Island Press is provided by The Nathan Cummings Foundation, Geraldine R. Dodge Foundation, Doris Duke Charitable Foundation, Educational Foundation of America, The Charles Engelhard Foundation, The Ford Foundation, The George Gund Foundation, The Vira I. Heinz Endowment, The William and Flora Hewlett Foundation, Henry Luce Foundation, The John D. and Catherine T. MacArthur Foundation, The Andrew W. Mellon Foundation, The Moriah Fund, The Curtis and Edith Munson Foundation, National Fish and Wildlife Foundation, The New-Land Foundation, Oak Foundation, The Overbrook Foundation, The David and Lucile Packard Foundation, The Pew Charitable Trusts, The Rockefeller Foundation, The Winslow Foundation, and other generous donors.

The opinions expressed in this book are those of the author(s) and do not necessarily reflect the views of these foundations.

Precaution, Environmental Science, and Preventive Public Policy

PRECAUTION

ENVIRONMENTAL

SCIENCE

AND

PREVENTIVE

PUBLIC

POLICY

Edited by
Joel A. Tickner

ISLAND PRESS

Washington • Covelo • London

#50448104

Library of Congress Cataloging-in-Publication Data
Precaution, environmental science, and preventive public policy
  / edited by Joel A. Tickner.
     p. cm.
  Includes bibliographical references and index.
  ISBN 1-55963-331-X (hardcover : alk. paper) — ISBN 1-55963-332-8
  (pbk. : alk. paper)
    1. Environmental policy—Case studies. I. Tickner, Joel A.
  GE170.P74 2002
    363.7'056—dc21   2002012499

British Cataloguing-in-Publication Data available.

Book design by Brighid Willson

Printed on recycled, acid-free paper ✪

Manufactured in the United States of America
10 9 8 7 6 5 4 3 2 1

CONTENTS

Part VII. Conclusions and Afterword   369

ACKNOWLEDGMENTS

The completion of a volume such as this can rarely ever be attributed to a lone editor. Rather, this book is the product of many dedicated scientists, policy analysts, and advocates. It is the outgrowth of several years of discussions among of wide range of experts from around the world about how science and policy can more effectively support preventive, precautionary decisions in the face of uncertain and complex risks. In the United States, these discussions began with the Wingspread Conference on the Precautionary Principle held in January 1998. I had the great privilege of co-coordinating that landmark conference and co-editing the book *Protecting Public Health and the Environment: Implementing the Precautionary Principle* that followed. Discussions about the precautionary principle have evolved to a great degree since 1998. The present volume reflects the discussions that have taken place with regards to the relationship between science and precaution.

I would like to personally thank the many individuals who have helped shape the book's content and who have ultimately given it life. I am completely indebted to my colleague Sara Wright for the countless hours she put into the book—editing, communicating with authors, and dealing with the often frustrating task of formatting—all with a wonderful sense of humor and dedication. Drs. David Kriebel and Kenneth Geiser have served as great mentors and colleagues, providing me with ideas, inspiration, and moral support. The staff of the Lowell Center for Sustainable Production at the University of Massachusetts, Lowell, particularly Cathy Crumbley, its program director, have been an important source of support, especially when working on controversial issues around science and precaution.

Rachel Massey and Nancy Myers lent critical editing support to this volume. With so many authors from so many disciplines and countries, it would have been impossible to have taken on all of the editing tasks myself.

Many thanks go to those who helped in planning the International Summit on Science and the Precautionary Principle—the conference that provided the background to this volume. They helped shape the discussions, questions, and topics that were the book's foundations. These individuals include Richard Clapp, Ted Schettler, Molly Anderson, Paul Epstein, Colin Soskolne, Andrew Stirling, and Finn Bro-Rasmussen. I would like to extend special thanks to Lee Ketelsen from the Clean Water Fund who has worked tirelessly to ensure that our work on science and precaution serves an important role in informing preventive policies. Of course, I greatly appreciate ideas and vision that each of the participants at the summit contributed.

I am deeply grateful to all of the chapter authors for their contributions and their dedication to reevaluating science and policy. Authors graciously spent large amounts of time answering questions and clarifying issues in their chapters. As a result, we have compiled a book with sufficient detail for experts in particular subjects, but simple enough to engage a broad audience. It was important to learn from the authors' experience in scientific research and policy analysis from throughout the world. The breadth of expertise and disciplinary perspectives contained in this volume is its greatest asset.

Heather Boyer at Island Press has been exceptionally supportive of this work from the beginning and has helped me work through the various "crises" that have occurred through the editing process. Her patience and input are greatly appreciated.

I would like to acknowledge generous grants from the V. Kann Rasmussen Foundation and the New York Community Trust that supported the research and editing of this book.

Finally, I would like to thank my wife, Judit, and sons Ariel and Maximilian for being there for me and for putting up with my periodic frustrations, time away from home, and frequent lack of attention to small details around the house. It is such a gift to come home to a loving family.

INTRODUCTION

Joel A. Tickner

Policy makers often mistakenly view science as an incontrovertible source of knowledge on which to base policy decisions. However, in the context of complex environmental and health risks, it is much more useful to think of science and policy as dynamically informing each other: science provides important information on which to base policy, and public policy outlines critical societal research and knowledge needs. Science is critical to solving some of our most pressing environmental and health problems. As environmental science faces the increasing challenges of more complex risks with greater uncertainty and ignorance, the nexus between science and preventive policy becomes even more important. As an applied science, it seems appropriate to understand how it can better inform policy, without sacrificing objectivity.

The precautionary principle was established in the 1970s in response to concerns about the limitations of science and policy structures to adequately address complex and uncertain risks and in recognition of the severe consequences to health and the economy from damage to health and ecosystems. The precautionary principle bridges the gap between science and policy by encouraging policies that protect human health and the environment in the face of uncertain risks. In this broad sense, it is not new. Precaution is at the heart of centuries of medical and public health theory and practice. A part of the Hippocratic Oath, "First do no harm," underscores a duty to prevent damage to health as well as the concept of primary prevention in public health.

The precautionary principle was characterized in the 1998 Wingspread Statement on the Precautionary Principle as follows: "When an activity raises threats of harm to human health or the environment, precautionary measures

should be taken even if some cause and effect relationships are not fully established scientifically." The statement listed four central components of the
principle: (1) taking preventive action in the face of uncertainty, (2) shifting
burdens onto proponents of potentially harmful activities, (3) exploring a
wide range of alternatives to possibly harmful actions, and (4) increasing public participation in decision making (see Raffensperger and Tickner 1999).

As a principle of decision making, the precautionary principle has its roots
in the German word *Vorsorgeprinzip*. An alternative translation of this word is
"foresight or forecaring principle," which emphasizes anticipatory action. It is
more proactive than "precaution," which to many sounds reactive and even
negative. Although it has its roots in German environmental policy, over the
past twenty years the principle has served as a central element in international
treaties addressing North Sea pollution, ozone-depleting chemicals, persistent
organic pollutants, genetically modified organisms, fisheries, climate change,
and sustainable development. The European Union has espoused precaution,
along with prevention of pollution at source and the polluter pays principle,
as a central element of environmental health policy.

Discussion about the role of the precautionary principle in the development of environmental policy has greatly intensified during recent years, in
part due to science and policy debates over the complex, global health risks
such as those posed by beef hormones, genetically modified organisms, toxic
chemicals, fisheries, and global climate change, among others. Much of the
recent debate about the precautionary principle has focused on the questions
of whether precaution poses a barrier to trade and of what specific level of evidence is sufficient to justify action to prevent harm.

Science and Precaution

When the precautionary principle has been discussed in the context of its relationship to science, it has often been portrayed either as antiscience (inconsistent with the norms of rational, science-based decision making) or as a risk-
management principle that is implemented only after objective, expert
scientific inquiry takes place. Both of these claims appear to be based on a mis-
understanding of environmental science.

There are ways in which the methods of scientific inquiry can implicitly
impede precautionary action, making it more difficult for policy makers to
take action in the face of uncertainty. Too often scientific research focuses on
narrowly defined issues, whereas the problems we face are complex and
require interdisciplinary research methods. Current scientific practice also
often attempts to minimize uncertainties and focus on those aspects of a problem that are quantifiable. This may mean narrowing the research focus so much
that important aspects of the problem are missed. Our current selection of

scientific tools and the ways they are used may not be sufficiently refined to deal with some of the complex, multigenerational environmental issues, such as global warming, that decision makers face today.

Yet, environmental science has a critical role in implementing the precautionary principle, by providing insights into the normal functioning of natural systems, the ways they are disrupted by technologies, opportunities for prevention and restoration, and gaps in our understanding of phenomena. To support the precautionary principle, science and policy must be able to identify and anticipate harm to health or the environment and support the development of options for precautionary action. This requires scientific methods, tools, and institutions that are adequately adapted to decision-making problems that policy makers face.

A shift to more precautionary policies creates opportunities and challenges for scientists to think differently about the way they conduct studies and communicate results. If the precautionary principle is presented to environmental scientists as an opportunity for more and better science—rigorous and transparent about uncertainties—it may be possible to find support from researchers who are presently unaware of such developments, or even hostile to a perceived "attack" on science. Scientists are also needed to respond to critiques of precautionary decisions, particularly when the uncertainties in science are misrepresented.

Background to This Volume

While the precautionary principle and its background, history, and implementation have been discussed at length in a variety of books, journal articles, and other publications (see Raffensperger and Tickner 1999; O'Riordan et al. 2001), this is the first compilation of essays dedicated entirely to the role of science and the science-policy interface in implementing the precautionary principle. Although the relationship between environmental science and precaution—referred to as precautionary science or the "ecological" paradigm (see chapter 8) (Barrett and Raffensperger 1999; Wynne and Mayer 1993)—has been discussed by numerous scientists during the past ten years, this volume provides an in-depth, multidisciplinary examination of ways environmental science can limit or better support precautionary, preventive decisions. It extends the often political discussion of precaution into the realm of science. It is meant to expand upon existing literature on the precautionary principle by broadening the range of disciplines and individuals thinking about its implications for science.

This volume begins with the presumption that the precautionary principle, as defined in the Wingspread Statement, is an important principle of environmental and health policy under uncertainty. Our bias is toward protection of health and ecosystems under uncertainty. The merits of the precautionary principle or critiques of it are not discussed. Rather, the focus of this book is on how science can more effectively address complex, uncertain environmental risks and lead to precautionary actions.

Some analysts have claimed that the health and ecosystem risks we face are exaggerated and that precautionary policies are unnecessary (Lomborg 2001; Bailey 1995). We disagree. Although industrialized activities have certainly improved health in many ways throughout the world in the past century, these human activities have also brought with them substantial costs to health and the environment that could have been avoided through better foresight. These include devastation of some fish stocks, deforestation, species loss, global climate change, ozone depletion, and disease in humans and wildlife caused by toxic substances. We believe that science, technology, and policy can be harnessed more effectively for prevention and innovation in safer, cleaner technologies and practices.

The book builds on more than two years of discussions held with an interdisciplinary group of scientists to understand how the conduct of environmental science can hinder or facilitate the implementation of precautionary policies (see Kriebel et al. 2001).

The underlying theme of the book is that the role of science must be a central aspect of discussions about the precautionary principle. Even if the reader disagrees about the merits of precautionary principle, we believe that most scientists can agree that there are important issues to discuss about whether the tools and methods we currently employ in environmental science and the ways we characterize uncertainty and ignorance are adequate to address the complex, sometimes global risks we face. Thus, the precautionary principle is relevant to environmental science. We have found that even among scientists who support the concept of precaution, issues of hypothesis generation, scientific method, and the use of science in policy are highly challenging and debatable. There are no simple answers.

The primary objective of this book is to build better understanding of the role of science in implementing the precautionary principle. To achieve this, the chapters in this book (1) outline how the current practice of science can both support and limit precautionary decision making, (2) envision and explore changes needed in the practice and application of science that would better support the precautionary principle, and (3) discuss cultural and sociopolitical differences across nations that affect how science is applied in decision making.

Obviously, the question of what to do in the face of uncertainty is a political question and must be left to the political process, with a multitude of societal stakeholders involved. The questions of science and precaution cannot be separated from questions of economics, political power, and institutional capacity and will. These are important considerations that in some cases may be far more important than science in determining whether precautionary decisions occur or do not occur. Nonetheless, we have purposely chosen to avoid those questions to the degree possible and focus primarily on science and the science-policy nexus.

Structure of the Book

Through case studies, analyses, and commentaries, written by an international, interdisciplinary group of leading scientific, legal, and policy scholars, the book provides an overview of some of the limitations in current approaches to science for policy and tools and a vision for a more forward-looking, anticipatory environmental science. We have found that case studies are a critical educational tool for scientists and others. They provide concrete examples of problems and solutions as well as lessons learned when damage does occur to health or ecosystems. Such concrete examples are even more important when discussing a broad, difficult-to-conceptualize principle like precaution and its application in science and policy.

The book is divided into six parts that frame the various aspects of the implications of the precautionary principle for science and science policy. In general, the chapters in this volume provide overviews to particular limits in science, methodological changes, and ways to improve the science-policy interface. It would be impossible to take any one scientific issue—for example, genetic engineering—and give it the thorough scientific treatment it deserves. It would also be impossible to engage in a detailed discussion on particular methods in science—for example, differences between frequentist and Bayesian statistics.

Our goal with this book is to allow scientists to step back from the detailed, issue-specific debates in which they engage daily and to consider the broader implications of their work and environmental science in general. Many of the issues discussed throughout the book—how we characterize uncertainty, how cumulative and interactive effects are examined, the integration of qualitative data into science, and the role of interdisciplinary collaborations—are not limited to single disciplines. They are part of a broader debate on the evolution of environmental science and its role in policy. One novel issue that the book brings to the fore is the need to build links between scientists studying risks and those studying and developing solutions. A more precautionary approach to policy requires both.

The authors in this volume represent a wide range of disciplines and cultures. Many have been active participants in the evolution of the science and policy issues they address in their chapters. As such, they bring vast practical experience into the evaluation of complex environmental and health risks—

and the often controversial nexus of science and policy. Although the precautionary principle has often been portrayed as a concept of Western, industrialized countries, it is important to note that some of the authors in this volume are from developing countries.

Part I presents reflections and analyses from the perspective of three scientists on the importance and role of the precautionary principle in science and policy.

Part II discusses the implications of the ways we choose to act or not act in the face of uncertainty. A starting point for exploring science and the precautionary principle is to discuss the foundations of science and changes in its underlying assumptions and practice that would be needed to more effectively deal with complex risks.

Part III is the heart of the book. A number of case studies—on fisheries, chemicals, climate change, genetically modified organisms, and biodiversity—explore the various types of uncertainties in studying complex risks and how they are currently addressed in science, as well as the implications of uncertainty for public policy. The case studies provide concrete examples of how the conduct of science can inhibit precautionary policies and how it can support them.

Part IV explores the complexities of the science-policy interface and how that interface differs between regions. The section also examines lessons for improving the way science is integrated into public policy.

Part V outlines a new, forward-looking role for science in implementing precautionary policies—that of developing alternatives and solutions to harmful activities and technologies. Many environmental scientists have been trained to avoid intervening in production systems and products: that is the realm of policy and engineering. The authors in part V argue that science can be a much more effective tool if we use it to achieve positive goals for society, while avoiding the unintended consequences of human activities, although these can never be risk free.

Part VI builds on the earlier case studies and analyses to begin outlining a road map for a scientific method that more effectively addresses uncertainty, has a more dynamic interface with policy, and can lead to early warnings and anticipatory, preventive actions.

We believe that the chapters in this volume can inform critical reflection on environmental science and its application in policy. Such reflection is necessary if we are to avoid the problems of past and leave our children a diverse, healthy world.

References Cited

Bailey, R., ed. 1995. *The True State of the Planet*. New York: The Free Press.

Barrett, K., and C. Raffensperger. 1999. "Precautionary Science." In *Protecting*

Public Health and the Environment: Implementing the Precautionary Principle, edited by C. Raffensperger and J. Tickner, 106–22. Washington, D.C.: Island Press.

Kriebel, D., J. Tickner, P. Epstein, J. Lemons, R. Levins, E. Loechler, M. Quinn, R. Rudel, T. Schettler, and M. Stoto. 2001. "The Precautionary Principle in Environmental Science." *Environmental Health Perspectives* 109: 871–76.

Lomborg, B. 2001. *The Skeptical Environmentalist: Measuring the Real State of the World.* Cambridge: Cambridge University Press.

O'Riordan, T., A. Jordan, and J. Cameron. 2001. *Reinterpreting the Precautionary Principle.* London: Cameron and May.

Raffensperger, C., and J. Tickner, eds. 1999. *Protecting Public Health and the Environment: Implementing the Precautionary Principle.* Washington, D.C.: Island Press.

Wynne, B., and S. Mayer. 1993. "How Science Fails the Environment." *New Scientist,* 5 June, 33–35.

Precaution, Environmental Science, and Preventive Public Policy

PART I

Scientists' Perspectives on Precaution, Science, and Policy

Through case studies, analyses, and commentaries, this volume presents a series of perspectives on the role and implications of the precautionary principles in the conduct of environmental science and policy. In this section, three scientists analyze the influence of the precautionary principle on science, policy, and their own work.

In chapter 1, Joel A. Tickner examines some of the limitations and opportunities for environmental science in supporting precautionary decisions. He argues that the principle presents environmental scientists with a unique opportunity to help refine scientific tools and structures and to participate in a more proactive use of science that not only anticipates harm but contributes to the integrity of health and ecosystems.

Based on his involvement in environmental science and policy debates at home in the Philippines and internationally, Romeo F. Quijano has developed a conceptual framework for thinking about the precautionary principle in science and policy. In chapter 2, he argues that precaution requires substantial changes to the way science is conducted for policy and to the policy structures themselves.

Barry Commoner, considered by many to be the father of the American environmental movement, reflects in chapter 3 on his experience addressing risks where precaution was not taken. He notes that because so little is known about human and ecological systems, the sensible approach to complex risks, such as chemical and biotechnology, is to be cautious.

CHAPTER 1

The Role of Environmental Science in Precautionary Decision Making

Joel A. Tickner

> The future is quite likely to involve increasing rates of change;
> greater variance in system parameters; greater uncertainty
> about responses of complex biological, ecological,
> social, and political systems; and more surprises.
>
> —JANE LUBCHENCO, zoologist and former
> president of the American Association for
> the Advancement of Science

This chapter provides an overview of the role of science in implementing the precautionary principle. Its goal is to frame the discussions and issues examined throughout this volume. Following a presentation of some of the limitations of the current uses of science in environmental decision making, this chapter describes a framework for science to support precautionary decision making for complex, uncertain risks, including what is studied and how. It then describes some of the policy and institutional structures needed to institute such a vision and concludes that implementation of the precautionary principle demands a "defragmentation" of science and policy.

The recommended changes in this chapter and throughout the book do

3

not require discarding the variety of scientific tools and methods that have been used to date. These tools have played an important role in the detection, understanding, and prevention of many environmental health risks. These changes do demand, however, that we continuously refine and improve upon them as our knowledge of exposures, hazards, and disease progresses. The precautionary principle presents environmental scientists with a unique opportunity to help refine scientific tools and structures and to participate in a more proactive use of science that not only anticipates harm, but contributes to the integrity of health and ecosystems.

The Complexity of Contemporary Environmental Risks

Contemporary environmental and health risks such as climate change, loss of biodiversity, and biotechnology present unique challenges to the ability of science to anticipate and prevent harm. If science is unable to predict, quantify, or characterize these risks, then precautionary actions may be precluded while greater evidence of potential harm is sought. Thus, implementation of the precautionary principle requires not only changes in environmental policy but also a reexamination of the limits of science in the face of complex risks as well as steps toward addressing those limitations.

The complexity of contemporary ecosystem problems presents the following challenges for science.

Scale. Many current environmental problems pose hazards that affect wide geographic areas and can affect humans and the environment for long periods of time. Long latent periods may occur between the onset of stress and detectable change (Rapport et al. 1998), increasing the likelihood that intervening factors may play a role in causing an effect (Krimsky 2000) and decreasing the ability of scientists to characterize causal relationships.

Context. Many of these contemporary risks are best characterized as the result of the cumulative and interactive effects and multiple stressors. Humans and ecosystems are often exposed to more than one hazard at a time, and the interaction among these hazards is relatively unstudied. For example, the root causes of the collapse of marine fisheries may be the result of disease caused by pathogens, chemical insults, *and* loss of diversity through overfishing (Epstein 1998). Some of the effects of human activities on health and the environment are subtle and not highly visible at the level of the individual or a particular ecosystem, but become evident at the population level. Large-scale variability in both exposures to and effects of some hazards, combined with the unique susceptibility of some individuals and populations, complicates efforts to anticipate and prevent impacts.

Uncertainty. Many contemporary environmental problems are characterized by extreme uncertainty, resulting from both limitations in current scientific tools and the nature of complex systems. There is uncertainty about not only such large-scale effects as global change but also smaller impacts, such as the toxicity of a particular chemical and its effects on adult humans or a developing fetus. While some of this uncertainty may be reduced through more information and scientific inquiry, some results from the complex nature of natural and social systems. This type of uncertainty, called indeterminacy, discussed throughout this volume, cannot be readily reduced.

Limitations of the Current Practice of Science for Policy

Limits on the conduct of environmental science have important implications for the capacity of science to inform precautionary decision making in the face of uncertain and complex risks.

Limited hypothesis formation. The formulation of hypotheses largely defines the types of results that can be found in environmental research. Generally scientific hypotheses are formulated so that they can be answered with the time and resources available, leading to reduced, narrowly defined research with a limited temporal and geographic scope. This "reductionist" approach attempts to understand complex systems by isolating one or a few factors under controlled and reproducible conditions (Lemons et al. 1997). This approach works well in experimental science, where the goal is to develop predictive theories to explain specific cause-and-effect phenomena. But in environmental science, it may cause specific hazards or impacts to be missed.

Epidemiologist Neil Pearce observes, "We seem to be using more and more advanced technology to study more and more trivial issues while the major population causes of disease are ignored" (Pearce 1996: 682). The attempt to translate problems into manageable research questions means that we may be finding extremely precise answers to incomplete or incorrect questions (Schwartz and Carpenter 1999). The narrowness and specificity of hypotheses and research also frequently lead to a compartmentalization of disciplines and knowledge. Such a fragmented approach often precludes an integrated understanding of the root causes of disease and degradation. This compartmentalization of disciplines extends to a separation between qualitative and quantitative researchers. While there is a strong bias among many environmental researchers toward quantitative methods and evidence, qualitative information is often necessary (and sometimes the best data available) to gain a more comprehensive understanding of complex environmental risks. Thus, scientists and decision makers frequently lack the widest array of tools, methods, experience, and knowledge to make precautionary decisions.

An emphasis on main effects and not interactions or cumulative impacts. Biological and ecological systems are complex, with multiple feedback loops and interactions. Nevertheless, there is a tendency in environmental health research to focus on the more quantifiable causal factors and to assume that they act independently. Agencies do not generally examine cumulative and interactive effects because they are difficult to study and broadly accepted methodologies are lacking. Yet, difficulties in studying these effects do not mean that they are absent. The deemphasis of interactions and cumulative effects can inhibit broad efforts to examine the root causes of environmental risks.

Limited treatment of uncertainty and errors. Uncertainty is inherent in most environmental health research and a primary rationale for invoking the precautionary principle. Uncertainty is frequently considered as a temporary lack of data that can be quantified, modeled, and controlled through additional scientific inquiry (Barrett and Raffensperger 1999). The formal evaluation of uncertainty in environmental science research is generally limited to a narrow discussion of errors in main results that indicate the magnitude of error in the outcome variable. Sometimes more comprehensive probabilistic or quantitative uncertainty analyses are undertaken to present a distribution of uncertainties for various independent variables and the outcome variable. But these may leave out potentially more important sources of uncertainty, such as errors in the model used to analyze and interpret data, variability and susceptibility of specific populations, systemic uncertainties, interactions of variables, and biases from limitations in the conduct of the study. Most uncertainty analyses leave out qualitative uncertainty information such as interpretations of what is known or not known or what is suspected. There is still little consensus on how uncertainty should be characterized.

Errors in scientific research are treated asymmetrically. While conventional methods focus on avoiding the error of observing an effect that really is not there—a type I error—this increases the probability of committing a type II error: not observing an effect that exists. This is particularly true when sample size is not controllable. Statistical significance is viewed as central to the acceptance of a study or hypothesis. If the results do not demonstrate an effect with a high enough degree of significance, that effect is considered not to have occurred. Often if the effect does not diverge much from background, even if it is statistically significant, it is not considered important. Statistical significance may be difficult to achieve if the effect is rare or the number of subjects studied or magnitude of effect is small.

The convention of guarding against type I errors can create a bias toward the null hypothesis, usually stated as "no effect." While the type I error rate is traditionally set at 5 percent (the finding is so strong that there is less than a 5 percent probability that the result would have been seen by chance alone), the type II error rate is often set at 20 percent. This means that 20 percent of the time a real phenomenon will be missed because the data were not strong enough to convincingly demonstrate its existence (Kriebel et al. 2001). While having certainty before accepting a causal hypothesis or theory is a central and necessary tenet of experimental science, this level of certainty is often difficult to achieve in the complex, uncertain world of environmental science. It may work against precautionary policies. Sir Bradford Hill noted that while significance serves as a guide to caution before drawing conclusions, there is no "correct" level of significance, for every situation. He concluded that uncertainty *"does not confer upon us a freedom to ignore the knowledge we already have, or to postpone the action that it appears to demand at a given time"* (Hill 1965: 299–300, emphasis added).

Limitations in the Policy Structures Incorporating Environmental Science

Environmental law and policy in the United States and elsewhere in the world have for the most part institutionalized these constraints in the conduct of environmental science. Pollution and natural resource policies are largely fragmented, often into individual media—air, soil, water, and waste, leading to narrowly framed policy questions and answers. Discussions among agencies—worker health, environmental, public health, natural resources, atmospheric—are even more rare. Government decision makers generally ask questions about direct, quantifiable impacts, those that are easily regulated or controllable and defensible in public and the courts (Clark 1994). They demand precise answers such as point estimates or risk figures and tend to avoid the muddling effects of uncertainty or qualitative information. Krimsky (2000: 206) concludes that "like the science that informs it, the process of regulation has taken a reductionist approach, seeking chemical by chemical solutions; focusing on too few etiological outcomes; neglecting additive, cumulative, and synergistic effects; and allowing a balkanization of regulatory authority."

Defining the Ways in Which Environmental Science Can More Effectively Support Precautionary Policies

This section begins to outline the principles and practice of an approach to environmental science to "facilitate the investigation of complex, interdisciplinary problems that span multiple spatial and temporal scales, to encourage interagency and international cooperation of societal problems, and to construct more effective bridges between policy, management and science" (Lubchenco 1998: 495). This approach is examined from two important perspectives: subject matter and methods. Subject matter, or what is studied, defines what will be observed or what results can be found. Methods define how the problem will be studied.

Under this vision of science and precaution scientists need not understand every mechanistic detail before reaching conclusions (Chan et al. 1999). Protection and restoration of health and ecosystem integrity are important goals. The practice of science for precautionary decision making is not just diagnostic and curative, but also preventive (Rapport et al. 1998). Science, in this context, serves a powerful role in identifying quality objectives and options for prevention (see chapters 18 and 19).

Subject Matter

To support more precautionary decision making, scientists and policy makers need to ask multiple questions at different scales of inquiry. No single question can usually adequately characterize the complex etiology of environmental degradation. What is needed is a "broad lens" through which to view problems as well as more narrow hypotheses to examine or understand parts of that problem (Krimsky 2000). Through such questions, scientists can examine both the broader, complex root causes of degradation and proximate ones.

Asking a wider array of questions can increase the likelihood of earlier detection of environmental and health hazards. However, widely framed questions and hypotheses are more likely to be criticized by the scientific and policy establishment as providing evidence of only "weak causality" (Krimsky 2000). Further, the scientific tools for examining such widely framed questions, compared with those for more narrow questions, are much less refined and developed, and thus open to criticism.

Systems and Parts of Systems

To support more precautionary decisions, questions and inquiry at various system and molecular levels are needed. Susser and Susser (1996b) conclude that both reductionist and systems-level thinking are necessary for anticipatory, preventive action and sound public policy. Examination of the full range of exposures and hazards throughout the entire life cycle of a potentially harmful activity, such as extraction, manufacturing, use, and disposal, can create a more comprehensive picture of risk to individuals and ecosystems. In a broader approach to studying environmental impacts, system–level effects are analyzed, the components of the system are disentangled and studied individually, and the entire puzzle is then reassembled. The more traditional approach to studying environmental health risks would omit the last step.

Populations and Individual Level Effects

Environmental research has moved away from studying effects on populations to focusing more on factors that raise risks for individuals and small parts of systems, such as behaviors and genetic makeup. This change in focus has had profound impacts on the types of responses that policy makers seek. Rather than changing social conditions, they seek to change individual behavior.

However, there are important influences on health at the population level that cannot be observed at the individual level (McMichael et al. 1999; McMichael 1999). Subtle impacts of exposures to environmental hazards may only be visible at the population level, for example, in changes in the IQ of children. Differences in susceptibility to disease also may not be captured through an examination of the makeup and behaviors of individuals. This means that certain adverse effects of environmental degradation may not be captured, and sensitive populations may not be adequately protected.

Numerous epidemiologists (see Susser and Susser 1996a) have concluded that current methods are inadequate to address the complex etiology of human disease and are exploring new methods based on the concept of "ecoepidemiology." Susser and Susser (1996b) describe this as the Chinese nesting-box model, in which the etiology of disease is examined at the population, individual, and molecular levels, in addition to factors that influence disease within and between each level of organization. This model incorporates those intervening and interactive factors that are often excluded in traditional epidemiology.

Interactions and Cumulative Effects

Levins and Lopez (1999) find that it is at the level of complex interactions where science has been least successful. A failure to comprehensively examine such interactions may result in overlooking or underestimating environmental risks. Research must more effectively examine synergies, additive effects, and antagonistic effects. Such interactions may include chemical exposures, physical stressors, and socioeconomic factors and may occur at the cellular, organismal, and community levels.

Environmental scientists also need to more thoroughly examine the cumulative effects of multiple exposures over time. Frequently, causes of disease and degradation are examined as if an individual or population had not experienced any previous exposure. Yet, accumulation is one of the most important sources of adverse effects (Lane 1998). For example, the impacts of global change are likely the result of multiple, cumulative stressors, including chemical emissions, greenhouse gas emissions, and destruction of natural resources (see chapters 9 and 10). The "tyranny of small decisions," made independently, can slowly build up to a threshold and result in large-scale damage by affecting the resistance and resilience of systems (Epstein 1998).

Broader Definition of Disease and Effects

Environmental disease and degradation are generally defined in dichotomous terms, such as diseased/not diseased or impacted/not impacted, and health is frequently defined using the biomedical concept of "absence of disease." Dichotomous definitions may not adequately account for small, cumulative impacts that may eventually result in disease, indirect effects, nonmedically diagnosed illness, or effects that cannot be quantified.

A more comprehensive definition of health, such as the World Health Organization's notion of "the complete state of physical and social well-being" (see IJC 1995), would be more helpful in preventing harm. Such a definition takes into consideration the adaptability of systems to new or changing circumstances. It would frame the role of environmental science as proactive maintenance of the integrity of health and ecosystems (see chapter 18). This requires indicators and techniques to determine how far a certain sector is from a healthy state, as well as to issue early warnings and signals for intervention to prevent degradation. Adopting an ecosystem health perspective (Rapport et al. 1998) can focus concern on what types of dysfunction exist, how systems entered that state, and what can be done to prevent further harm.

Methods for Science to Support Precaution

To answer these bigger questions, scientists need methods that are not yet well developed in many fields (Schwartz and Carpenter 1999) and may not be considered appropriate in some other scientific endeavors.

A broader variety of methods will need to be adopted within individual disciplines, and new cross-disciplinary approaches will need to be developed. Such reforms are already beginning in disciplines such as marine science, climate science, and epidemiology. Lubchenco (1995: S8) outlines a set of key features of scientific inquiry that should form an integral part of these new methods. They should delineate:

• what is known and the certainty with which it is known;
• what is not known;
• what is suspected;
• the limits of the science;
• probable outcomes of different policy options;
• key areas where new information is needed; and
• recommended mechanisms for obtaining high-priority information.

Cross-Disciplinary Approaches and Multiple Lines of Evidence

The most decisive characteristic of a science that more effectively addresses complex and uncertain risks and informs precautionary decision making is the plurality of methods, disciplines, approaches, and evidence that it incorporates (Stern et al. 1992). The fragmentation of knowledge, disciplines, and institutions that characterizes the conduct of environmental science to date has limited the ability of scientists to understand the impacts of humans' industrial activity on the environment. While changes are needed in each individual scientific discipline to address the challenges of large-scale environmental risks, a more important need is the integration of knowledge across disciplines. To support more precautionary decisions, a wider array of evidence from various disciplines and constituencies (quantitative and qualitative, observational and experimental, empirical and theoretical) needs to be considered. The use of multiple sources of evidence can lead to scientific conclusions that are more robust and more explicit about uncertainty.

The emerging complexity of environmental hazards has already forced a reintegration of scientific knowledge from various disciplines to address such hazards as endocrine disruption, global change, and the challenge of ecosystem health.

For example, a comprehensive analysis of different types of evidence on the effects of persistent pollutants on wildlife in the Great Lakes led Dr. Theo Colborn to hypothesize that a common mechanism of action may be causing the effects. The organization of the now-famous Wingspread sessions on endocrine disruption provided a multidisciplinary opportunity to review similarities in research, test hypotheses, and weigh evidence. The first Wingspread conference organizers noted the importance of the event: "So shocking was this revelation that no scientist could have expressed the idea using only the data from his or her discipline alone without losing the respect of his or her peers" (Colborn and Clement 1992: xv).

Transdisciplinary approaches integrate the knowledge and contributions of different disciplines—medicine, toxicology, epidemiology, ecology, veterinary science, meteorology, engineering—into a base of knowledge that can lead to the development of novel questions, hypotheses, and ideas, and more efficient problem solving. Engineering and other design fields, not traditionally considered in the realm of "science," play an important role in the development and understanding of preventive options (see chapter 19).

Reviewing evidence from various disciplines enhances the ability of scientists to identify potentially harmful substances and effects. For example, cross-disciplinary scientific analysis, including scientists from different constituencies and backgrounds, played a critical role in understanding the sometimes subtle impacts of human industrial activities on the health of the Great Lakes ecosystem. These formed the basis for the U.S.-Canada International Joint Commission's recommendation to phase out discharges of persistent and bioaccumulative substances in the region (Durnil 1999).

Entirely new scientific collaborations and disciplines, such as earth systems analysis, ecosystem health, conservation biology, climate science, and conservation medicine, have been founded on interdisciplinary analysis. Conservation medicine links human and animal health with ecosystem health and global change, bringing together interdisciplinary teams of veterinary and medical health professionals to develop a greater understanding of the ecological context of health (Tufts Center for Conservation Medicine 1999).

Fuller and More Explicit Discussion of Uncertainty

Uncertainty is an unavoidable, but often underappreciated, aspect of environmental decision making involving complex causal relationships and social interactions. Acknowledging uncertainty is a positive part of scientific inquiry: It clarifies what is known and not known and, thus, stimulates further investigation. Yet, *uncertainty is often confused with low-quality information* (Funtowicz and Ravetz 1992). Uncertainty is often a reason for undervaluing research results, denying the existence of a problem, or minimizing its importance. Bureaucracies tend to avoid uncertainty because it muddles decision-making processes that are based on rules and absolutes. However, as biologist John Cairns aptly concludes: "Unrecognized risks are still risks; uncertain risks are still risks; and denied risks are still risks" (Cairns 1999: A57).

A first step toward a science that supports precaution is an admission that science cannot deliver certainty in matters regarding environmental health. Environmental scientists must be more explicit about uncertainty and potential errors. This means preparing for uncertainty in scientific assessment and anticipating it (Levins and Lopez 1999). Under a precautionary approach, uncertainty includes both areas of the unknown that can be filled in through further analysis and quantitative modeling as well as the irreducible uncertainty (including variability) that is inherent to complex systems. Scientists need to consider and characterize, both qualitatively and quantitatively, what they *do not* know as well as what they *cannot* know with current tools.

Scientific methods to support precautionary decision making would include a more comprehensive quantitative and qualitative (descriptive) analysis and discussion of uncertainty, examining:

- the sources of uncertainty (exposure estimation, confounding, bias, variability);
- the type of uncertainty (in measurement, model, or systemic uncertainty resulting from complex conditions);
- the degree of uncertainty and certainty in conclusions;
- how much uncertainty can or cannot be reduced through additional research; and
- the implications of uncertainty.

Environmental scientists should also be more explicit about potential errors and the ability of the research to detect an adverse effect. A clearer explanation of statistical significance—its meaning, factors affecting it, potential for errors—is a first step. When public and ecosystem health is at stake, it is appropriate to pay greater attention to avoiding a type II error, that is, not observing an effect when it exists.

Environmental scientists should be open about the predictive capacity of research to detect an effect, including what magnitude of effect the study might be able to predict. When a conclusion of "no adverse effects" is made, environmental scientists must be clear about whether the conclusions reflect a true lack of effect or the lack of evidence, understanding, or ability to detect that effect.

Incorporation of Qualitative and Quantitative Knowledge

In the conduct of environmental health research, there is a general bias toward quantitative methods. Qualitative information is frequently viewed as less valid, and often conclusions of "no adverse effects" are reached if effects cannot be quantified. Qualitative research questions, such as whether there is exposure or a hazard, what patterns exist in the data, or whether there are alternatives, are rarely asked. Scientists may discard large quantities of useful qualitative information that do not fit current tools or models or cannot be integrated into quantitative estimates. Quantitative estimates, on the other hand, may not reflect the nuances, context, and implications of adverse environmental effects on individuals or populations. The messy interpretation of the particulars of a population can seldom be reduced to numbers, and this description is usually relegated to the discussion section of research papers. Some types of knowledge, such as complexity and interrelatedness, experience, and explanation, simply cannot be numerically represented (see chapter 21).

Methods need to be developed to more effectively incorporate qualitative descriptions and observations of context, impacts, and interactions. This includes processes to incorporate observational, experiential, and historical knowledge collected from "extended peer communities," individuals or groups who have a stake in the results of scientific research (Funtowicz and Ravetz 1992).

Expert Judgment and Panels

Expert judgment and panels can provide an important supplement to "hard" data in precautionary decision making. They allow the introduction of experience and multiple viewpoints in analyzing scientific information. An everyday counterpart to the expert panel is the way physicians care for their patients. They rely on screening test data and observation but also on their prior experience and the entire base of knowledge about a patient's condition, including the patient's history (Sox et al. 1988). Further, medical doctors often communicate with other doctors who have been involved in the care of the patient. Environmental science needs to incorporate not only what is known about a problem as defined by conventional scientific techniques but also what scientists believe about the problem based on their own scientifically derived data, other information, experience, and suspicions.

Expert panels can evaluate all relevant information, quickly update knowledge, and maintain flexibility in analysis. They can examine strengths and weaknesses of evidence and set priorities for research and policy actions (IJC 1995). For example, the health-related impacts of climate change have been documented and supported by the Intergovernmental Panel on Climate Change (see chapters 9 and 10). Convening scientists from numerous disciplines and backgrounds to share information and reach consensus is an important mechanism for putting together the available data on threats to environmental health.

Iterative Approaches to Scientific Knowledge

Environmental science is more than a body of facts and understandings; it is also a continuous process for discovering new information. It can best inform precautionary policies if it reflects the state of knowledge and provides mechanisms for incorporating new information. The interaction between science and policy must be dynamic (Lubchenco 1995). Iteration can involve initially evaluating hazards using screening techniques that tend to err on the side of caution and then using more resource-intensive methods to gather data necessary to reduce uncertainties (National Research Council 1994).

Two techniques, Bayesian statistics and adaptive management, provide some direction for updating and integrating past and future knowledge. In Bayesian statistics, prior probabilities of harm are influenced by new data and become posterior probabilities. This method allows researchers to cyclically learn and systematically factor in prior knowledge in weighing evidence, incorporating everything from hunches and experience to hard data. While traditional scientific methods frequently leave out past knowledge, Bayesian methods encourage scientists to observe how conclusions change with increasing information, allowing a more rapid reaction as new information is acquired (Malakoff 1999). They require a much more explicit acknowledgment of uncertainty, assumptions, and bias because values and training can influence prior beliefs (Robins et al. 1985). Such methods are commonplace in clinical decision making, where rapid decisions are needed to prevent adverse effects in patients (Sox et al. 1988). In 2001, the State of California utilized the qualitative Bayesian approach to understand the highly uncertain risks of human exposure to electromagnetic fields from power lines (see Neutra and Delpizzo forth-coming and www.dhs.ca.gov/ehib/emf).

Monitoring, Surveillance, and Mapping for Feedback and Early Warnings

Monitoring systems can provide early warnings of potential effects, so that additional research and policy decision making may be quickly put into place. Ongoing monitoring and evaluation are necessary not only to detect hazards but also to understand the effectiveness and potential unintended conse-quences of precautionary actions (Goldstein 1999). Various scientific methods provide such early warning and feedback. Monitoring tools include use of indicators to measure degradation, the impacts of policy interventions, and progress toward environmental health goals (Rapport et al. 1998); human health surveillance (Thacker et al. 1996); and integrated modeling (Haines and McMichael 1997). It is important to point out that monitoring should not be used to promote a "wait and see" approach to hazards (Haines and Epstein 1993).

Presenting Scientific Information in the Best Way to Inform Decision Making

Because it will be used in decision making, environmental science research results and conclusions should be presented in the most user-friendly, policy-relevant format so that results are not misunderstood (Lubchenco 1995). This will require that environmental scientists more explicitly describe uncertain-ties, scientific disagreements, limits in the research, what is suspected, and future research needs (see chapter 22).

Public Policies to Support Novel Approaches to Environmental Science

Debates in various fields, such as epidemiology, conservation biology, and cli-mate science, as well as the emergence of new fields of inquiry indicate that the environmental sciences are beginning to adapt their subject matter and methods to address complex environmental hazards. Policies and institutions must also be adapted to incorporate a broader, cross-disciplinary scientific base.

Changes in policy and funding that are flexible and adaptive to emerging knowledge, encouraging broader analysis of problems, greater characterization of uncertainties, and more thorough analysis of prevention opportunities, would support more precautionary decision making. Such changes include more effective coordination and communication across agencies and increased government and private funding for long-term, speculative, uncertain, complex, and cross-disciplinary research that incorporates novel methods and broader hypotheses. Changes to policy and funding must be accompanied with changes in science education to encourage interdisciplinary learning, communication, and consideration of the impacts of research on health and ecosystems (see chapter 19). Education must be supplemented by venues for interdisciplinary professional interaction and debate.

Changes in environmental science should encourage scientists to incorporate their experience and informed judgment into conclusions and to be aware of their social responsibility to do science that protects human health and the environment. As citizens and purveyors of knowledge, environmental scientists have a social responsibility to sound an alarm when irreversible or serious effects may occur but are not immediately apparent (IJC 1995). It is the role of the scientist to help define the research questions, frame the issues, synthesize complex information, evaluate the consequences of different options, and develop recommendations for fundamental prevention.

Conclusions

The remaining chapters in this book discuss novel scientific methods and policy structures necessary to address contemporary environmental hazards that are serious, global, multigenerational, and potentially irreversible in nature. Several chapters note that current structures seldom cast a sufficiently wide net to anticipate harm, determine the root causes of effects, or identify far-reaching, preventive solutions across disciplinary boundaries. To support precautionary decision making, the current fragmentation and narrow focus of science and policy will need to be dissolved, allowing a much broader framing and examination of questions. It will require cross-disciplinary collaboration, flexibility, willingness to experiment with new approaches, and creative thinking. While there is no one ideal structure for environmental science research, *its foundations should be dictated by the complexity and nature of the scientific challenge* (National Research Council 1999). A broader approach to environmental science requires creative public policy structures that support and integrate it. The precautionary principle highlights the tight, problematic linkage between science and policy, encouraging environmental scientists to address the challenges of complex environmental hazards head-on and actively participate in discussions to develop more precautionary environmental policies.

References Cited

Barrett, K., and C. Raffensperger. 1999. "Precautionary Science." In *Protecting Public Health and the Environment: Implementing the Precautionary Principle,* edited by C. Raffensperger and J. Tickner. Washington, D.C.: Island Press.

Cairns, J., Jr. 1999. "Absence of Certainty Is Not Synonymous with Absence of Risk." *Environmental Health Perspectives* 107, no. 2.

Chan, N, K. Ebi, F. Smith, T. Wilson, and A. Smith. 1999. "An Integrated Framework for Climate Change and Infectious Disease." *Environmental Health Perspectives* 107(5): 329–38.

Clark, T. 1994. "Creating and Using Knowledge for Species and Ecosystem Conservation: Science, Organizations, and Policy." In *Environmental Policy and Biodiversity*, edited by R. E. Grumbine, 335–64. Washington, D.C.: Island Press.

Colborn, T., and C. Clement, eds. 1992. *Chemically-Induced Alterations in Sexual and Functional Development: The Wildlife/Human Connection*. Princeton: Princeton Scientific Publishing.

Durnil, G. 1999. "How Much Information Do We Need Before Exercising Precaution?" In *Protecting Public Health and the Environment: Implementing the Precautionary Principle*, edited by C. Raffensperger and J. Tickner, 266–76. Washington, D.C.: Island Press.

Epstein, P. 1998. "Integrating Health Surveillance and Environmental Monitoring." In *Ecosystem Health*, edited by D. Rapport, R. Costanza, P. Epstein, C. E. Gaudet, and R. Levins. Malden, Mass.: Blackwell Science.

Funtowicz, S., and J. Ravetz. 1992. "Three Types of Risk Assessment and the Emergence of Post-Normal Science." In *Social Theories of Risk*, edited by S. Krimsky and D. Golding, 251–74. Westport, Conn.: Praeger.

Goldstein, B. 1999. "The Precautionary Principle and Scientific Research Are Not Antithetical." *Environmental Health Perspectives* 107, no. 12: A594–95.

Haines, A., and P. Epstein. 1993. "Global Health Watch: Monitoring Impacts of Environmental Change." *Lancet* 342: 1464–70.

Haines, A., and A. McMichael. 1997. "Climate Change and Health: Implications for Research, Monitoring and Policy." *British Medical Journal* 315, no. 7112: 870–75.

Hill, A. B. 1965. "The Environment and Disease: Association or Causation." *Proceedings of the Royal Society of Medicine* 58: 295–300.

IJC-U.S.-Canada International Joint Commission. 1995. *1993–95 Priorities and Progress Under the Great Lakes Water Quality Agreement*. Windsor, Ont.: U.S.-Canada International Joint Commission.

Kriebel, D., J. Tickner, P. Epstein, J. Lemons, R. Levins, E. Loechler, M. Quinn, R. Rudel, T. Schettler and M. Stoto. 2001. "The Precautionary Principle in Environmental Science." *Environmental Health Perspectives* 109: 871–76.

Krimsky, S. 2000. *Hormonal Chaos: The Scientific and Social Origins of the Environmental Endocrine Hypothesis*. Baltimore: Johns Hopkins University Press.

Lane, P. 1998. "Assessing Cumulative Health Effects in Ecosystems." In *Ecosystem Health*, edited by D. Rapport, R. Costanza, P. L. Epstein, C. Gaudet, and R. Levins: Malden, Mass: Blackwell Science.

Lemons, J., and D. Brown. 1995. "The Role of Science in Sustainable Development and Environmental Protection Decision-Making." In *Sustainable Develop-*

ment: *Science, Ethics, and Public Policy,* edited by J. Lemons and D. Brown, 11–38. Boston: Kluwer Academic Publishers.

Lemons, J., K. Schrader-Frechette, and C. Cranor. 1997. "The Precautionary Principle: Scientific Uncertainty and Type I and Type II Errors." *Foundations in Science* 2: 207–236

Levins, R., and C. Lopez. 1999. "Toward an Ecosocial View of Health." *International Journal of Health Services* 29, no. 2: 261–93.

Lubchenco, J. 1995. "The Role of Science in Formulating a Biodiversity Strategy." *BioScience,* Supplement, S7–S9.

———. 1998. "Entering the Century of the Environment: A New Social Contract for Science." *Science* 279: 491–97.

Malakoff, D. 1999. "Bayes Offers a New Way to Make Sense of Numbers." *Science* 286: 1460–64.

McMichael, A. 1993. *Planetary Overload: Global Environmental Change and the Health of the Human Species.* New York: Cambridge University Press.

———. 1999. "Prisoners of the Proximate: Loosening the Constraints on Epidemiology in an Age of Change." *American Journal of Epidemiology* 149, no. 10: 887–97.

McMichael, A., B. Bolin, R. Costanza, G. Daily, C. Folke, K. Lindahl-Kiessling, and B. Niklasson. 1999. "Globalization and the Sustainability of Human Health." *BioScience* 49, no. 3: 205–11.

National Research Council. 1994. *Science and Judgment in Risk Assessment.* Washington, D.C.: National Academy Press.

———. 1999. *Global Ocean Science: Toward an Integrated Approach.* Washington, D.C.: National Academy Press.

Neutra, R., and V. Delpizzo. Forthcoming. "How Scientific Judgments Can Be Packaged to Facilitate Democratic Foresight Strategies: Lessons from the California EMF Program." *Public Health Reports.*

Pearce, N. 1996. "Traditional Epidemiology, Modern Epidemiology, and Public Health." *American Journal of Public Health* 86, no. 5: 678–83.

Rapport, D., R. Costanza, P. Epstein, C. Gaudet, and R. Levins, eds. 1998. *Ecosystem Health.* Malden, Mass: Blackwell Science.

Robins, J., P. Landrigan, T. Robins, and L. Fine. 1985. "Decision-Making Under Uncertainty in the Setting of Environmental Health Regulations." *Journal of Public Health Policy,* September, 322–28.

Schwartz, S., and K. Carpenter. 1999. "The Right Answer for the Wrong Question: Consequences of Type III Error for Public Health Research." *American Journal of Public Health* 89, no. 8: 1175–80.

Sox, H., M. Blatt, M. Higgins, and K. Marton. 1988. *Medical Decision Making.* Boston: Butterworth-Heinemann.

Stern, P., O. Young, and D. Druckman, eds. 1992. *Global Environmental Change: Understanding the Human Dimensions.* Washington, D.C.: National Academy Press.

Susser, M., and E. Susser. 1996a. "Choosing a Future for Epidemiology. I. Eras and Paradigms." *American Journal of Public Health* 86, no. 5: 668–73.

————. 1996b. "Choosing a Future for Epidemiology. II. From Black Box to Chinese Boxes and Eco-Epidemiology." *American Journal of Public Health* 86, no. 5: 674–77.

Thacker, S., D. Stroup, R. G. Parrish, and H. Anderson. 1996. "Surveillance in Environmental Public Health: Issues, Systems and Sources." *American Journal of Public Health* 86: 633–38.

Tufts Center for Conservation Medicine. 1999. *Introducing the Center for Conservation Medicine.* North Grafton, Mass.: Tufts University.

CHAPTER 2

Elements of the Precautionary Principle

Romeo F. Quijano

The main objective of a scientific exercise to determine the potential threats of harm from chemicals is to protect health and the environment. Yet this fundamental objective is often forgotten or ignored in the appraisal of risks inherent in the production, distribution, and use of potentially harmful chemical products. Dominant forces in the scientific community and regulatory agencies impose an evaluation system that relies heavily on numerical data and on the "smoking gun" type of evidence of harm that presumes the chemical to be innocuous until proven otherwise. This supposedly "science-based" risk assessment methodology has proven to be more effective in protecting vested interests rather than protecting health and the environment. In fact, risk assessment is not the decisive factor in determining the regulatory status of a toxic chemical. The reality is that economic interests and political expediency are generally the dominant considerations influencing regulatory decisions pertaining to toxic chemicals, especially in southern countries where financial, technical, human, and other resources are sorely lacking and where sociopolitical circumstances are particularly conducive for powerful chemical companies to exert influence and manipulate public policy (Quijano 2000).

In a country (the Philippines, for example) where thousands of toxic chemical products are imported under a liberalized economic regime and where the income of a foreign chemical company from just one chemical product exceeds many times over the entire budget of the regulatory authority, it is not surprising that health and environmental protection are at the mercy of the market economy. This situation is aggravated by the fact that only a

21

handful of medical toxicologists, most of whom have to contend with numerous other responsibilities as faculty, researchers, clinicians, and so forth are available to provide technical expertise to the government that, for some reason, is even reluctant to tap that expertise.

The unequal power relations between the strong and the weak, between the rich and the poor, and between the First World and the Third World are very much in the decision-making processes of government. Decisions that tend to protect health and environment are allowed only insofar as they do not significantly threaten the dominant economic interests, and only when strong public pressure is exerted on government. It is not unusual, for example, for bureaucrats to ignore the recommendations of a government-appointed toxicology committee or even abolish the committee itself rather than ban or restrict the toxic chemicals that the committee has deemed to be too dangerous to be allowed into the market. It is also not unusual for a technical expert in that committee who criticizes government inaction and bad corporate practices to be marginalized, harassed, threatened, and sued in court for publicizing the dangers presented by toxic chemicals to human health and the environment.

Even intergovernmental bodies are not immune to corporate influence, as technical committees are often packed with scientists with vested interests in the outcome of reviews (Castleman and Lemen 1998). The chemical producers, regulatory authorities, and other decision makers often rely on the "risk assessments" done by these supposedly objective "scientific" bodies as the basis for concluding a certain toxic chemical is "proven" to be safe (Rampton and Stauber 2001; Fagin and Lavelle 1999). To them, this is sound science.

True science, however, is about truth and the search for new knowledge in a systematic and logical manner so that people may benefit from it. True science involves astute observation of objects and events, careful formulation of hypotheses, unbiased experimentation and analysis, and logical conclusions. On the other hand, the pseudo-science of economic interests is frequently characterized by manipulation of objects and events, vested interest-driven formulation of hypotheses, biased experimentation and analysis, and market-directed, predetermined conclusions. In this distorted kind of science, data are collected, generated, or even fabricated to support economic objectives and achieve marketing targets. Information in this context is not something that may be true or false but something that is created and packaged to sell a product. Thus, science and scientists, all too often, have become effective tools of economic interests at the expense of public health and the environment. The case histories of endosulfan, methylene chloride, and asbestos illustrate this unfortunate development (Quijano 2000; Rampton and Stauber 2001; Fagin and Lavelle 1999).

The obfuscation of science is in no small measure due to the inherent reductionist, disciplinary character of most scientists. By the very nature of their training and work, scientists often develop tunnel vision. Too much specialization, coupled with increasing dependence on narrowly defined research funding (often from industry), tends to make them miss social realities and the bigger-picture implications of their work.

Scientists need to get out of this trap. The first step, perhaps, is to think like an ordinary human being (we are all citizens in the end) and recognize that human-made chemicals are in many cases inherently hazardous and should generally be presumed harmful unless proven otherwise. This is the precautionary approach (Tickner et al. 1998; Smith 2000; Raffensperger et al. 2000). It recognizes the fact that, historically, most hazardous chemicals have been shown to cause serious and irreversible damage to human health and the environment. It accepts the reality that the long-term impacts of toxic chemicals are difficult to predict and often impossible to prove. It is not dependent, as is risk assessment methodology, on a system of decision making that demands generation of extensive scientific data and requires exhaustive and quantitative analysis of risks as preconditions to policy formulation and action. This is particularly relevant to Third World countries where the resources needed to characterize the risks are not readily available.

From my perspective, the precautionary principle has several essential elements:

1. *Prevention:* Prevention is the first essential element of the precautionary principle. Prevention should be the focus of decision making, not mitigation after damage is done. Avoidance of exposure is the major concern, not defining the limits of exposure, as in the risk assessment approach. The question asked is not how much exposure is allowable but whether the exposure is necessary in the first place. Very often, it is claimed that the precautionary approach is already applied in risk assessment, and the use of "safety factors" to allow for uncertainties in limits of exposure is cited as an example. This is not correct. This is not precautionary but a "reactionary" measure. True precaution does not expand the dragnet to capture an escaped convict but takes immediate preventive measures so that the convict does not escape in the first place and addresses, on the long term, the sociopolitical factors that tend to create the "criminal."

2. *Reverse onus:* Reverse onus means placing burdens and responsibilities for safety and understanding on producers and not putting the burden of proof of harm on the potential victims. Too often, toxic chemicals cannot be restricted or banned because current laws and international agreements, like those of the World Trade Organization, mandate liberalized entry and persistence of toxic chemicals in the environment unless proven harmful beyond reasonable doubt. Efforts to regulate, restrict, or prohibit the pro-

duction, sale, and distribution of toxic chemicals to protect health and the environment are often considered "trade restrictions" and are challenged by the chemical companies or by countries in which these companies are based. This situation is obviously biased in favor of business interests and can be disadvantageous to people's health and the environment. The precautionary approach attempts to remedy this unjust situation.

3. *Elimination:* The ultimate goal under the precautionary principle is the elimination of harmful chemicals, not just the management of risks. Especially for persistent organic pollutants (POPs), elimination is the only long-term option because risks are considered unmanageable. The recent discovery that very low levels of POPs can cause significant reproductive, developmental, neurological, immunological, and other disorders directly or indirectly because of endocrine disruptive effects reveals that previous assumptions about tolerable levels based on risk assessments are incorrect. Increasingly, toxic chemicals characterized as persistent, bioaccumulative, and transported over long distances are now beginning to be allocated a zero level of tolerance, which means that for such chemicals, there is no safe level at all.

4. *Community orientation:* The health of communities is a primary concern of the precautionary principle. The basic human right to health and to a healthful environment takes precedence over economic and proprietary rights. The right to engage in a profit-making venture (like selling a chemical) is a derogable, conditional right, while the right to health is a nonderogable, fundamental human right. Corporate "rights" are ascribed rights and nonhuman, whereas community right to health is a basic human and social right. Any potential threat of harm from chemicals must be dealt with in a precautionary manner that protects basic human rights using the best available knowledge. We should not wait for rigorous scientific studies to provide proof of harm. Evidence of harm in preclinical studies must be presumed to be evidence of potential harm to humans. Community monitoring data and people's testimonies of harm must be given due importance and should inform the basis of a precautionary action.

5. *Alternatives assessment:* Assessing the alternatives to address the needs that the toxic chemicals are supposed to fill is another important element of the precautionary principle. This is not even usually considered under the risk assessment paradigm. More often than not, the need that chemicals are supposed to address can be addressed more effectively and safely over the long term by nonchemical alternatives. For example, the use of highly toxic pesticides is often justified in terms of increasing crop yields. However, a closer study of factors that contribute to sustainable crop yields would reveal that pesticides are not really necessary and that an integrated,

ecological approach to plant, soil, and pest management would be the better option for a sustainable crop production that would not endanger health and the environment.

6. *Uncertainty is a threat:* Unlike in risk assessment where uncertainty is given the benefit of the doubt, the precautionary principle considers uncertainty as a potential threat. While those fixated with the risk paradigm often consider absence of evidence as evidence of absence (of harm), precautionary principle advocates consider absence of evidence as no evidence of absence (of harm). Infinitesimal uncertainty factors often preclude demonstration of cause-and-effect relationships and probabilistic characterization of risks. To be meaningfully protective, therefore, an assessment process looking into the potential environmental and health impacts of a chemical should consider uncertainties as a warning signal. Addressing the reasonable knowledge gaps pertaining to that chemical, to the people's satisfaction, should be made obligatory for the chemical manufacturer before chemicals are allowed to be released into the environment.

7. *Technically/scientifically sound:* Contrary to what the critics often say, the precautionary principle is scientifically and technically sound. The evaluation process using a precautionary approach is not just an arbitrary procedure based on mere speculations and unfounded fears. It is based on the best available scientific evidence and guided by technically sound analytical procedures. For example, the potential toxicity and kinetic disposition of many chemicals in human populations and the environment can be assessed by analyzing structure-activity relationships, physicochemical characteristics, molecular mechanisms of action, animal toxicologic and ecotoxicologic data, and other types of information relevant to the chemicals in question, in much the same way as in the risk assessment methodology. There is a wide array of available scientific data that could provide sufficient basis to make a sound judgment as to the potential risks that a chemical poses to human health. However, for existing chemicals in commerce where scientific data are lacking or are inappropriate or impractical to generate (such as direct experimentation on humans), precautionary action protective of human health and environment should be taken even if there are doubts that the chemical in question poses unacceptable risks, making use of the best available knowledge and taking into account not only scientific but also sociocultural factors.

8. *Information unrestricted:* A key element in the practice of the precautionary principle is access to information. While the risk assessment paradigm accepts confidentiality of information to protect corporate proprietary rights, the application of the precautionary principle requires full disclosure and accessibility of information relevant to the appraisal of potential threats that a chemical brings to human health and the environment. Since

the protection of health and the environment is the paramount objective, all relevant information should be made available and accessible; otherwise, the appraisal process would be made subordinate to corporate interests. This would be tantamount to the violation of the people's fundamental right to health and to a healthful environment. The right to information is an extension of the right to health, and any abridgement or restriction of the right to information would violate the nonderogable nature of the right to health.

9. *Open:* A risk appraisal system based on the precautionary principle is an open, democratic, and participatory process. It is not the exclusive domain of elite scientists. It is not just a matter between industry and the regulatory authorities. The main objective of the exercise is protecting people. Therefore, people have the right to look into and scrutinize what steps are being done to protect them from hazardous chemicals. It is the people's right to participate in the decision-making processes relevant to the protection of their health and their environment. The right to participate in decision making is an extension of the people's right to self-determination. The people have the right to determine for themselves which chemicals they need and which they don't need, what risks are acceptable and what are not acceptable. This right is also an extension of the right to health, since without it, the right to health is unattainable.

10. *Need based:* Most of the chemicals that were introduced into the market after World War II were not the outcome of mission-oriented researches directed toward fulfilling particular human needs. They were by-products of the oil industry and the war machinery of the industrialized countries. To maintain profitability after the war, the corporations that created by-products began searching for "needs" and began creating demands that would be filled by these chemicals. Thus, the synthetic dye industry, the solvent-based manufacturing process, pesticide-dependent agriculture, the chemical preservative–dependent food products, and many more chemical-dependent activities were created. The demand for synthetic chemicals became phenomenal; today hardly anyone can pass the day without consuming or using a synthetic chemical-dependent product. The need for synthetic chemicals has been automatically presumed in the current regulatory system under the risk assessment paradigm. This system, however, has led to disastrous consequences and has now put the living organisms in this planet, including humans, in danger. Clearly, our "need" for synthetic chemicals, especially pesticides and other intrinsically hazardous chemicals, must be reassessed. This is precisely what the precautionary approach does: assess the need for the chemical as part of a comprehensive and integrative approach to risk appraisal before allowing its release into the market and throughout its life cycle. The benefits that the chemical

brings to people must be reasonably clear and more important than the potential threats of harm.

In the end, the precautionary principle reorients the way science is done for policy, inherent presumptions about chemicals and their risks, and ultimately the policies necessary for protecting health and ecosystems. Precaution must rely on science to identify potential risks to health and the environment. But we need not wait for elusive exhaustive data before preventive actions are taken.

References Cited

Castleman, B., and R. Lemen. 1998. "Corporate Influence at International Science Organizations." *The Multinational Monitor,* January/February.

Fagin, D., M. Lavelle, and Center for Public Integrity. 1999. *Toxic Deception.* Monroe, Maine: Common Courage Press.

Quijano, R. 2000. "Risk Assessment in a Third World Reality." *International Journal of Occupational and Environmental Health* 6: 312–17.

Raffensperger, C., T. Schettler, and N. Myers. 2000. "Precaution: Belief, Regulatory System, and Principle." *International Journal of Occupational and Environmental Health* 6: 266–69.

Rampton, S., and J. Stauber. 2001. *Trust Us, We're Experts.* New York: Center for Media and Democracy, Jeremy P. Tarcher/Putnam.

Smith, C. 2000. "Introduction. The Precautionary Principle and Environmental Policy." *International Journal of Occupational and Environmental Health* 6, no. 3: 263–64.

Tickner, J., C. Raffensperger, and N. Myers. 1998. *The Precautionary Principle in Action: A Handbook. Written for the Science and Environmental Health Network.* http://www.sehn.org/rtfdocs/handbook-rtf.rtf.

CHAPTER 3

A Cautionary Tale

Barry Commoner

The precautionary principle may be a relatively new term, but the idea has been part of environmental campaigns for a long time. In this chapter I examine some of the ways in which the environmental movement has struggled with the issues encompassed by the precautionary principle. Two major factors and their interaction must be taken into account. One is the scientific basis of uncertainty—the inability to predict—that should lead to caution. The other element is the role of the public. In this chapter I consider how these factors illuminate the experience of the environmental movement with uncertainty, precaution, and public action.

Four examples from the twentieth century show how precaution has not been incorporated into the decisions made on hazardous technologies. They illustrate the lessons we have still failed to learn from these repeated mistakes.

Nuclear Testing

The modern environmental debates started with the public campaigns to limit fallout from nuclear weapons testing. A huge concern arose with the development of nuclear weapons and the fact that we had not only exploded weapons that killed many thousands of people in Hiroshima and Nagasaki, but then repeatedly tested them in the open. If anything called for the precautionary principle, that did.

The existence of radioactive fallout affecting the general population came as a total surprise to the public. It emerged from military secrecy after a severe thunderstorm over Troy, New York, that caused some physicists' counters to go wild. The physicists went out, wiped material off cars, and realized that radioactivity was coming out of the sky in the rain. That made news. Until then, no one had been willing to say that radioactive fallout could spread through the air. A series of battles ensued among scientists, the concerned public, and the government about what was going on. Hiding behind official secrecy, the government revealed very little, so members of the scientific community took it on upon themselves to learn about fallout. In the process, they found it was critical to work with the public.

Nuclear testing in the open air produced enormous quantities of radioactive material and created a significant biological insult to the food supply. Very little was known about its consequences. The hazard was disseminated, through the air, all over the Northern Hemisphere. In the United States, people had heard about fallout and were worried, but had not the foggiest idea what it was. In St. Louis, people would come to me with bits of grass with mold spots on them, asking if that was fallout. Those of us who were concerned with fallout were faced with the necessity of finding out what was really known so that we could talk to people about it. For example, when it was determined that radioactive iodine, along with strontium 90, was present in milk, people asked what they could do about it. Physicians began advising people to give their kids nonradioactive iodine (potassium iodide) in order to dilute the effect of the radioactive iodine in the milk. This is now well-recognized as an essential preventive step in case of a nuclear accident (Wurtz 1962).

In 1958 an official United Nations science committee was set up to evaluate the problem. The committee concluded that radioactive fallout "has new and largely unknown hazards for present and future generations." I think it was clear, at least from this point on, that the precautionary principle ought to have been applied to nuclear technology from the onset.

One outcome of the evolving movement of scientists and the public against nuclear testing was that the process of putting radiation in the air was stopped. I would call the Nuclear Test Ban Treaty, which ended the surface explosion of nuclear weapons, the first success of the precautionary principle. The premise of the treaty was that because we did not understand the dangerous consequences of this activity, we should stop it (United Nations 1959).

The fundamental science behind radiation—the science of dosage, the relation between dose and response—was problematic. Ever since Hiroshima and Nagasaki, there had been debate about whether the relationship was linear or whether there was a threshold below which radioactivity was safe. Some nuclear scientists argued for the existence of a threshold. They maintained not only that humans could tolerate a certain amount of radiation exposure, but that a little radiation might even be good for us. They believed the threshold was a result of cellular mechanisms that repaired the damage to DNA. It took many years before it became clear that there is no threshold, that exposure to radiation should be avoided or, if necessary, kept to a minimum (Commoner 1973).

The use of radioactivity in dentistry and in medicine was common since well before World War II. But exposure standards were lax; when I was a child in the 1920s and 1930s, when you went to the store to buy shoes, you put your foot in a fluoroscope to see if the shoe fit, giving your foot an enormous exposure.

Things changed after the antinuclear campaign. Foot fluoroscopes were banned and Eastman Kodak came out with a very sensitive X-ray film, so that dental and medical X-rays required much lower exposures than before. The standards of acceptable exposure to radiation were brought sharply down (Commoner 1973). So, the antinuclear fallout movement, driven by concern over a major public policy, raised debate over a new, esoteric matter of science—dose-response curves and thresholds— that in turn created improved public policy on radiation exposure.

History tells us that if you can find the proper linkage between a serious, widely disseminated problem and the public, you can raise the possibility of doing something about it, even when the scientific questions have not been resolved. In fact, stimulating public action will lead to demand for more and better science. The demand for action then intensifies, for in many environmental issues, the more we learn, the worse the problem appears. We still have a lot more to learn about nuclear radioactivity.

DDT

The next big issue after nuclear fallout was pesticides. I learned from a recent biography that Rachel Carson was motivated by the fallout issue to do her own investigations into pesticide toxicity and to write about it for the general public (Lear 1998). She knew about public concern and activism around fallout in St. Louis, Detroit, New York, and elsewhere. She saw the role that scientists were playing both in gathering information and bringing it to public attention, and she saw pesticides as an analog to what was happening with fallout. *Silent Spring* was a brilliant success. It changed the entire public attitude toward DDT, which until then had been worshiped as a wonder of modern science. And the intense public concern was followed by a sharp increase in research on DDT and related pesticides.

But before Rachel Carson, the environmental hazard of DDT was neither a scientific nor public issue. I was in the Navy during World War II. One of my assignments was to design and build equipment to put on torpedo bombers to spray DDT on the beaches of the Pacific islands that we were going to invade. I was told that DDT—its insecticidal property only recently discovered—was a magic chemical that would kill insects at extraordinarily low concentrations. You could put a relatively weak solution into a vaporizer and kill a mosquito with a single microscopic drop. That is, a very small amount of this substance would have a big biological effect. That was all I was told.

The spray plane was built, and we tested it over a jungle in Panama. That is when we learned that snakes are irritated by DDT. I do not like snakes. I was in the jungle when we sprayed, and when I came out, there were snakes everywhere I walked. We once sprayed a beach where troops had been bothered by flies while they were testing rockets. The plane came along, sprayed, and did a marvelous job—no flies. Three days later, we got an urgent call to come back. Flies were everywhere, feasting on piles of dead fish.

Whenever you used DDT, there were a lot of surprises and unexpected results. This was the first widespread use of a chemical that was extremely potent at very low concentrations. People should have taken notice and thought more carefully what the surprises meant. But they did not.

Then, Rachel Carson pointed out that bird populations were falling because their eggshells were cracking. DDT was barging into the delicate hormone systems that govern sexual development—and the thickness of eggshells. Hormones and their receptors operate at very low molecular concentrations, in ways that are still poorly known. It was a triumph that DDT was banned; the public, alerted and informed by *Silent Spring*, demanded it. *We should have learned from the DDT experience that any substance potent enough, at very low concentrations, to impede a poorly understood biological system that also operated at very low concentrations should be kept out of the environment.* But this precautionary warning went unheeded, and toxic pesticides proliferated in the environment.

Such low-concentration effects lead to uncertainty and controversy of a simpler kind as well. Those of us in toxics activism have had experience with the problem of "nondetects." Somebody analyzes an environmental sample and says, "I didn't detect any, so there is no effect to worry about; it's safe." This brings up the question of the sensitivity of the analytical method. The Environmental Protection Agency loves to say that a nondetect means that there may be zero effect. That is true; but it is also true that the actual concentration may be just under the detection limit and so might well have an effect. In the argument that follows, a compromise is often reached that interprets "nondetect" as half the detection limit. When you deal with such very low concentrations, there may be built-in uncertainties in the analytical technology that foster controversy about low-level effects. A poorly interpreted "nondetect" may cloak an important precautionary warning.

Petrochemicals

Another major environmental issue that I'd like to discuss grew out of the concerns over pesticides. Petrochemicals in general—not only chemicals made to kill something, but those sold in seemingly benign consumer products as well—should have been introduced with a great deal of precaution. These chemicals, which are organic carbon compounds, have proliferated enormously by invading existing markets for inorganics or naturally occurring organics. Glass and wood have been supplanted by plastic, soap by detergent, and in ball fields, real grass by fake grass. These big existing markets were taken over without any consideration of what would happen in the environment.

The overall output of the petrochemical industry is about 500 billion pounds a year of synthetic organic compounds. One by one, these chemicals have been shown to have an array of adverse biological effects, such as hormone disruption, birth defects, and cancer (Commoner 1992). As a scientist, you have to ask, what's going on here? Why is this particular sector of manufacturing generating one problem after another, each one a surprise: dying birds in the Great Lakes, sex-reversal in frogs, tumors in fish, not to speak of the deaths of chemical workers.

No one stopped to think about a very simple thing. For billions of years, the only organic compounds on the earth were manufactured by living things—not in a laboratory, and not in a factory. The biosphere, living things, had a complete monopoly on organic chemistry. Then, in 1828 a chemist synthesized the first natural organic compound, urea. By the end of that century chemists were making many *unnatural* organic compounds, and fifty years later the petrochemical industry was in full sway. I remember how DuPont used to boast that they tailor-made new, synthetic molecules to suit any particular purpose. The petrochemical industry massively invented new organic chemicals—a job previously reserved for living things—and put them into the environment, where, finding their way into birds, frogs, fish, and people, they caused toxic surprises. Why?

Synthetic organic molecules are often structurally similar to natural ones because in both the basic structure is built around linear arrays or rings of carbon atoms. After all, whether it occurs in a protein or a plastic, a benzene ring is a benzene ring.

One of the remarkable features of organic chemicals is their enormous variety; a current compilation of the properties of chemical compounds contains over 12,000 organic compounds, the great majority of them made of only four elements: carbon, oxygen, hydrogen, and nitrogen. In comparison, the list contains only one-fifth that number of known inorganic compounds that are made of all the rest of the elements. But the range of possible kinds of organic compounds that are actually produced by living things is enormously restricted. For example, a very simple protein might consist of one hundred amino acid units, linked up in a particular order. Since there are twenty different kinds of amino acids, there is a huge number of different sequential orders in which they can be arranged. In fact, if you made one 100-unit protein molecule with each possible sequence and added them all together, you would get a weight larger than the weight of the known universe (Elsasser 1966). That tells you immediately that the proteins made in living things are a fantastically small fraction of what is possible.

In the long course of evolution and in the huge numbers of organisms that have participated in it, some of these possible kinds of organic compounds must have arisen—and because they were incompatible with the complex cellular system, caused it to fail. Any synthetic organic chemical that does not now occur in living things is likely to be an evolutionary reject. A billion years ago, some cell might have taken it upon itself to make DDT or dioxin—and has not been heard from since. In a way, by going into the business of synthesizing organic chemicals, the petrochemical industry was plagiarizing the chemistry of life. But like all plagiarists, they didn't quite get it right, and most of what the industry makes is incompatible with life. The creation of the petrochemical industry violated the principles that have governed the evolutionary development of the chemistry of life. Its products, launched into the biosphere, have become a huge, unpredictable, trial-and-error experiment—a casualty of a dangerously incautious industrial technology.

Genetic Engineering

Looking at this history, we can see that the precautionary principle is the only way to deal with such problems. Lately, a new problem has come up. The petrochemical companies, once content with producing artificial forms of life's organic chemicals, have now begun to produce artificial forms of life itself. And once again, on a massive scale, they are substituting their new, synthetic products for existing natural ones. In the United States, most of the acreage of our major crops—corn, soybeans, and cotton—is now planted with genetically modified seeds. The reincarnated petrochemical industry—biotechnology— claims that their new methods are "precise, specific, and predictable" (Masse 2000). They claim that they can avoid the uncertainty of natural breeding by manipulating genes to produce precisely the proteins they want, with no risk of unexpected consequences (Commoner 2002).

According to Francis Crick's "central dogma" pronounced in 1958, genes, made of DNA, are transcribed into RNA, which in turn determines the structure of proteins. The nucleotide sequence of the gene specifies the amino acid sequence of the protein, which in turn specifies its structure, the resulting biochemical activity, and the inherited trait that it engenders. One consequence is that for every protein and the resultant inherited trait, there is just one particular gene.

In February 2001, after less than a decade of intensive research, the $3 billion Human Genome Project, designed to identify and enumerate all the genes in the human body, was completed. The main result was "unexpected." Based on the estimated number of human proteins—guesses were in the range of hundreds of thousands to a couple of millions—only thirty thousand genes were found.

Even more embarrassing, a mere weed and a lowly worm have nearly as many genes as you and I. Apparently genes alone cannot account for inheritance. Yet this is the essence of the central dogma, in which the gene is enthroned as the sole source of the genetic information that is transferred unaltered to a protein and the inherited trait. In fact, the massive amount of research inspired by the central dogma over the past forty years has produced a number of results that the Human Genome Project describes as "discordant." They show that although proteins, in their amino acid sequence, contain genetic information derived from the nucleotide sequence of a particular gene, they may also contain in their folded-up, tertiary structure genetic information that is *not* derived from the gene. A notorious example is the infectious nucleic acid–free prion that causes Mad Cow disease and related human diseases, in which genetic information is passed from protein to protein. Moreover, by correcting errors in the nucleotide sequence of newly replicated DNA, protein enzymes contribute to the genetic information of the gene itself.

Finally, some of the discordant results show that specialized proteins participate in processes that dismember gene-derived genetic information and reassemble it into a range of different protein products. All this tells us (well, at least some of us) that the DNA gene is not the exclusive molecular agent of inheritance, operating through the rigid, linear DNA-RNA-protein line of command. Instead, the molecular processes that participate in the system of inheritance, which remains a unique property of the living cell, transfer genetic information from DNA to proteins and vice versa. The molecular system of inheritance is circular rather than linear, and deeply complex rather than seductively simple.

Such conclusions, which profoundly question the central dogma, could have been inferred from the data available many years ago (and some of us did so). But the prevailing wisdom has insisted that DNA does it all, that if the appropriate gene, alone, taken from any organism or even made synthetically is inserted into any other living thing, it will do exactly—and no more—that it is supposed to. The petrochemical industry's "life scientists" heard the message and put it to work. Now each year trillions of plants are grown, encumbered with segments of DNA that are alien to the plants' own complex, tested-by-evolution genetic system. But the hazards of these attempts to overcome the rules of natural selection have been hidden in the biotechnology industry's laboratories. It is widely known that most transgenic experiments fail to produce satisfactory, or even surviving, plants before the very few that can be commercialized are in hand. The industry boasts that its marketed plants are safe and effective because they have been "winnowed" from a much larger population of failures. Even so, in at least one case, Monsanto's herbicide-resistant transgenic soybean, the presence of the alien gene is accompanied by a section of disrupted host DNA sufficient to give rise to novel proteins.

The biotechnology industry's transgenic crops present us with a new twist in the tragic history of precautionary neglect. This time the fault lies not so much in the area of technological application, but more in the realm of the basic science. By neglecting to consider the impact of "discordant" experimental results on the conveniently simplistic central dogma, molecular geneticists have legitimated the invasion of a massive uncontrolled transgenic experiment into U.S. agriculture. Biotechnology and the precautionary principle, as the saying goes, "were made for each other," but the match itself has yet to be made.

Conclusion

In this essay, I have tried to show how time and again decisions about new technologies have been made without regard to their environmental consequences. Each of these examples presents a case in which unintended consequences of some new technology led to serious risks to the biosphere. The importance of a precautionary principle is very clear. It is time, now, to put it into action.

References Cited

Commoner, B. 1973. *Closing Circle*. Beekman Publishers.
———. 1992. *Making Peace with the Planet*. Peter Smith Publishers.
———. 2002. "Unraveling the DNA Myth." *Harpers Magazine*, December 18.
Elsasser, W. 1966. *Atom and Organism: A New Approach to Theoretical Biology*. Princeton: Princeton University Press.
Lear, L. J. 1998. *Rachel Carson: Witness for Nature*. New York: Henry Holt.
Masse, A. 2000. *Guide to Biotechnology*. Biotechnology Organization, Washington, D.C. 10 May. www.bio.org/aboutbio/guide2000/whatis.html.
United Nations. 1959. *Report of the Scientific Committee on the Effects of Nuclear Radiation*.
Wurtz, R. H. 1962. "The Iodine Story." *Nuclear Information* 4, no. 9 (September): 62.

PART II

Precaution, Ethics,
and the Philosophy of Science

This section explores the underlying ethical, scientific, and philosophical basis for a more precautionary approach to science and policy. The chapters in this section examine the nature of the problems and types of uncertainty that demand new approaches to science.

Matthias Kaiser outlines in chapter 4 several efforts to understand the ethical implications of precaution in Norway. He argues that the precautionary principle is based on a moral responsibility to reduce and act on our ignorance of possible future harm and that it requires a broader perspective for decision making, including democratic processes for participation under uncertainty.

In chapter 5, Juan Almendares discusses the difficulties of applying precaution in science and policy in Honduras, a poor country with limited public health infrastructure and where both human rights and democracy are lacking. He argues that any discussion of science and precaution must consider the context under which risks occur.

CHAPTER 4

Ethics, Science, and Precaution: A View from Norway

Matthias Kaiser

This chapter examines the question of how applications of the precautionary principle are interwoven with values and ethics. The importance of these concerns has often been neglected in formulating policy. I argue that in applying the precautionary principle, it is vital that value and ethical judgments be explicitly addressed in suitable democratic fora.

To illustrate the importance of ethics, this paper examines several recent policy discussions in Norway. The precautionary principle enjoys wide public acceptance in Norway, and it is already relatively well entrenched in the legal framework. The principle has been actively discussed and accepted by scientists and government officials. Norway has also more than ten years of experience with national advisory bodies in the area of scientific ethics. These bodies work with a definition of scientific ethics that makes it natural to include issues of applying the precautionary principle among the scope of their deliberations. While there is consensus on the validity of the precautionary approach, no unified methodology for implementing that approach in science and technology has been established. Instead of interpreting this as an argument against the principle, I claim that the ethical and value dimensions that are part and parcel of the precautionary principle are, as yet, not adequately accounted for and that the integration of those concerns will help in establishing a model for decision making.

Expert Commission Reports

In Norway, the precautionary principle played a crucial role in the reports of two recent government-appointed expert commissions in 2000 and 2001. The first of these analyzed the health consequences of genetically modified (GM) products (NOU 2000: 29; hereafter referred to as the "Walløe report") and the other one examined xenotransplantation (NOU 2001: 18, hereafter referred to as the "Gjørv report"). While both are unusual in their explicit and lengthy discussion of the precautionary principle, they come to roughly opposite conclusions with regard to the conditions necessitating the application of the principle.

It is important to note the distinction between (a) checking whether conditions for applying the precautionary principle are fulfilled at all (these conditions are outlined in later chapters) and (b) determining what kinds of precautionary measures are indicated once it is agreed that the principle should be applied. Norwegian expert commission reports are written for the government on the basis of an explicit mandate, are published and accessible for everybody, and include both a presentation of the relevant scientific findings and the recommendations of the commission with regard to suitable regulatory measures. Normally the commissions work to maximize consensus on specific topics, but it is not unusual that a commission has a split vote. The final reports, normally drafted by a secretariat or by individual members of the commission, thus represent the viewpoints of all commission members.

The Walløe report considers various potential health risks of GM food. For almost every health risk examined in the report, all but one member of the commission concluded that existing evidence to date provides no reason to apply the precautionary principle. For instance, concerning the question of whether a new gene can be transferred from food plants to mammalian cells through the intestines, the commission was unanimous that this is normally not the case, according to generally accepted knowledge. However, the commission also mentioned that a number of apparently well-documented new studies seem to suggest significant exceptions to this rule. The causes behind these exceptions were, however, not known.

The group then mentioned that the very few animal studies based on long-term feeding with GM food had not considered the question of gene-transferal to body cells. Thus, there is a clear lack of direct scientific evidence to show that genes from GM food could enter mammalian body cells. The majority then concluded from these observations that there is no reason to apply the precautionary principle based on concerns about genetically altering mammalian body cells; the few possible exceptions to the complete breakdown of DNA in intestinal tracts that we know of would equally apply to nonmodified foods. Only one member of the commission warned that we do not know the mechanisms behind these exceptions and that GM food might indeed result in gene transfer, so the potential health risks are too significant to be neglected. This member opted for applying the precautionary principle.

The philosophy behind the majority in the Walløe report was apparently that we need specific scientific studies that establish through accepted methods that the risks are real. As a second step we may then consider whether the risks are too great, and whether we should apply the precautionary principle. It may be noteworthy that the report repeatedly criticizes the lack of independent studies in this area, but this had apparently no effect on the committee's conclusion that risks were nonexistent or too small to justify applying the precautionary principle.

The commission preparing the Gjørv report had to address uncertainties similar to those involved in the Walløe report. The main risks of xenotransplantation derive from the possibility of xenosis, the transferal of infectious diseases from animals to humans. The porcine endogene retrovirus (PERV) is of particular concern. Similar to the case of GM food, to date no studies have directly demonstrated any transferal of PERV from primary (i.e., not grown in the laboratory) pig cells to primary human cells. The report also identifies seven steps necessary for PERV infections to become a health risk to humans (e.g., that infectious PERV must be able to infect human body cells, that the transplant organ might excrete PERV, etc.). The fact that four of the seven steps were already shown to occur in laboratory studies, even though some of them only under idealized circumstances (e.g., with immune-deficient mice), was taken as indication that the risk might be real. Furthermore, the report referred to the development of HIV infections as a relevant model for predicting potential xenosis. HIV infections have developed as zoonosis from apes to humans. *Thus, the report concludes that there is a scientifically based scenario of possible harm, though there is no indication about its likelihood.* On this basis, the commission was unanimous in concluding that the precautionary principle should be applied to xenotransplantation.

These examples illustrate how the precautionary principle has entered debates about scientific and technological developments with potential major social or environmental effects. In essence, they show that it is virtually impossible to separate science and its use in policy when dealing with complex, uncertain systems. There are significant inconsistencies in different people's understanding of how to interpret the conditions for the use of the precautionary principle in specific cases. Judgment is only in part a matter of science.

In Europe, and at least in Scandinavia, discussions are not about whether or not to endorse the precautionary principle in policy. Rather, the questions relate more to the issues of how radical a break the precautionary principle is from standard procedures of risk assessment, what kind of information it demands from science, what truly precautionary measures are, and who is to be consulted in these matters.

Proponents of a narrow interpretation of the precautionary principle, like the Walløe commission, typically base their arguments on what kind of "facts" have been proven by "sound science." They ask for scientific evidence of the reality of probable harm based on indisputable direct studies. Others find it paradoxical that the very core of the precautionary principle, that is, the existence of scientific uncertainty, needs to be described by "hard scientific facts." For these people, arguments from analogy and from model scenarios often seem sufficient ground to apply the principle. In the face of uncertainty and high-decision stakes, all available evidence, including indirect evidence, needs to be considered and taken into account.

Scientific Ethics: Broadening the Scope

In Norway there are three national committees for research ethics, established in 1991: one for medicine, one for social science and the humanities, and one for science and technology (abbreviated as NEM, NESH, and NENT, respectively; see www.etikkom.no). These committees use a broader definition of research ethics than is usually the case in the Anglo-American countries. First, the field of research ethics is not limited to research with human subjects, but includes the environment. Second, research ethics are not limited to individual scientists' behavior in a research setting, but also include macro reflections on the behavior and policies of larger entities such as scientific institutions. Third, research ethics are considered not only with regard to the process of doing research, but also with regard to the responsible handling, interpretation, and communication of research results. In fact, this last area has been dominant in Norwegian discussions about ethics of science and technology.

The rationale for this wide use of ethics is twofold:

1. Science is an important actor in translating knowledge into social reality.
2. Knowledge implies moral responsibility. In matters of policy all involved parties have some responsibility, but those with the most information on and best insights in possibilities and dangers have a special co-responsibility to utilize their knowledge (see Mitcham and Schomberg 2000).

Given this understanding of research ethics, the Committee for Science and Technology (NENT) undertook a close examination of the precautionary principle. The committee produced a comprehensive report in 1997 (NENT 1997). The reason for the committee's interest was that the precautionary principle might face the same fate as the concept of sustainability: policy makers wanted scientists to say something about it, and scientists felt this was exclusively the decision maker's responsibility. In such a situation there is a real danger that both parties might overlook the novel challenges posed by the precautionary principle.

What is the relationship between the precautionary principle and ethics? NENT argued that the principle relates to ethics and ethical values in several ways (NENT 1997).

First, the precautionary principle is to a large degree based on or justified by the moral principle of *culpable ignorance,* which has a long history in ethical theory. The committee discovered Ian Hacking's discussion of this principle (Hacking 1986) and found it very much in line with legal thinking. The law of many countries provides conditions for punishing individuals when it can be shown that they were in a position to prevent an accident and failed to act upon some available information that warned of the danger ahead. This is usually referred to as negligence.

The concept of culpable ignorance is easy to grasp by way of example. Suppose you own a very old car. If you have an accident because another car suddenly pulls to the left without warning and hits you, then you were simply ignorant of the other driver's intentions and not to blame for the accident. However, if you have an accident because your brakes fail, then the case might look differently. Knowing that brakes in old cars tend to become faulty, you should have had them checked at regular intervals. Of course, you will be ignorant about the possible damage of the system until the brakes finally fail. But if you have also failed to check them regularly, then you will also be *culpably ignorant* and bear some moral (and legal) responsibility for the resulting accident.

In general, one is culpably ignorant if on the basis of some general knowledge, one recognizes the need for certain supplementary specific information or measures in order to avoid harm, but fails to do so. This general knowledge will typically be such that it comprises types of situations rather than the specifics of a given situation. In other words, it reasons by analogy rather than proof. In the example above, this general knowledge would relate to what one should expect from a technical system after long-term use. Old cars should be checked for proper functioning; failure to check is negligent.

If this reasoning is sound for everyday behavior, then there is good reason to assume that it also should hold for science and policy making. As discussed below, the question then is to define what types of general knowledge place this burden of responsibility on us.

This is not the only way the precautionary principle relates to ethics and ethical values. In fact, when weighing uncertainties and risks, when deciding what methodological approach befits a subject, when evaluating several precautionary strategies, and when deciding who has a say in the decision process, values of some kind (other than purely epistemic values) play an important role. Since these values tend to be neglected by both scientists and policy makers, NENT believed it was important to examine and point out these value-related issues.

As an illustration of this interdependence of science and ethics one may consider the discussion about statistical type I versus type II errors (e.g., Lemons et al. 1997; Fjelland 2002). Normally there is a certain trade-off between type I and type II errors: increasing one means decreasing the other. In standard significance tests researchers more often than not only control for type I errors. But when dealing with decision making about possible future harm, for example, about some environmental matter, the really crucial information may be what the chances are of overlooking a real effect. In overlooking this statistical insight in standard significance tests, science makes a value assumption not adequate for preventive measures in the light of possible harm.

Of course, the first result of such an inquiry into the relation of values, ethics, and the precautionary principle could be considered negative: applying the precautionary principle is not simply a matter of "objective" standard science or routine administrative work. There is no straightforward objectivity attached to the principle. No science can deliver the "right" answers. One cannot avoid some kind of explicit stand on basic values (see also JAGE 2002).

Applying the Precautionary Principle

Some analysts often criticize the notion of precaution as being too imprecise; that there is no definition available that allows an immediate operationalization of the principle (Sandin 1999; Graham 2001; Goklany 2001; Morris 2000). This is, of course, true for all the diverse definitions and formulations that this principle has undergone over the years. All need interpretation. This skepticism seems to persist in many quarters of science and policy, despite the many academic efforts to clarify precaution further (see, for example, Foundations of Science 1997; JoRR 2001; JAGE 2002; Cottam et al. 2000; Freestone and Hey 1996; see also Lemons and Brown 1995; Lemons 1996).

It has been pointed out (O'Riordan and Cameron 1994) that the vagueness of the principle is by no means surprising, nor is it a drawback. Jordan and O'Riordan (1999) state that "the application of precaution will remain politically potent so long as it continues to be tantalizingly ill-defined and imperfectly translatable into codes of conduct, while capturing the emotions of misgivings and guilt." The precautionary principle has a similar semantic status to moral norms or ethical principles (like human dignity, equity, and justice). It needs to be interpreted and specified on a case-by-case basis, and it will sometimes change its specific content according to the available information and current practices. It is well recognized that ethical principles, such as the protection of human dignity, sometimes call for a certain measure of paternalism (e.g., when institutionalizing certain patients) that may in other cases be disrespectful of human dignity. This is quite similar to precaution. In order to protect, for instance, the biodiversity of a given region it may be wise simply to leave a disturbed or polluted river leading into the region to its natural course and stop all kinds of human interaction with the river. But in some cases it may be indicated to take active steps to bring this river back into a quasi-natural state again, such as by restocking fish species, reducing its salinity, and so forth.

We need to look at the case at hand in order to find out what precaution means in that specific case. Partly this is due to the complexity of the scientific facts to which we need to relate. But partly this is also due to the varying interests and values that enter such a case. Typically there will be competing interests (e.g., besides biodiversity) at stake, some of which deserve special attention (e.g., to preserve cultural diversity by providing the economic basis for human settlements). While the precautionary principle can remind us of our moral duty to prevent harm in general, it cannot prescribe what kind of sacrifice we should be prepared to make in each and every case. Thus, the precautionary principle has the semantic status of a general norm rather than that of a detailed step-by-step rule of operation. It follows from this that it may make its occurrence in the guise of a multitude of different formulations and goal expressions.

The lack of clarity in definitions should not prevent one from spelling out some of the crucial conditions for the principle's application. The conditions NENT (1997) embraced are the following:

1. There exist considerable scientific uncertainties.
2. There exist scenarios (or models) of possible harm that are scientifically reasonable.
3. Uncertainties cannot be reduced without at the same time increasing ignorance of other relevant factors (i.e., attempts to reduce uncertainties by model building or laboratory studies typically imply abstractions that lead away from the real system under study, and there is no "adding back" to real conditions) (Fjelland 2002).
4. The potential harm is sufficiently serious or even irreversible for present or future generations.
5. If one delays action, effective counteraction later will be made more difficult.

While the NENT conditions for the application of the precautionary principle do not necessarily represent a widespread consensus on the principle, it is noteworthy that the European Union communication on the precautionary principle (EU 2000) seems to express a similar spirit when it states that "recourse to the precautionary principle presupposes that potentially dangerous effects deriving from a phenomenon, product or process have been identified, and that scientific evaluation does not allow the risk to be determined with sufficient certainty."

It was on the basis of these criteria that the commission on xenotransplantation concluded that the precautionary principle should be applied in this

case. It was on the basis of point 2, the lack of existence of scientifically reasonable scenarios of harm, that the majority in the Walløe report rejected
application of the principle. The difference lies in the fact that the xenotransplantation group was willing to accept reasoning by analogy (such as the HIV
experience or basic research in laboratories under idealized circumstances),
whereas the group on GM food wanted these scenarios or models to be based
on feeding experiments directly relevant to the issue of GM food. In other
words, they operated on a narrow interpretation of what could fall under "scientifically reasonable scenarios of possible harm."

The reason the Gjørv group included research results not directly related
to xenotransplantation was that they deemed the risks to be too great for the
general public to overlook or disregard evidence that signaled caution. They
were looking for some positive scientific evidence, but the existence of one
model-scenario of one kind of zoonosis, combined with partial laboratory evidence, was considered sufficient scientific reason to warrant caution. When the
Walløe group concluded differently, it made a value judgment of significant
importance.

Precautionary Measures

Once one has established that the precautionary principle has to be applied,
one normally faces the question of what to do about it. How shall we act
(including refraining from acting at all)? What measures should be counted as
precautionary in some sense? Any action that can be assumed to effectively
reduce the risks in question and that prepares us for handling future crises
could be counted as a precautionary strategy. However, choosing a strategy
invariably involves taking a stand on basic value issues and, thus, ethics again.

The xenotransplantation group was most explicit about different kinds of
measures that could reduce risks. In this report four such possible strategies
were mentioned explicitly:

1. A moratorium until further research can be conducted.
2. A step-by-step strategy with predefined targets for research before development is brought another step forward.
3. A go-slow strategy where practical use is restricted to few applications over
   a longer time.
4. A monitoring strategy where a system is set up to report on potential
   problems immediately and where possibly affected individuals are con-
   tacted and isolated.

In the end, the committee opted for a combination of all four strategies,
with a limited moratorium in the beginning in order to establish a body of
oversight and then a combination of the other strategies.

The NENT report (1997) and Kaiser (1997) discuss possible precautionary strategies on a more general level via selected case studies, in particular, the question of fish escapes from farms. Four types of possible responses are presented and discussed. Each of these strategies is based on implicit value assumptions. These have to do with the degree to which individuals tend to be risk averse or risk taking. To the extent that people believe that nature is very robust to change, people tend to be more willing to take certain risks with nature. To the extent that people think nature is in a rather delicate balance, people tend to become risk averse.

The same can be said about society. Society, just like nature, has evolved over a long time, and its institutions (e.g., its economic or institutional operations) are tuned to each other. One may then believe that society is a very robust entity, which would incline one for risk taking. Or one can believe that society with all its subordinate functions and workings is a rather delicate affair, inclining one to be risk aversive. Thus, when combining these attitudes, one ends up with what could be regarded as four ideal types of risk handling. It is important to note that none of the attitudes described above is directly inferred from science. They are prescientific matters of belief. They indicate where we might want to put our values when dealing with risks. This is not to say that people in general do this in a very consistent manner. In fact, there may often be bias regarding the risks from which we ourselves benefit as contrasted to the risks from which others benefit but that might affect us. None of these attitudes is based on hard science.

Once it has been established that the precautionary principle should be applied, we are faced with a multitude of possible precautionary strategies to reduce risks. There is no one "right" strategy in any objective sense. We sometimes have to make trade-offs, for example between effects on nature and effects on society. If we decide, as the Walløe report implicitly suggests, to go on with developing GM food and simply to monitor its further development carefully, then we put a much larger value on maintaining certain socioeconomic processes than on protecting health or protecting the environment. This is certainly legitimate, but it is not a question of straightforward science. It is a value decision. In the xeno case the situation is different. While the risks to nature are of minor importance in this case, it is the balancing of individual health benefits with public health concerns that seems to be the overall consideration.

Precaution and Participation

Once one accepts that the precautionary principle implies value decisions of some kind or another and rests on basic ethical intuitions, what does one do with regard to the decision process pertaining to it? Who is in a legitimate position or in authorized office to make these decisions? In principle the answer could be easy: the democratically elected representatives in parliaments are justified to make these decisions. This is why they are elected in the first place.

However, in practice the situation may be, and normally is, much more complex. First, many of these decisions will involve long-term strategies, extending beyond the terms of individual elected decision makers. Second, many of these decisions are quite complex and need scientific expertise to be sorted out. That is why authorities and politicians expect scientists to play an active role in decision making. Very often, however, this means that scientific expertise provides *all* the information that goes into the decision process. Third, democratic societies are pluralistic societies, with a multitude of values and individual preferences. There are no expert shortcuts to this multitude of values. No ethicist has a higher authority in basic value matters than the people themselves. Fourth, those who are directly or indirectly affected by certain hazards should have a say in the management of these hazards. This is particularly relevant when risks affect minorities or other small groups without formal power. Risk-cost-benefit analysis in its standard forms, with its summation over individual preferences, is defective with regard to principal considerations of justice. When large burdens on a few are contrasted with small benefits to many, very often the minority will lose out. Thus, it has been suggested that if risk-cost-benefit analysis is to be employed at all, it needs to be supplemented by ethical weighting techniques (Shrader-Frechette 1991).

These considerations provide the basis for arguing that decision making with regard to the precautionary principle should be supplemented by policy tools that are participatory in character. This does not mean that our democratic institutions lack the legitimate formal powers to decide for or against a precautionary measure. Nor does it mean that the decisions should be made by referendum instead of the established institutions. The final decisions will always rest with institutional decision makers. The claim is rather that from a moral point of view the decision makers should engage in a process where the public is extensively consulted and where input from this process is important for the final decision.

When values and ethics enter the picture, pluralistic societies tend to oppose expert-based decision making. This is the basic challenge in what is sometimes referred to as the new governance of science and technology. Apart from expanding the inputs on relevant values and evaluation of knowledge claims, such a new form of governance also promises to provide for more robust long-term strategies. If the public is given a voice in the final decisions, then the public is also willing to share responsibility for the outcome.

In this respect, the application of the precautionary principle differs from traditional risk assessment methods. Very often risk assessments are conducted on a routine basis, in order to supply standardized information needed for decision making. A new drug, for instance, does not necessarily involve problems of the kind that would call for a large public consultation. It is, however, worthwhile noticing that even risk assessors are now arguing for some form of public consultation as a supplement to risk analysis (JoRR 2001), and they realize that both value-based motivations and interest-based motivations play a crucial role in implementations of the precautionary principle (Tait 2001; Stirling et al. 1999).

There are a wide variety of participatory tools for decision making in public policy. What kind of tool, what kind of procedure one adopts, is obviously dependent of the kind of problem one faces. In Norway, we have some positive experience with the use of consensus conferences (also known as lay panels) (Joss and Durant 1995). For these conferences a panel of lay people is selected (by advertisement in newspapers or by random draw from population register) and given a general theme for deliberation. They meet in two preparatory weekends and define specific issues to address and select the experts they want to hear at the subsequent three-day conference. At the conference, the lay panel hears presentations by experts and is allowed to question them before making recommendations to the public and press. These conferences seem to be an adequate tool to deal with large-scale decisions affecting all society, such as GM food. In fact, NENT cosponsored two such consensus conferences on GM food, one in 1996 and one in 2000 (see www.etikkom.no and www.teknologiradet.no). Both provided an interesting footnote to the Walløe report. The 1996 lay panel expressed concern that we know so little about the environmental effects of GM food and about the few independent studies done in this field. Many scientists complained about this lack of evidence to the panel. The second panel was even more troubled by the fact that the scarcity of studies on impacts had not changed significantly by the year 2000, despite undiminished assurances of safety by the same scientists. No wonder the lay panel recommended a strong precautionary strategy.

One might also meet well-defined and conflicting group interests, rather than general worries about fundamental values at risk. In such cases NENT has used scenario workshops with interested parties (Kaiser and Forsberg 2000).

Despite growing interest, we still know too little about the possible uses and the impacts of participatory policy tools. In Europe we have started to assess some of them in larger European studies (Kluver et al. 2000; Fixdal 1998). All these participatory decision tools that make up the new governance of science and technology seem culturally dependent and topic sensitive. It may not be possible to specify the best participatory tools in advance, and it may be that some of these approaches have shortcomings in handling the complex issues involved. This should not distract from the observation that the precautionary principle, because of its inherent value dependency and ethics dimensions, calls for a consultation with a wide section of the general public to supplement institutional decision-making processes.

It is perhaps also noteworthy that the xenotransplantation report explicitly demanded that some form of public consultation should take place before the moratorium on xenotransplantation may be lifted. The commission preparing the report argued that the possible risks involved demanded an active public dialogue from the beginning that could enlighten decision makers about the kind and level of risks people find acceptable, given the positive potential of the treatment. It is then also noteworthy that the Walløe report provided no such recommendation about consulting the larger public about GM food.

Conclusion and Recommendations

The precautionary principle is based on an ethical conviction that sometimes our ignorance of possible future harm makes us morally culpable. It assumes that we should actively seek information and knowledge that may have a bearing on uncertain, complex issues and that in looking for that information we cannot apply a narrow perspective. We need to consider and weigh knowledge and information that comes in from the sideline or enters by analogy. The precautionary principle is not based on pure imagination of remote and speculative possibilities. The principle, when placed in a realistic decision context, refers back to some scientifically relevant information. But where we find this information, and how we decide what is more or less relevant, may have to be decided on a case-by-case basis. Values will implicitly guide us when we pick the information we consider relevant.

To take the further step of precautionary action, we again have to make reference to values, since there is normally a multitude of possible strategies available to address suspected hazards. And often we have to make trade-offs between considerations concerning nature and considerations concerning society. Where we strike the balance in the end is dependent on what we value and what we believe about various trade-offs.

In order to translate these value dimensions into policy, especially long-term and socially robust policy, new forms of governance are called for. The characteristics of such governance are the wide use of participatory decision tools in one form or another. While these decision tools are normally not expressed as part and parcel of the precautionary principle, they should be seen as a direct consequence of the value base on which any application of the principle will be dependent.

In the end, the precautionary principle emerges as more closely related to other moral principles than we might have thought in the beginning: it is not the facts or the science or finding the range of possible strategies that poses the

---

Box 4.1. Changes to Environmental Science to Support Precaution

• Perform a critical evaluation of the *research problem* at hand, with special attention to the possible fallacy of so-called type III errors (i.e., "good research, but wrong problem altogether").
• Do not uncritically accept common perceptions of who the *experts* are or where they are to be found, but look actively for alternative or supplementary expertise.
• Try to be maximally inclusive rather than exclusive in your discussions of what the relevant *options* for actions are.
• Try actively to develop new platforms for *dialogue* between science and the public.
• Seek to *cooperate* as much as possible with relevant partners from governmental authorities, industry, and nongovernmental organizations.
• Establish routines for critical *self-reflection and evaluation* of your own activities, including reflection on the extent to which funding sources may have colored your research.

---

greatest problems. It is rather the consensus on the more or less implicit values that poses the greatest challenge. One may ask what the consequences of this interrelatedness of the precautionary principle and ethics are for responsible science. What change is actually implied for scientific practice? The list of six points (box 4.1) presents a kernel of potential precautionary science and relates to some of the points made in other contributions to this volume.

When science is not self-conscious about its own potential pitfalls and shortcomings, when science assigns to itself a better track record than is justified by history, when science forgets the many idealizations and abstractions that are prerequisites for its model-building and testing procedures, and finally, when science portrays itself as unaffected by large commercial or political interests, it stands in grave danger of becoming socially irresponsible. Science, and in particular the scientific expertise that we might utilize in policy making, is much more pluralistic, uncertain, divided, interest-based, and value-based than often appears in the picture scientists present of their activity. Failing to concede this much may be seen as ethically defective. A truly precautionary science does not refrain from entering discussions and arenas where values and ethics are at stake, but contributes to these in a balanced and self-reflective manner. The new objectivity in science is not an attempt to stick to "hard facts" alone. It is the "hard decisions and soft facts" that pose the challenge of our times. Thus, the new objectivity of precautionary science amounts to new modes of organizing and managing research, including new forms of quality control (extended peer reviews), with the aim of providing relevant information for policy making that only the scientific method can reveal and that can effectively contribute to a balanced picture of the various options that society has to consider. Present-day science does not in general fit this picture, but luckily it contains all the intellectual resources necessary to become precautionary science to the benefit of society and nature.

References Cited

Cottam, M. P., D. W. Harvey, R. P. Pape, and J. Tait, eds. 2000. *Foresight and Precaution,* vols. 1 and 2. Rotterdam: A. A. Balkema.

EU. 2000. *Communication from the Commission on the Precautionary Principle COM (2000) 1.* Brussels: Commission of the European Communities.

Fixdal, J. 1998. "Public Participation in Technology Assessment." *TMV skriftserie* nr. 37. Oslo: University of Oslo.

Fjelland, R. 2002. "Facing the Problem of Uncertainty." *Agricultural and Environmental Ethics* 15, no. 2: 155–69.

Foundations of Science. 1997. Special issue. "The Precautionary Principle and Its Implications for Science." *Foundations of Science* 2, no. 2.

Freestone, D., and E. Hey, eds. 1996. *The Precautionary Principle and International Law: The Challenge of Implementation.* The Hague: Kluwer Law International.

Goklany, I. M. 2001. *The Precautionary Principle: A Critical Appraisal Environmental Risk Assessment.* Washington, D.C.: Cato Institute.

Graham, J. D. 2001. "Decision-Analytic Refinements of the Precautionary Principle." *Journal of Risk Research* 4: 127–41.

Hacking, I. 1986. "Culpable Ignorance of Interference Effects." In *Values at Risk,* edited by D. MacLean. Savage, Md.: Rowman and Littlefield Publishers, Inc.

JAGE. 2002. Special issue. "The Precautionary Principle." *Journal of Agricultural and Environmental Ethics* 15, no. 1.

Jordan, A., and T. O'Riordan. 1999. "The Precautionary Principle in Contemporary Environmental Policy and Politics." In *Protecting Public Health and the Environment: Implementing the Precautionary Principle,* edited by C. Raffensperger and J. Tickner. Washington, D.C.: Island Press.

JoRR. 2001. Special issue. "The Precautionary Principle." *Journal of Risk Research* 4, no. 2.

Joss, S., and J. Durant, eds. 1995. *Public Participation in Science: The Role of Consensus Conferences in Europe.* London: Science Museum.

Kaiser, M. 1997. "Fish-Farming and the Precautionary Principle: Context and Values in Environmental Science for Policy." *Foundations of Science* 2: 307–41.

Kaiser, M., and E. M. Forsberg. 2000. "Assessing Fisheries: Using an Ethical Matrix in a Participatory Process." *Journal of Agricultural and Environmental Ethics* 14: 191–200.

Kluver, L., M. Nentwich, W. Peissl, H. Torgerson, F. Gloede, L. Hennen, J. van Eijndhoven, R. van Est, S. Joss, S. Bellucci, and D. Butschi. 2000. *European Participatory Technology Assessment: Participatory Methods in Technology Assessment and Technology Decision-Making*, edited by L. Kluver et al. Copenhagen: The Danish Board of Technology. www.tekno.dk/europta/2000.

Lemons, J., ed. 1996. *Scientific Uncertainty and Environmental Problem Solving*. Cambridge, Mass.: Blackwell Science.

Lemons, J., and D. A. Brown, eds. 1995. *Sustainable Development: Science, Ethics, and Public Policy*. Dordrecht: Kluwer Academic Publishers.

Lemons, J., K. Shrader-Frechette, and C. Cranor. 1997. "The Precautionary Principle: Scientific Uncertainty and Type I and Type II Errors." *Foundations of Science* 2: 207–36.

Mitcham, C., and R. Schomberg. 2000. "The Ethic of Scientists and Engineers: From Occupational Role Responsibility to Public Co-responsibility." In *The Empirical Turn in the Philosophy of Technology*, edited by P. Kroes and A. Meijers. Research in Philosophy and Technology series, volume 20. Amsterdam: JAI Press.

Morris, J. 2000. *Rethinking Risk and the Precautionary Principle*. Oxford: Butterworth/Heinemann.

NENT. 1997. *Føre-var Prinsippet: Mellom Forskning og Politikk*. Oslo: De Nasjonale Forskningsetiske Komiteer. English translation forthcoming in 2002 at www.etikkom.no.

NOU. 2000. *2000: 29, GMO-mat; Helsemessige konsekvenser ved bruk av genmodifiserte næringsmidler og næringsmiddelingredienser*. Oslo: Statens forvaltningstjeneste.

———. 2001. *2001: 18, Xenotransplantasjon; medisinsk bruk av levende celler, vev og organer fra dyr*. Oslo: Statens forvaltningstjeneste.

O'Riordan, T., and J. Cameron, eds. 1994. *Interpreting the Precautionary Principle*. London: Earthscan Publications Ltd.

Sandin, P. 1999. "Dimensions of the Precautionary Principle." *Human and Ecological Risk Assessment* 5, 889–907.

Shrader-Frechette, K. 1991. *Risk and Rationality*. Berkeley: University of California Press.

Stirling, A., O. Renn, A. Klinke, A. Rip, and A. Salo. 1999. *On Science and Precaution in the Management of Technological Risk: An ESTO Project Report*. Sevilla: European Commission, Institute for Prospective Technology Studies, Sevilla. The full report can be downloaded from ftp://ftp.jrc.es./pub/EURdoc/eur19056en.pdf.

Tait, J. 2001. "More Faust Than Frankenstein: The European Debate About the Precautionary Principle and Risk Regulation for Genetically Modified Crops." *Journal of Risk Research* 4, no. 2: 175–89.

CHAPTER 5

Science, Human Rights, and the Precautionary Principle in Honduras

Juan Almendares

Is it possible to apply the precautionary principle in the less industrialized countries? One way to answer this question is to observe that it can be particularly difficult to establish scientific arguments in poor countries because there are not many scientists, and they are not well equipped to produce new knowledge. In addition, economic and political dependence can contribute to unethical scientific approaches. The majority of scientific work is from borrowed ideas, and the financial assistance for research is framed by the globalization politics of the industrialized countries. Our governments spend less than one percent of their budgets for investigation and development. The tendency is toward privatization of universities and less support for investigation and access to secondary education, leading many scientists to leave for the developed world.

The precautionary principle offers advantages for Third World countries because it is simple and has some precedence in the international legal system. It is possible to share information from other countries that have similar experience and have taken preventive action in advance of scientific proof of causality. More importantly, in countries where scientists and medical professionals can appear as elites separate from the rest of society, the precautionary principle encourages them to be more aware of and participate with members of civil society in making decisions and taking actions on environmental problems.

In this chapter, I argue that application of the precautionary principle in countries with economic dependence requires consideration of history, politics, science, and ethics. I outline some of the barriers to precaution in developing countries and provide two examples of how technologies and activities from more developed countries are exacerbating the environmental health problems of Honduras and making precautionary action and policies impossible. I conclude that until we address major problems of democracy and human rights, applying the precautionary principle will be difficult in developing countries, even though it may present some opportunities to avoid problems that the northern countries have created. At the same time, the precautionary principle may guide scientists and civil society toward establishing the democratic strength in these countries that precaution requires.

My perspective on the precautionary principle is the result of years of work as a medical doctor, educator, ecologist, and human rights advocate in Honduras. Despite being born into a poor family, I had the chance to study. I developed a great passion for serving poor people. Over time, I have learned that there is a critical interconnection between philosophy, health, science, ethics, and respect for human rights.

The Context: A Lack of Democracy and Human Rights

Since the "discovery" of America by European colonizers, true democratic process has not existed in Honduras. Honduras, in the heart of the Americas, remains hidden from sight in the international political arena. One only hears its name linked to tragedies: Hurricane Fifi, Hurricane Mitch, starvation in the twenty-first century, drug trafficking, prostitution, the killing of children, corruption. Honduras has been known as a "banana republic," a "country for sale," a "U.S. airstrip," the "Green Prison." Recently, Honduras got its best press by defeating Brazil in the Copa America soccer tournament.

Our borders with Guatemala, El Salvador, and Nicaragua inspired the dream of our illustrious leaders, Francisco Morazàn (Central America Federation) and Jose Cecilio del Valle, of forming a "Grand Nation" (Becerra 1995). It is that same geographic location that has caused Honduras to be converted into a strategic geopolitical, economic, and military entity by northern countries and multinational corporations. This has happened, of course, without Hondurans' democratic participation in the decision process.

Wars among and within Central American countries in recent decades have been due more to alliances with foreign interests than to civil conflicts originating in these nations. Honduras was occupied by the U.S. military and Nicaraguan contras during the 1980s. U.S. troops occupied Honduran territory before the National Congress of the Honduras Republic gave its approval. In this period, Honduras served as a center for espionage and training for torture practices. Later, it became one of the United States' most important military bases in Latin America and was placed in the middle of multilateral wars, particularly the conflict in Colombia.

The impacts of the 1980s war are still felt by the friends and relatives of the "disappeared," as well as in the fear and terror that reverberate in the population. The conflict has also caused mistrust of those who govern, the proliferation of corruption, confrontations among religious groups, and ever-increasing racism.

Hurricane Mitch, in 1998, generated and exacerbated physical, social, and spiritual tragedies in Honduras. The hurricane destroyed more than 70 percent of the infrastructure of the country, killed 5,000, and injured 12,000 people. Military maneuvers have increased in the areas where the indigenous people live in extreme poverty. Based on the hostile advice of international economic policies—"kick them while they're down"—governments and companies took advantage of the Honduran people and environment during the crisis following Hurricane Mitch. One month after the disaster, under pressure from multinational mining companies, the Honduran government covertly changed mining laws to allow free exploration throughout Honduras. Companies obtained concessions harmful to the national interest.

The Denver Agreement, signed by governments of the United States and Honduras and approved by the National Congress in 1998, after Hurricane Mitch, established more protections for foreign investment with tax exemptions and decreased environmental controls. The agreement permits any U.S. citizen or company to freely negotiate for or purchase any part of Honduran territory. Furthermore, it makes possible the privatization of basic services such as water, electricity, health, and communication. It has been operative since July 2001 (La Gaceta 2000).

Before and since the tragedy of Hurricane Mitch, the country has been for sale, for either military or economic objectives. The present privatization of the economy has produced a dramatic increase in poverty, external debt, and failures of small businesses, and it has strengthened multinationals. This has led to higher rates of consumption of alcohol, tobacco, and other drugs in the population (Castro and Carranza 2001). Poverty and violence have also increased in the population (UNDP 2000; Arroyo and Espinoza 2001).

Violence: Overt and Covert

Unfortunately, experts in Honduras have used the media to create the erroneous notion that the root of the country's violence rests in children and youngsters. The purported violence of young groups, who are labeled "gangs," has received much attention. The solution has been to eradicate the gangs.

Casa Alianza (Silva 2001), an organization for the well-being of children, has bravely denounced the killings of children: 94 in 1998, 286 in 1999, and

221 in 2000. In her report on the extrajudicial and arbitrary executions in Honduras, UN Special Consultant Asma Jahangir confirmed the existence of "hunters of children" when she stated: "Documented reports exist in regard to 66 minors that died in the first six months of the year. The government files show that some of these children were killed by security forces, and have been assured that the accused will be brought to justice." Every day, death squads operate with impunity in Honduras.

Meanwhile, repressive authorities expand their use of violence against indigenous peoples and prosecute human rights leaders. For example, they beat women and children from the peasant community of Gualaco, who struggled against the prevailing terror of the armed group that assassinated the environmentalist Carlos Flores. Although substantial gains were made throughout the 1990s in upholding human rights, the situation has worsened steadily in recent years, especially since Hurricane Mitch, with rising homicide rates, along with torture and organized violence, which often accompany poverty. These two forms of violence claim nearly equal numbers of victims in Honduras. For example, in 1998 there were 2,490 homicides and 2,003 victims of torture and organized violence (UNDP 1999).

That same year, 63.7 percent of households were below the poverty line. Some children eat from dumpsters; their best friends are mangy dogs and buzzards. Some retrieve cigarette butts, sniff glue, or get what they can from discarded bottles of alcohol, all of which cause brain damage and worsen the effects of the malnutrition from which they are suffering. While children are being killed deliberately or by neglect, the crimes of white-collar delinquents, who are almost never arrested or imprisoned, are ignored.

In this context we must consider whether and how the precautionary principle might be applied. The precautionary principle emanates from the desire to protect humankind and planet, even if there is no certain scientific evidence of the extent and cause of the environmental problem. The following examples, dibromochloropropane (DBCP) and mining, represent the kind of abuse of the environment and public health that is often considered in making the case for precaution. If the government acted with transparency it might be possible to apply the precautionary principle and avoid the kind of problems described below, in which companies avoid paying damages to workers and communities, claiming that there is no scientific proof of harm.

However, the whole context of violence, poverty, and international collusion in Honduras is relevant to applying precaution. If silence is kept in the face of intentional murder and neglect of children, the spirit of humanity ceases to exist. We must ask ourselves not only how we can dine in peace while thousands of children are dying of hunger, and sleep peacefully when children are killed in the streets, but also whether we contribute to the masking of the violence. Are we complicit in maintaining the inhuman policies of national and international systems?

Dibromochloropropane

Dibromochloropropane is a highly toxic nematacide used in banana plantations. It was manufactured in the U.S. by Shell Oil (as Nemagon) and by Occidental Chemical (as Anvac). Early laboratory evidence suggested it caused testicular atrophy. It was used widely during the 1960s and 1970s by Costa Rican and Honduran banana growers. In the 1970s, some plant workers at Occidental Chemical Company's facility in California discovered that they were sterile. In 1977, the Environmental Protection Agency banned all food-related use of DBCP in the United States (Audilett 2001).

Even before DBCP was banned in the United States, workers on Honduran banana plantations suspected that this chemical was causing infertility. But the observations of the workers suffering from these problems were discounted as not credible. In other words, the precautionary principle was not applied. It was necessary for data from the First World to demonstrate the impacts. In a 1998 letter, the Standard Fruit Company in Honduras states that it stopped using DBCP in 1980.

According to the World Health Organization, DBCP is absorbed into the body by inhalation, skin contact, and ingestion. It decomposes at high temperature and produces toxic fumes: hydrogen bromide and hydrogen chloride and carbon monoxide. DBCP causes acute effects including skin, eye, and lung irritation. It can affect the lungs, liver, and kidneys. Chronic effects include those to the kidneys and testicles, reproductive toxicity, and possible carcinogenicity.

During the past two decades we have learned through scientific studies and journalistic reviews about the impacts of the use of pesticides by banana companies. There was no response to these studies from the government or worker representatives until 1998, when I was invited by the labor unions of Standard Fruit Company to a conference on the impact of pesticides on the workers. (The day after the conference, for unknown reasons, one of the workers who participated with me was assassinated.)

Early in 2001, the unions of the Standard Fruit Company invited me to represent workers on a medical commission. The goal of the commission was to evaluate damages caused by DBCP. The other representatives on the commission were a company doctor and a government doctor. Before accepting, I set forth the conditions that I could not receive any compensation by the company, the government, or workers and that I wanted to evaluate impacts on both men and women.

There were two controversial points as we began deliberation. The first had to do with the evaluation of infertility in men based on not only a clinical exam but also on hormonal studies and studies on morphology and motility of sperm. The company doctor did not accept evaluating morphology or motility because studies in other countries had focused only on sperm count. The second point of controversy was about the evaluation of impacts on women. The company doctor said that since gynecologists had not reviewed the evidence, he would not offer an opinion. The government doctor said that because impacts in women had been described in terms of nonspecific outcomes in the literature, more research was needed before any conclusions could be made.

Nonetheless, my position as doctor representing the workers was that this was an occupational health problem, that there was sufficient technical evidence of impacts on women as well as men, and that this evidence should not be excluded from the review. The commission met on various occasions with the company, workers, and the government. The more than 1,000 women who worked in the industry were not evaluated for effects of DBCP. Epidemiologic studies on hormonal effects, cancer, malformations, or stress on couples associated with infertility were not conducted. However, according to the California Code of Regulations, both men and women should be evaluated (California Code 2000).

My concerns were sufficient to cause the government and the company to disband the commission. I do not know how they are now evaluating impacts, but I know that women are not being evaluated.

Two decades have now passed and more than 5,000 workers have yet to be compensated for damages. There is an obvious economic advantage to postponing compensation, because hundreds of workers will die in this period or they will be old and unable to claim compensation. The 1998 letter from Standard Fruit explains that although the company began a medical evaluation of exposed workers in 1992, this effort was suspended in 1993 because various workers had sued the company in the United States. However, it claimed that its research found that male reproductive capacity—as measured in sperm counts—was what would be expected in a normal male population. Its research thus showed that there was no infertility caused by DBCP in Honduras.

Mining

The second case I present is the mining industry in Honduras, a country that lives in political and natural uncertainty. The situation would be much better today if the precautionary principle had been applied several centuries ago to the impacts of mining. Mineral toxins from mining spill into the soils and rivers. Tropical storms and hurricanes regularly occur in Honduras, exacerbating this pollution, along with spilling the chemicals used in mining. Mining companies have insurance for their own benefit, but the Honduran rivers and soils are frequently polluted. In this situation of uncertainty, application of the precautionary principle is crucial.

Some of the pollution is more than a century old. Historically, Honduras has been a mining country. The principal mineral resources have been lead, zinc, gold, silver, antimony, mercury, and iron. One of the important centers of the New York and Rosario Company, which operated in the nineteenth century, was San Juancito, where mercury and copper were mined. Though it has been ninety years since the mines have closed there is still a dead area named *las animas* (souls of death) where mercury mined by the company is still found. In this area it is impossible to grow animals or plants. During Hurricane Mitch, the old mining areas were flooded. The floods moved sediments containing mercury and copper, polluting water and soils not only in San Juancito but also in surrounding areas.

Most of the problems are more current. Without any scientific and ethical principles, the State of Honduras authorized the exploration of new mines in the wake of Hurricane Mitch. The mining industry took advantage of this liberality, and now the government has approved licenses for exploration and exploitation on about 35,359 sq km, about one-third of the Honduran territory (ASONOG 2001). Environmental impact assessments are conducted by the mining companies themselves. Civil society participation is negligible. Deforestation, pollution, and heavy-metal contamination are increasing in soils and rivers. For example, arsenic in the principal rivers of Honduras exceeds 50 micrograms per liter. Lead and cadmium levels in some rivers exceed those identified as safe by the World Health Organization (Secretaría de Salud 2001). Companies are subject to requirements on managing the toxic residues they generate, but these are not enforced. Even since Hurricane Mitch, there has been no consideration of the impact of hurricanes and other natural disasters in the design of these projects.

Mining is not even always profitable to the country. Honduras received more income than mining companies did from mining operations before Hurricane Mitch. However, during the year 2001, Honduras' income from mining was zero, while the mining companies obtained approximately $79.4 million (Almendares 2001).

Conclusions

Scientific work usually tries to separate object from subject, quantity from quality, and so forth. But a deeper examination would show that the way science is defined and used depends on the context. It is influenced not only by the philosophical basis, history, and development of the scientific community, but by the full range of social, political, and economic factors that operate in a society.

In order to apply the precautionary principle we need to educate and create the conditions for genuine democratic participation in the decision-making process. From an ethical point of view, however, we should not allow the participation of persons and organizations who have been responsible for human and environmental abuses and corruption. But these groups and individuals are often difficult to identify and track down, or they are protected by laws and sophisticated technology.

The precautionary principle should be applied in our societies with a broader interdisciplinary and historical scope. More science is required that is based on social, historical, political, and ethical aspects, instead of knowledge that concentrates only on empirical facts and inductive methods.

The best way to begin this process in less industrialized countries is for scientists to begin cooperating with community members and civil society in evaluating the ecological impacts of development projects as well as in developing legislation and polices on health, environment, and human rights issues. Even in a country like Honduras, alliances between scientists and communities are possible. We have created such an alliance in the Mother Earth Movement (Movimiento Madre Tierra). Forward-looking policies in other countries can help move our policies in Honduras along.

In the end, the precautionary principle has practical, ethical, and solidarity value in less developed countries. It can help to create a more transparent and democratic process in these societies as a whole. Thus, the application of the precautionary principle has relevance not only for science but also for the social context.

References Cited

Almendares, J. 2002. "Situación Ecológica de Honduras." *Semana Científica*. Tegu-cigalpa, Honduras: Universidad Nacional Autónoma de Honduras.

Arroyo, V., and A. Y. Espinoza. 2001. Armas y Violencia Social en Centroamerica. In *El Arsenal Invisible. Armas Livianas y Seguridad Ciudadana en la Postguerra Centroamericana*, 357–64. San Salvador: Fundacion Arias para la Paz y el Progreso Humano.

Associación de Organizmos no Gubernamentales (ASONOG). 2001. *Porque Honduras vale más que oro!* Mega Print Honduras Centroamerica.

Audilett, T. 2001. *International Law Society Newsletter* 1 (September), no. 1.

Becerra, L. 1995. *Evolución Histórica de Honduras.* Honduras: Editorial Batkun.

California Code. 2000. http://www.dir.ca.gov/title8/5212c.html. Subchapter 7. General Industry Safety Orders Group 16. Control of Hazardous Substances Article 110. Regulated Carcinogens Section 5212. 1,2 Dibromo-3-Chloropropane (DBCP), Appendix C California Code of Regulations, Title 8, Section 5212. 1,2 Dibromo-3-Chloropropane (DBCP), Appendix C.

Castro, M., and M. Carranza. 2001. *Maras y Pandillas en Centroamèrica. Volumen I.* Erci, IDESO, IDIES, IUDOP. UCA Managua. Las Maras en Honduras, 221–324.

La Gaceta, Diario Oficial de la República de Honduras. 2000. Poder Legislativo Decreto 207–98. Tegucigalpa, Honduras.

Mattison, D. R., D. R. Plowchalk, M. J. Meadows, A. Z. al-Juburi, J. Gandy, and A. Malek. 1990. "Reproductive Toxicity: Male and Female Reproductive Systems as Targets for Chemical Injuries." *Medical Clinics of North America* 74, no. 2: 391–411.

Raffensperger, C., and J. Tickner, eds. 1999. *Protecting Public Health and the Environment: Implementing the Precautionary Principle.* Washington, D.C.: Island Press.

Secretaría de Salud de Honduras. 2001. *Informe de avance sobre calidad del agua por metales pesados en el sur de Honduras.* Tegucigalpa, Honduras.

Silva, C. I. 2001 ¿Hasta cuándo? Ejecuciones extrajudiciales en Honduras 1998–2000. Costa Rica: Color Tech S.A.

United Nations Development Program (UNDP). 1999. *Informe sobre desarrollo humano.* Honduras.

———. 2000. *Honduras: Informe sobre el desarrollo humano.*

PART III

The Implications of Uncertainty on Science and Public Policy: Case Studies from the Field

The precautionary principle is about scientific uncertainty and the appropriate actions to take to reduce that uncertainty and protect health. Uncertainty is an inevitable condition underscoring almost all environmental and health decision making, but because it complicates decision making, it is often played down or ignored (see chapter 13). Uncertainty is inevitable because humans operate in complex, unpredictable, and uncertain systems that are difficult to control and may produce consequences that are unpredictable, irreversible, and very costly. Uncertainties are also introduced into analyses as a result of our choices of assumptions, our disciplinary perspectives and backgrounds, and the questions asked or not asked in an analysis.

Uncertainties make it possible for both proponents and opponents of regulation to interpret the scientific basis for regulations in ways that advance their particular policy objectives. They are often viewed as temporary gaps in information, which can be controlled, reduced, or eliminated through more research. Under this view, precaution is often viewed as an interim measure until more scientific data is accumulated.

Because of uncertainties, environmental decision making can hardly ever be based purely on "sound" science, probabilities, and certainty. The case studies in this section analyze various types of uncertainties involved in characterizing hazards to health and the environment. These include simple parameter uncertainties; uncertainties in models used to bridge gaps in knowledge; broader, "systemic" uncertainties; and what might be termed "political uncertainties," where risks are not studied, where uncertainties are ignored, or where uncertainties are created to make it more difficult to reach causal judgments and policy action.

Several chapters examine two types of more profound uncertainties—

indeterminacy and ignorance—that are generally not considered in decision making. The condition of indeterminacy reflects the lack of direct causal linkages in open-ended systems with multiple influences. Ignorance is the state of not knowing what we do not know.

The case studies in this section provide an in-depth analysis of uncertainties associated with a number of ecosystem and health risks. They examine three particular issues: the types of uncertainties involved in studying complex risks, how uncertainties have been addressed in science and policy, and the influences of uncertainties on public policy. They examine the question of whether the ways in which science is currently conducted for policy limit our ability to address the problem of uncertainty and the need to make decisions. They provide recommendations on how uncertainties can more effectively be characterized, communicated, and addressed in science and science policy.

In chapter 6, in which Boyce Thorne-Miller examines the impacts of human activities on fisheries, it is noted that while there is always uncertainty in examining marine ecosystems, fisheries managers have focused most of their attention primarily on establishing quantitative "conservation goals." She argues that this approach neglects systemic uncertainties and indeterminacy in understanding the interconnectedness of the marine environment as well as the human pressures on marine systems.

Finn Bro-Rasmussen examines the types of uncertainties in examining chemical risks, including trivial scientific uncertainty, model or extrapolation uncertainty, ignorance, and indeterminacy of complex systems. In chapter 7, he outlines several ways in which they can be more effectively addressed in science and policy.

Using endocrine-disrupting chemicals as a case study in chapter 8, Joe Thornton examines the myriad uncertainties involved in understanding the impacts of toxic substances on human and ecological health. He argues that due to these uncertainties and the technical and resource limitations in studying chemical risks, a new "ecological" approach to chemical assessment and management is needed.

In chapter 9, Alistair Woodward examines major sources of uncertainty concerning links between climate change and mosquito-borne disease as well as the reasons behind divergent scientific opinions regarding the nature these links. He finds that arguments about the quality of the science will not resolve differences if the explanation lies fundamentally in the way the problem has been conceived.

Donald A. Brown notes that enough is understood about the risk of global warming to trigger an ethical responsibility to act, despite uncertainty about impacts. In chapter 10, he notes that in a more holistic approach to uncertainty, scientists should make clear the importance of risks they are unable to quantify, yet recognize as plausible.

Katherine Barrett and Stuart Lee examine in chapter 11 the dynamic and *creative* tension between claims of certainty and uncertainty with regard to approval processes for genetically modified organisms in Canada. They argue that the recognition of certainty and uncertainty is significantly shaped by the institutional, political, and scientific context.

Reinmar Seidler and Kamaljit Bawa examine in chapter 12 uncertainties involved in protecting biodiversity, including the many unanswered questions about ecosystem function. They note that biodiversity protection belongs to a category of problems that have at their heart a level of scientific uncertainty that is effectively irreducible, and they examine approaches conservation biologists can adopt to conserve biodiversity and effectively guide policy within this context.

CHAPTER 6

Fisheries and the Precautionary Principle

Boyce Thorne-Miller

In 1995, the Food and Agriculture Organization of the United Nations (FAO) adopted a Code of Conduct for Responsible Fisheries that recognized the precautionary approach as one of its guiding principles. In 1996, the Magnuson Act regulating marine fisheries was reauthorized and amended by the Magnuson-Stevens Fisheries Conservation Management Act, which adopted the precautionary approach in spirit if not in name. Leading up to and following those actions, fisheries scientists have developed detailed guidelines for implementing the precautionary approach and have worked on numerous fisheries models and decision frameworks in that context.

Yet these efforts have been based on poor assumptions, narrow views, and an "obsession with quantitative methods" (Rose 1997) that have prevailed for years in fisheries science. What is happening in marine fisheries illustrates the pitfalls of a limited precautionary approach, one that interprets precaution simply as a more cautious application of risk assessment. This narrow view allows managers to continue to look at small, fragmented pictures, such as the protection of particular regional fisheries or the status of a single species or population, without considering the larger picture: the ultimate well-being of marine ecosystems and also of the people who depend on them for their livelihoods. By separating issues that are interconnected, it is possible to avoid confronting some of the most fundamental causes of the problems.

The only hope for long-term sustainable fisheries is a much broader precautionary approach that incorporates ecological, social, and economic sustainability, including intergenerational equity. This would involve shifting marine fishing from its current industrial status to an activity that is in harmony with natural ecosystems and economies. Fish should not be viewed simply as commodities, of value only when dead. They are critical components of living systems. It makes little sense to treat fisheries as profit centers in an industrial economy, when maximizing profits spells doom for the fishery and the ecosystem (and those who depend on it for their livelihoods).

"Precautionary" Fisheries Management Today

International fisheries bodies and treaty organizations pride themselves on having taken the initiative to adopt, define, and implement the precautionary approach. Even U.S. law, the Magnuson-Stevens Fishery Conservation and Management Act, and regulations, while carefully avoiding mention of the word *precautionary*, have more or less followed the lead of the FAO in seeking sustainable fisheries (Darcy and Matlock 1999; FAO 1995a,b; Restrepo et al. 1998). This all seems quite promising until one closely examines the various recommendations for implementing the precautionary principle in the context of fisheries management. While they certainly represent improvements in the system, they do not address fundamental causes and broad impacts of fish depletion.

The ability to look at the larger ecosystem picture and address fisheries on appropriately large temporal and spatial scales has been hampered by institutional traditions and jargon that perpetuate unsustainable patterns of fishing and fisheries management. A number of terms are common language of fisheries management, but the terms themselves have perpetuated nonprecautionary thinking. Nevertheless, they continue to be used and play an important role in decisions. Here are some of the most important:

Maximum sustainable yield: Maximum sustainable yield (MSY) is "the largest long-term average yield (catch) that can be taken from a stock (or stock complex) under prevailing ecological and environmental conditions" (Restrepo et al. 1998). The idea is to take as many fish as possible without hampering a species' ability to reproduce. MSY is typically estimated by risk assessment techniques at about 80 percent of the population biomass of reproductively mature fish (Kirkwood and Smith 1995).

Optimum yield: Not to exceed MSY, optimum yield (OY) is "the amount of fish that will provide the greatest overall benefit to the Nation, particularly with respect to food production and recreational opportunities and taking into account the protection of marine ecosystems" (Restrepo et al. 1998). OY is generally less than MSY but may be well above 50 percent of the reproductive biomass (Kirkwood and Smith 1995).

Yield or harvest, stock, and underutilized stocks: These terms and others perpetu-

ate the attitude that wild fish are primarily crops and exist for the sole purpose of being captured for human benefit (NMFS 1999; Safina 1995).

Fisheries production: This implies that the catch is actually produced by the fishers and fuels the idea that fisheries are industries. The FAO, which is responsible for monitoring fisheries worldwide, reports its catch data as "fisheries
production" (FAO 1999). The data are based on counts of dead fish (i.e., the
catch of targeted fish species) rather than the populations of live fish.

By-catch: This term includes all fish, invertebrates, birds, mammals, and rep-
tiles that are caught in a fishery that is targeting some other species. It may
also include juveniles of the targeted species. The implication is that these
are useless animals and with a few exceptions, they are neither counted nor
directly regulated but are estimated at 25 percent of the global catch (Safina
1997).

The widely accepted template for implementation of the precautionary
approach may be summarized as follows:

Quantitive conservation goals: Managers establish quantitative conservation goals,
which are estimated reproductive populations needed for the sustained production of the MSY of fishery stocks. It aims to conserve the fishery. This is
a goal for attaining MSY, not for the maintenance of species' functions in
their ecosystem nor the conservation of marine ecosystems.

Target reference points: Quantitative target reference points are determined by
applying risk assessment that somehow takes uncertainty into account. Target reference points refer to the maximum number or biomass of particular
fish species that can be taken in a given season to achieve the desirable outcome (OY or MSY)—usually determined by risk assessment.

Acceptable risk: The risk of not achieving conservation goals is calculated, and
target reference points are chosen that are consistent with acceptable risk.
Acceptable risk, in fisheries, is the chance managers are willing to take of not
attaining sustainability, and thus possibly eventually driving a species to
extinction or commercial extinction (below commercially exploitable levels).

The types of uncertainty usually recognized include measurement errors,
natural population variability, model errors (errors in assumed values), estimation errors in reference point values, and inability to achieve targets (Rosenberg and Restrepo 1994). While these are anticipated sources of uncertainty,
the uncertainties associated with unexpected sources—for example, unanticipated ecosystem or global changes—are not easily taken into account. It is
widely accepted that uncertainty should be a basis for precaution, and the
amount of caution in establishing targets should be proportional to the uncertainty, but guidelines for implementing the precautionary approach in fisheries

seem to avoid explaining just how uncertainty should be estimated and quantified (FAO 1995b; Restrepo et al. 1998), although one approach might be to assign numerical factors to qualitative data (Restrepo and Powers 1999). Presumably, the greater the uncertainty, the lower (or more cautious) the targets (Garcia 1995; Restrepo et al. 1999).

Risk assessment carries with it an aura of quantifiability, even though both MSY and uncertainty are estimates generally based as much on qualitative as quantitative information. The mistake is not the use of qualitative information but the pretense of quantitative precision in an enormously complex and poorly understood system. Scientific validity need not rest solely on quantification (Rose 1997). The challenge of the precautionary approach is to incorporate all types of sound scientific information toward a better understanding of the options, in this case for managing fisheries in complex ecosystems.

Precaution has been typically triggered at the stage of deciding regulatory actions appropriate for specific fisheries as they exist, that is, traditional, single-species fisheries. In this way, it is prevented from playing an overarching role in guiding broader fishing patterns over time on local, regional, and global scales. Consequently, numerous alternatives for redesigning fisheries are never considered. The precautionary principle has been seen primarily as a mandate to change regulatory decisions without changing fundamental goals.

Why This View of Precaution Is Not Big Enough

As a result of this limited risk-focused approach, managers frequently end up arguing over the microregulation of catch levels for maintaining fish populations that are already in unstable states of decline, even though—because of increased fishing effort and gear "improvements"—the catch may be near the estimated MSY. Graphs of FAO data depict the production of world fisheries (i.e., quantities of dead fish) as rising over the period between 1950 and 1990, then leveling off, and for some species, dropping slightly over the following decade (FAO 1999). Fisheries science assumes that the catch data reflect fish availability, or living fish populations, but that relationship is clouded by changes in gear efficiency and misreporting of catch data. Graphs depicting reliable estimates of populations of *living* fish would produce a much different visual image showing that fish populations have steadily declined over the same period, in large part because of the effects of fishing. In addition, the FAO estimates of the status of fish and fisheries have recently been challenged by the discovery that China, with the world's largest fishery, has apparently been systematically overreporting its catches to FAO by as much as two times the actual catch (Watson and Pauly 2001).

Even when the precautionary principle is applied, fisheries management typically persists in single-species approaches where targeted species are assessed and regulated independent of other species in the ecosystem (Charles 2001). It fails to address the more complex problem of species interactions and the impacts of the removal of individual and multiple species on overall ecosystem integrity (indicated by biological diversity and species interactions or functions).Yet these considerations are fundamental to maintaining sustainable fisheries over many generations of fish, people, and ecosystems.To be sure, there is an increased awareness and concern about multiple species catch and by-catch and the importance of maintaining essential fish habitat, but research and regulation have generally lagged far behind.There seems to be an increasing interest in developing multiple-species models to study the effects of removing significant portions of target and nontarget species from their ecosystems (Charles 2001), and a new legal focus on essential fish habitat in the United States has stimulated ecological fisheries research (Fluharty 2000).

On the regulatory side, attention to by-catch and habitat destruction tends to focus on gear problems, and managers are still grappling with how to regulate such destructive technologies as long lines, giant trawling nets, and certain gill-net practices. Giant drift nets were so offensive that they were banned by the United Nations in 1989 (UN Resolution 44/225). Other examples of gear regulations to reduce by-catch include turtle exclusion devices mandated for the U.S. shrimp fishery and required "dolphin-safe" tuna fishing gear and practices. But on the whole, by-catch and physical habitat destruction remain unsolved problems that only increase with the growth of industrialized fishing (Dayton et al. 1995; FAO 1999; Safina 1997).

A new management option that is being heavily promoted by ecologists is the establishment of marine protected areas that are no-take zones for fisheries (Dayton et al. 1995; Lauk et al. 1998; Pauly et al. 1998). In this case, the goal of restoring a particular fishery is used to identify areas where fishing should be prohibited, thus providing an area for unhampered reproduction and development and allowing the population to restore itself both in numbers and age structure. It is thought that the healthy population within the protected area will spread out and thus enhance the stock available for fisheries in surrounding areas. Even though each protected area would be selected on the basis of its value as habitat for one or perhaps a few fishery target species, it would benefit the complement of species that share the habitat.

Experience has shown that a number of pleasant surprises arise when a marine habitat is protected—unexpected species find niches in the community so the species diversity generally increases, and the interactions among species often become quite different than they were when the ecosystem was stressed by fishing. Recent research based on historical and paleoecological records suggests that overfishing has been the primary cause of degradation of marine ecosystems and that other common stresses, such as pollution, habitat alteration, introduced species, and climate change, have had the most significant impacts on systems after they are weakened by overfishing (Jackson et al. 2001). Therefore, a system of areas protected from fishing should have a significant positive effect over the long term and is consistent with the goals and scope of precautionary action.

The precautionary approach is typically applied to specific decisions within the existing regulatory framework: Should there be some limitation placed on the catch of this or that particular species? If so, should it be accomplished by individual catch limits, by the number of boats allowed to fish, by the length of time each year that fishing is allowed, and so forth (applied to fishing of one or another particular population of a species)? Should a species fishery be closed down temporarily to allow some recovery of the stock? Should there be size or design restrictions on the gear used in a particular fishery? Confining the precautionary approach to these one-dimensional questions limits information and choices and leads to piecemeal decisions. Some of these may be effective in stopping or reversing the decline of individual fished populations in the short term, but a more comprehensive approach is needed to reverse the decline of marine ecosystems over the long term.

Ultimately, we place the implementation of the precautionary principle in the hands of our economic, political, regulatory, and educational institutions. Fisheries are regulated by institutions that are informed by fisheries science and directed by international treaties and national laws and policies. Both science and our legal and regulatory structures tend to fragment the issues into isolated entities that are addressed separately (Thorne-Miller 1999). Thereby, they avoid considering the critical interrelations among a variety of factors affecting fisheries, including coastal development, property rights, climate change, international trade policies, and market competition, as well as fisheries regulation itself. Extensive institutional reform is essential for the successful implementation of a truly precautionary approach to fisheries management, and the public and fishers must be major participants in the reformation. A scientific reformation is also needed, and that is beginning to happen as ecologists, who traditionally have not been involved in fishery biology, insist that they have important knowledge and ideas to contribute to the fisheries debate (Dayton 1998; Dayton et al. 1995; Fluharty 2000; Hammel et al. 1993; Lauk et al. 1998; Ludwig et al. 1993; Pauly et al. 1998). What is ultimately needed is a broader application of the precautionary approach that considers the structure and functions of natural ecosystems, the long-term well-being of fishers and their communities, and the interplay between human and natural economies.

Ecologically Precautionary Fisheries

Some experts believe that it is impossible to hunt wild populations at commercial levels sustainably, especially to satisfy market demands of such large human populations as we have now (Earle 1995), and they may be right. Nevertheless, we are so far down the commercial fisheries road that management seems to be a more realistic option than termination of the activity. Without management, exploited populations are fished as intensely as possible until they are depleted to commercially extinct levels. Other species are then fished instead, or the depleted population recovers, only to be fished down once again (Earle 1995).

There have also been more serious attempts on the part of the industry to find new fish to exploit. The worst possible outcome is illustrated by the rapid development and decline of the orange roughy fishery. In recent years it was discovered that seamounts are natural harbors for dense congregations of large demersal (bottom-dwelling) fish that take advantage of the food that falls out of the water when deep-water currents are interrupted by the protrusions from the ocean floor (Koslow 1997). Orange roughy and Patagonia tooth fish (a.k.a. Chilean sea bass) are two such fish that quickly became popular in restaurants and grocery stores as consumers were trained to want them. But these deep-water fish grow very slowly and may not begin reproducing before they are one or more decades old. Because of their lengthy reproductive cycle that precludes rapid replenishment, populations crashed within a few years of the opening of the fisheries.

Fisheries, especially industrialized fisheries, tend to be conducted as if they are single-species operations (even though in reality many species may be captured, only one is the focus of the effort). Globally, just under two hundred fish stocks supply 90 percent of the world's reported fish catch (Earle 1995), but a mere six species account for about 25 percent of the catch (FAO 1999). Consequently, there has been little effort to move away from the single-species approach to fisheries management. In the context of ecological sustainability, there has been movement from an exclusive focus on short-term stock assessments to a larger concern about ecosystem health (Charles 2001). Fisheries management now recognizes three objectives of biological sustainability:

- a desire to avoid depletion of stocks in the short term;
- a desire to maintain populations of target fish and the species upon which they depend, so that the harvest can continue into the future for successive generations of fishers; and
- a basic need to maintain or restore the overall health of the fishery ecosystem.

Historically, the first of these has been the primary focus of fisheries management and remains a priority today. As concerns about overfishing and sustainability have grown, the second has sometimes been added to decision making, and the third has become part of the research agenda and is the basis of a new interest in marine protected areas or no-take zones (Charles 2001).

The new attention to essential fish habitats and no-take zones offers hope of more precautionary approaches to fisheries management, but even that approach may have a weakness. Even the most enlightened discussions of marine protected areas focus on habitat for select fisheries species, so that the ecosystem is to be managed for the benefit of the fishery, rather than the fishery being managed for the sustainability of the ecosystem. It may seem like a subtle difference, but it could be critical to long-term success of fishery management plans.

To achieve ecological sustainability in fisheries and a precautionary approach to fisheries management, the ecosystem should be at the center of decision making. Simply reversing the priority of the three objectives listed above could dramatically alter management priorities and establish goals consistent with a precautionary approach:

- *Maintain the ecosystem and design fisheries to accomplish that.* Placing the ecosystem first might change what we fish, when we fish, how we fish, and how much we take. The reward is that an ecosystem thus cared for will continue to be able to provide the gift of food: "We must take care of systems that take care of us, whether we are required to do so or not" (Earle 1995). This will require a new emphasis on marine ecology with fisheries as a subset of the research agenda. Establishing extensive marine protected areas would serve the dual function of providing havens for fish recovery and important areas for research on less stressed ecosystems. The question is, how large must these protected areas be to be effective enough to result in the recovery of fish populations regionally? Ecological research would also inform decisions about where the protected areas should be located for optimum effectiveness.

- *Maintain populations of fish and the species they depend on.* If fishers and fisheries managers are taking care of the ecosystem first, they will be sensitive to the species diversity and relative abundance that characterize a healthy ecosystem. Conservation goals should be defined in terms of the population essential to maintain a species' function in the ecosystem rather than to maximize the fisheries catch. Management decisions made on the basis of available information should be flexible enough to allow adjustments to avoid or correct large deviations from the conservation goals. Such fisheries management might involve new strategies, such as:
    - changing the fish targeted or areas fished from one year to the next—a fishing rotation;
    - using fishing techniques—sometimes old-fashioned ones—that avoid by-catch of juvenile fish or unwanted species; and
    - regulating, counting, and utilizing more or less everything caught.

- *Avoid depleting stocks.* Management adopted as a result of the first and second consideration should automatically avoid such depletions. However, there will always be uncertainty associated with management decisions, and unexpected events (human caused or natural) may change the dynamics of the ecosystem and its species. Fishers and managers must be alert to population fluctuations, and adaptive management plans should provide for adjustments in fishing patterns as soon as unanticipated declines are noticed.

One important criterion for ecosystem-based management is to recognize fishers as part of the ecosystem. Ecological research has studied and defined ecosystems as they are without humans, whereas fisheries science has defined fish and their habitat in terms of humans. The two sciences should be blended into the study of ecosystems that include fisheries. The science should then inform management decisions with the objective of maintaining a healthy, naturally functioning ecosystem that successfully supports its characteristic biodiversity, including the human component. It does seem unlikely, however, that an ecosystem perspective would lead to the conclusion that fishers can gobble up 80 percent of a reproductive population (a level of fishing that has typically been recommended by fisheries science in the past). Such levels of fishing on a species with a critical ecosystem role is bound to impoverish the ecosystem and, consequently, the fishery.

The greatest scientific challenge is to understand ecosystems well enough to know how to manage fisheries responsibly. In marine ecosystems, major obstacles may be encountered in gathering accurate information and understanding possible sources of uncertainty (Thorne-Miller 1999). *The fluid matrix of the ecosystem makes it unusually difficult to predict and susceptible to surprises. The opaque veil that hides the ecosystems from human view exacerbates the difficulties of assessing populations and communities of animals in motion.* Generally, there is not much historical data for a marine ecosystem being studied, so it is difficult to know its status with respect to potential biodiversity and natural functions. Given the difficulty of monitoring fish populations over their full ranges and the poor understanding of their recovery dynamics, it is always difficult to estimate fishery impacts. However, fisheries science could do a much better job of using the information that is available.

Given the high levels of uncertainty, it is often feared that the precautionary principle will only lead to closures (possibly permanent) of many fisheries. As a result, policies regarding precaution tend to perpetuate traditional risk-based fisheries' management decision-making processes but season them with extra caution (FAO 1995b; Restrepo et al. 1998).

Adaptive management is an approach that allows a more complex goal, a wider range of alternatives and innovation, and the flexibility to adjust decided actions if it is determined they are not working as predicted. It offers a precautionary solution to the dilemma of permitting an activity to occur under high levels of uncertainty. It is not a new concept, but, with a few exceptions, it has yet to be fully and appropriately implemented in the context of fisheries, and the fishing and management communities have not commonly been willing to be so flexible (Restrepo et al. 1999). The rapid response must work to close, open, or shift the fishing location; to change the species fished and their quotas; or to respond in other appropriate ways to monitoring. With adaptive management, precautionary decisions can be made regarding a fishery but then can be responsively modified as more scientific research and monitoring is done. Science and management should be continually engaged in a dynamic, adaptive interchange of information and ideas. One example of the adaptive management approach is the Alaska salmon fishery (see Tripp 1998; Marine Stewardship Council).

Sociologically Precautionary Fisheries

The effectiveness of fishery regulations and the willingness of the fishers and managers to be more flexible and responsive to the needs of the ecosystem that feeds them are ultimately related to socioeconomic and political factors in fishing communities. Often, short-term economic factors dominate the decision-making process. It is not sufficient to talk only about ecological sustainability of fisheries, because that is an unlikely accomplishment if sociological sustainability is not also addressed. Consistent with the need for institutional reform mentioned earlier, this may require changes in governmental and community institutions to address the "concomitant weakness of existing social institutions for handling relations between society and nature" (Gandy 1999).

It is too late to hope for the ideal action before damage is done with respect to marine fisheries. Precaution in this case is more likely to prevent further harm and sometimes to correct past harm and, thereby, help provide for the integrity of ecosystems so they can support future generations of people—on a limited scale.

Science can provide critical information about the status of ecosystems and can guide decisions regarding what actions would be most effective in protecting or restoring them. However, the decision to take precautionary action at all is a social one, and there are socioeconomic implications of the actions that are taken. In the case of fisheries, attempts to restrict fishing in one way or another often meet resistance from fishers on the grounds that their livelihoods are being compromised (Charles 1998). Fishers are suspicious that the precautionary approach is another way of reducing their income and making their lives miserable. However, when nonprecautionary decisions lead to fisheries collapse, the socioeconomic impacts are even greater. Both the fishers and their communities are affected, and there are impacts on markets and consumers outside the local community.

There seems to be a rising awareness of the importance of the social component of the fishery picture. Development of an enhanced sense of community and sense of place among fishers may hold the key to successful management (Charles 2001; Pauly 1997). Community management offers important precautionary options, at least for coastal fisheries (Charles 2001; Hilborn et al. 2001). Community-based management may originate with the fishers themselves, or it may involve cooperation among the fishers in response to governmental restrictions on total catch. In some cases it might be preferable to involve the whole community, so that fishers may productively (without hardship to themselves and with benefit to the community) shift between fishing and other occupations depending on the fluctuations in the condition of the populations of fished species. Hilborn et al. (2001) suggest that the precautionary approach might be applied to reduce risk to fishing communities through "portfolio management," in which fishers may choose each year from a diverse portfolio of harvestable resources. A certain level of responsibility for supporting fishers to manage their fisheries sustainably might also appropriately be assigned to the public, who benefit from the fisheries. The case of shrimp fishers of Suruga Bay in Japan, in which the fishers voluntarily restrict their fishing and share the profits evenly each year, demonstrates that this approach can work (Omori 2002).

Charles (2001) suggests that community-based management is most practical for fisheries that are small scale, backed by strong tradition, and characterized by strong connections between the fishers and their community, and that it may be more difficult to achieve sustainability in offshore and distant water fisheries. One approach that some consider precautionary is the use of individual transferable quotas (limited number of permits for individual boats to fish a set quota each year), which is the system established to rescue the Alaskan halibut fishery (Heinz Center 2000)

Pauly (1997) argues, however, that this privatization of fishing rights does not address the disconnection between human and natural time scales, and therefore it will fail to protect fish populations and marine ecosystems. He suggests that only a system of protected reserves that fishers themselves understand and willingly honor will adequately protect many fished species—especially those with long lives and slow reproduction. Consistent with the community-based approach, this approach provides a sense of place and pride that favors long-term outlooks as fishers "become the local guardians of the resources and not their roving executioners" (Pauly 1997). The challenge is to establish a greater connection among fishers, the open ocean ecosystems, and the global citizenry.

Economically Precautionary Fisheries

Economic considerations, especially short-term economic goals, often conflict with efforts to achieve ecological sustainability, and these differences become a source of social conflict. This is the result of a misdirected economic value system. Their industrial focus means that commercial fisheries are governed by an artificial industrial economy that is inappropriate for natural resources that are integral parts of living ecosystems. Economic sustainability not only requires major changes in the fishery but also in the economic system applied to fisheries.

Lichatowich (1999) has suggested a change of viewpoint, which would shift from an industrial system that is linear and extractive and emphasizes production to a more natural system that is circular and renewable and focuses on reproduction. In the case of fisheries, the shift would be from an industry that emphasizes the "production" of fish for market to an activity that provides food for people while allowing the maintenance and reproduction of healthy wild populations and ecosystems. It is hard to imagine that this could come about without partly or wholly moving fishing from private to public realms.

Table 6.1 illustrates some of the specific differences between the two economies. The character of the existing linear, profit-driven, industrial fishing economy is out of step with natural cycles and natural ecosystems. Regulatory boundaries and patterns of fishing fragment cohesive ecosystems, and as a result, natural interdependencies among different species and between species and their habitats are interrupted. Ecosystems are simplified and become unstable because of selective and excessive extraction of some species and of habitat degradation caused by destructive gear and by-catch.

The regulatory and operative isolation of industrial fisheries from other activities prevents a holistic approach to food production and jeopardizes food security for future generations. It also prevents holistic ecosystem management that would deal with linkages between declining fish populations and nonfishing activities.

The character of a natural fishing economy, examples of which are almost entirely limited to precolonial aboriginal fishing communities, favors long-term sustainability of fisheries and ecosystems. It offers a precautionary approach to designing management schemes.

- Management boundaries should coincide with natural ecosystem boundaries, so management regimes could take into account the complex interplay between species and habitats. The importance of nontarget species and the critical interactions among different species should be recognized and maintained.
- Fishers, as well as fish, should be recognized as integral parts of the fishery ecosystem.

Table 6.1. Fishing Economies.

Industrial Economy	Natural Economy
Landscape is organized into political hierarchies with artificial boundaries (national exclusive economic zone [EEZ], regional regulatory zones, states, private property) that result in conflict over ownership, use, and distribution of resources.	Landscape is organized into ecosystems that function as an integrated whole (e.g., watersheds, estuaries, wetlands, enclosed seas, coastal shelf habitats, deep sea mounts).
Political boundaries fragment natural ecosystems and may interrupt interdependencies through inconsistent management. Fragmentation also occurs as a result of fishing practices that destroy habitat, and overfishing interrupts interdependencies.	Species co-evolve with one another and with their natural habitat(s), and sustainability relies upon maintaining interdependencies.
Ecosystems are restructured, simplified, and controlled to meet human needs. Overfishing by-catch, stock enhancement, and predator control cause less stable ecosystems.	Ecosystems support a complex diversity of life that has adapted and co-evolved to survive together. Humans may be a part of natural ecosystems if they fish to maintain diversity and complexity as successful predators do.
Infrastructure is visible and discrete, with functions generally understood by the public (e.g., fishing boats and gear, processing plants, aquaculture structures). Management and public and private funding tend to focus on maintaining this infrastructure.	Infrastructure is not readily visible and is poorly understood by the public and incompletely understood by experts. Public education and precautionary management that takes uncertainty into account would help address the need to maintain this infrastructure.
Large centralized and specialized production facilities in aquaculture (including hatcheries) lead to monocultures and mass production of single products. "Enhancement" and introduced species for fisheries purposes reduce genetic diversity and may lead to unbalanced competition, which reduces species diversity.	Production (i.e., reproduction) is dispersed among many small units (species and populations) that are interspersed among one another, which favors diversity. Fisheries in the context of a natural economy would be sensitive to the need to maintain diversity.
In constructed biological systems (e.g., fish farms), productivity is maintained artificially by external processes (fertilizers, pesticides, antibiotics, bred stock, biotechnology, predator control). Productivity of wild fisheries is artificially affected by inputs from unrelated land-based activities.	Productivity is maintained by processes internal to the ecosystem, including cycling of nutrients through the biosphere, genetic diversity, and natural selection. A more holistic view of the impact of all human activities on marine ecosystems would help fisheries.
Economy is linear and extractive, emphasizing production.	Economy is circular and renewable, encouraging reproduction.
Economy is driven by fossil fuel and the need to accumulate capital.	Economy is driven by solar energy and the need to reproduce.
Waste is created by failing to fully use fished resources. Other waste is generated in the process of operating and maintaining boats, gear, and related facilities.	No waste is produced; everything is recycled.
Ecosystems are partitioned into economic sectors that are operated and regulated independently: e.g., shipping.	Ecosystems are coherent and complex mazes of interconnected habitats.

Source: Adapted from Lichatowich, 1999, p. 48.

- Fisheries management should incorporate public education about the fish, their habitats, and the interactions of a variety of human activities and marine ecosystem functions.
- Gear should be designed to be selective and nondestructive of habitats.
- Fishing or other activities should be designed and regulated so as not to harmfully alter species diversity, nutrient cycling, genetic diversity, and food webs.
- Fishing and fish processing should minimize waste by minimizing by-catch and by recycling.

While it is impossible for fisheries to have no impact, it is possible, with a great deal of effort, to "match our numbers and our demands with what our planet can provide" (Pauly 1997).

The Bigger Picture

Although fishing has gone on for the duration of humankind, and overfishing has been happening for centuries, the global scale of overfishing that has brought vast areas of the marine ecosystem to destruction during the past two or three decades demands an entirely new approach. If we want to be precautionary, we cannot merely tweak the existing management system that has failed.

A truly precautionary approach must address whole ecosystems, but it also cannot ignore sociological issues and the difficulty of establishing long-term sustainability when short-term socioeconomic impacts are inevitable. The benefits to fish and ecosystems of a precautionary approach to fisheries management are obvious and have been discussed thoroughly among policy makers and managers (FAO 1995b; Lauk et al. 1998; Myers 1993; Myers and Mertz 1998; Restrepo et al. 1998). But the numerous benefits to the fishers and their communities have not been given adequate attention, and the public has not been adequately brought into the discussion.

Furthermore, it is not enough to address fishing without looking at the entire food production system (Williams 1996)—agriculture, aquaculture, and fishing together are called upon to feed excessive and growing populations of humans. In an age of global trade and human-dominated global ecosystems, we have to start with new goals, new options, and new questions to lead us to effective management and monitoring of our food production system. It is essential that the populace become engaged in the discussion.

We have both the privilege and the responsibility to have such holistic discussions and to implement responsible decisions regarding the extraction, production, and trade of living resources, including wild sea life.

References Cited

Charles, A. T. 1998. "Living with Uncertainty in Fisheries: Analytical Methods, Management, Priorities, and the Canadian Groundfishery Experience." *Fisheries Research* 37: 37–50.

————. 2001. *Sustainable Fishery Systems.* London: Blackwell Science.

Darcy, G. H., and G. C. Matlock. 1999. "Application of the Precautionary Approach in the National Standard Guidelines for Conservation and Management of Fisheries in the United States." *ICES Journal of Marine Science* 56: 853–59.

Dayton, P. K. 1998. "Reversal of the Burden of Proof in Fisheries Management." *Science* 279: 821–22.

Dayton, P. K., S. F. Thrush, M. T. Agardy, and R. J. Hofman. 1995. "Environmental Effects of Marine Fishing." *Aquatic Conservation: Marine and Freshwater Ecosystem* 5: 205–32.

Earle, S. A. 1995. *Sea Change: A Message of the Oceans.* New York: Putnam.

FAO. 1995a. *Code of Conduct for Responsible Fisheries.* Rome: UN Food and Agriculture Organization.

————. 1995b. *Precautionary Approach to Fisheries. Part 1: Guidelines on the Precautionary Approach to Capture Fisheries and Species Introductions.* Elaborated by the Technical Consultation on the Precautionary Approach to Capture Fisheries (Including Species Introductions), Lysekil, Sweden, 6–13 June 1995. FAO Fisheries Technical Paper 350/1.

————. 1999. *The State of World Fisheries and Aquaculture 1998.* Rome: Food and Agriculture Organization of the United Nations.

Fluharty, D. 2000. "Habitat Protection, Ecological Issues, and Implementation of the Sustainable Fisheries Act." *Ecological Applications* 10: 325–37.

Gandy, M. 1999. "Rethinking the Ecological Leviathan: Environmental Regulation in an Age of Risk." *Global Environmental Change* 9: 59–69.

Garcia, S. M. 1995. "The Precautionary Approach to Fisheries and Its Implications for Fishery Research, Technology and Management: An Updated Review." In *Precautionary Approach to Fisheries. Part 2: Scientific Papers,* edited by FAO, 1–63. FAO Fisheries Technical Paper 350/2.

Hammel, M., A. Jansson, and B. Jansson. 1993. "Diversity Change and Sustainability: Implications for Fisheries." *Ambio* 22: 97–101.

H. John Heinz III Center for Science, Economics, and the Environment. 2000. *Fishing Grounds: Defining a New Era for American Fisheries Management.* Washington, D.C.: Island Press.

Hilborn, R., J. Maguire, A. M. Parma, and A. A. Rosenberg. 2001. "The Precautionary Approach and Risk Management: Can They Increase the Probability of Successes in Fishery Management?" *Canadian Journal of Fisheries and Aquatic Sciences* 58: 99–107.

Jackson, J. B. C., M. X. Kirby, W. H. Berger, K. A. Bjorndal, L. W. Botsford, B. J. Bourque, R. H. Bradbury, R. Cooke, J. Erlandson, J. A. Estes, T. P. Hughes, S. Kidwell, C. B. Langue, H. S. Lenihan, J. M. Pandolfi, C. H. Peterson, R. S. Steneck, M. J. Tegner, and R. R. Warner. 2001. "Historical Overfishing and the Recent Collapse of Coastal Ecosystems." *Science* 293: 629–38.

Kirkwood, G. P., and A. D. M. Smith. 1995. "Assessing the Precautionary Nature of Fishery Management Strategies." In *Precautionary Approach to Fisheries. Part 2: Scientific Papers.* Technical Consultation on the Precautionary Approach to Capture Fisheries (Including Species Introductions), edited by FAO, 141–58. Lysekil, Sweden, 6–13 June 1995. FAO Fisheries Technical Paper 350/2.

Koslow, J. A. 1997. "Seamounts and the Ecology of Deep-Sea Fisheries." *American Scientist* 85: 168–76.

Lauk, T., C. W. Clark, M. Mange, and G. R. Munro. 1998. "Implementing the Precautionary Principle in Fisheries Management through Marine Reserves." *Ecological Applications* 8, no. 1 (Supplement): S72–S78.

Lichatowich, J. 1999. *Salmon without Rivers: A History of the Pacific Salmon Crisis.* Washington, D.C.: Island Press.

Ludwig, D., R. Hilborn, and C. Walters. 1993. "Uncertainty, Resource Exploitation, and Conservation: Lessons from History." *Science* 260: 17.

Marine Stewardship Council. *Alaska Salmon FAQ.* http://www.msc.org/html/content_85.htm.

Myers, N. 1993. "Biodiversity and the Precautionary Principle." *Ambio* 22: 74–79.

Myers, R. A., and G. Mertz. 1998. "The Limits of Exploitation: A Precautionary Approach." *Ecological Applications* 8, no. 1 (Supplement): S165–69.

National Marine Fisheries Service (NMFS). 1999. *Our Living Oceans: Report on the Status of U.S. Living Marine Resources, 1999.* U.S. Department of Commerce, NOAA Technical Memo NMFS-F/SPO-41.

Omori, M. 2002. "One Hundred Years of the Sergestid Shrimp Fishing Industry in Suruga Bay: Development of Administration and Social Policy." In *Oceanographic History: The Pacific and Beyond,* edited by K. R. Benson and P. F. Rehbock. Seattle: University of Washington Press.

Pauly, D. 1997. "Putting Fisheries Management Back in Places." *Reviews in Fish Biology and Fisheries* 7: 125–27.

Pauly, D., V. Christensen, J. Dalsgaard, R. Froese, and F. Torres Jr. 1998. "Fishing Down Marine Food Webs." *Science* 279: 860–63.

Restrepo, V. R., P. M. Mace, and F. M. Serchuk. 1999. "The Precautionary Approach: A New Paradigm or Business as Usual?" In *Our Living Oceans: Report on the Status of U.S. Living Marine Resources, 1999,* edited by National Marine Fisheries Service, 61–70. NOAA Technical Memorandum NMFS-F/SPO-41.

Restrepo, V. R., and J. E. Powers. 1999. "Precautionary Control Rules in US Fisheries Management: Specification and Performance." *ICES Journal of Marine Science* 56: 846–52.

Restrepo, V. R., G. G. Thompson, P. M. Mace, W. I. Gabriel, L. L. Low, A. D. MacCall, R. D. Methot, J. E. Powers, B. L. Taylor, P. R. Wade, and J. F. Witzig. 1998.

Technical Guidance on the Use of Precautionary Approaches to Implementing National Standard 1 of the Magnuson-Stevens Fishery Conservation and Management Act. U.S. Department of Commerce, National Oceanic and Atmospheric Administration. Technical Memorandum NMFS-F/SPO-1.

Rose, G. A. 1997. "The Trouble with Fisheries Science." *Reviews in Fish Biology and Fisheries* 7: 365–70.

Rosenberg, A. A., and V. R. Restrepo. 1994. "Uncertainty and Risk Evaluation in Stock Assessment Advice for U.S. Marine Fisheries." *Canadian Journal of Fisheries and Aquatic Science* 51: 2715–20.

Safina, C. 1995. "The World's Imperiled Fish." *Scientific American* 273: 46–53.

————. 1997. *Song for the Blue Ocean: Encounters Along the World's Coasts and Beneath the Seas*. New York: Henry Holt and Co.

Thorne-Miller, B. 1999. *The Living Ocean: Understanding and Protecting Marine Biodiversity*, 2nd ed. Washington, D.C.: Island Press.

Tripp, A., ed. 1998. *Pacific Salmon, Alaska's Story*. Santa Barbara, Calif.: Albion Publishing Group.

Watson, R., and D. Pauly. 2001. "Systematic Distortions in World Fisheries Catch Trends." *Nature* 414: 534–36.

Williams, M. 1996. *The Transition in the Contribution of Living Aquatic Resources to Food Security*. Food, Agriculture, and the Environment Discussion Paper 13. Washington, D.C.: International Food Policy Research Institute.

CHAPTER 7

Risk, Uncertainties, and Precautions in Chemical Legislation

Finn Bro-Rasmussen

The Precautionary Principle in European Union Policies

The precautionary principle was officially adopted as a central tenet of European Union (EU) environmental policies by the EU treaty signed in Maastricht in 1992. The EU Commission, in its communication of 2 February 2000, interpreting the precautionary principle, emphasized that the principle should form part of the structured approach to the analysis of risk, which includes risk assessment, risk management, and communication of risk.

By its inclusion in the Maastricht treaty, the principle was given a rather wide scope, but it did not gain any immediate or judicially specified status. The commission's communication did not close this gap but expressed the intention that guidelines be established for the application of the precautionary principle by authorities and community regulators. In this process, the precautionary principle was to be different from the scientific extrapolation practice and to be used only to cover cases where scientific evidence is insufficient, inconclusive, or uncertain. In the concluding words of the commission, the decision concerning its use or nonuse, therefore, is political (EU Commission 2000a).

Following the communication, it was obvious that legislation on chemical substances, processes, and products would be among the fields first to draw upon the possibilities and to consider the practical application of the officially accepted principle. This is the background for the observations and comments in this chapter.

Making the Precautionary Principle Operational

As the EU precautionary principle concept was being developed, nongovern-mental organizations and occasionally also national and international experts gave voice to the expectation that the precautionary principle should be made operational. Risk analysis of chemicals was developing rapidly. These groups thought it important to introduce the precautionary principle into chemical laws and regulations in a systematic manner and to use it for guidance in data development, evaluation, and application (e.g., Bro-Rasmussen 1999; Copen-hagen Chemicals Charter 2000). Public confidence in safety regulations would likely increase if the precautionary principle were applied to both existing and future regulatory structures.

In this chapter I accept the Commission's conditional assumption that the precautionary principle should be tied to insufficiencies and inconclusive find-ings in the evaluation of chemicals, and should cover uncertainties in the risk assessment process. This makes it imperative, however, to delve into the very nature of uncertainties. It points to the need to interpret the precautionary principle not only to embrace policy but also scientific variability and uncer-tainties, and to bring certain scientific practices under critical scrutiny (see Wynne 1992).

Ideally, therefore, the precautionary principle should be interpreted as a tool, complementary to the risk assessment scheme, transparent and accepted by all stakeholders. This will logically lead to a reconsideration of both princi-ples and precautionary measures as these are presently practiced. At least three areas must be included, namely:

- The choice of regulatory strategy, for example, negative versus positive lists, approval or authorization schemes, and so forth;
- The process and schemes of hazard assessment, as practiced for individual chemicals based on toxicological and ecotoxicological assessments;
- The risk assessment, in which effects/hazard evaluations are balanced with exposure estimates or data.

Regulatory Strategies: The Choice

Negative Lists

Chemicals now circulating in society are generally counted in numbers vary-ing from 50,000 to 100,000. (In the EU, 100,106 chemicals were listed as "existing chemicals" in the EINECS register in September 1981. In the order of 30,000–50,000 are considered as "daily chemicals," marketed in quantities over 1 ton per year. See also Bro-Rasmussen et al. 1996.) Traditionally these have been dealt with individually. The presumption has been that everything is permitted until evidence indicates otherwise and unless restrictions are specifically stated. Primarily from empirical knowledge gained more or less accidentally, but also, gradually, based on experimental testing, chemical lists have established mandating warnings, classification, labeling, and occasionally also restrictive use regulations. These may be called *negative lists*—lists of chemicals that are restricted in some way. All chemicals not on the list are considered unrestricted.

Only about 2,500 to 3,000 chemicals, or 2 to 5 percent of the universe of existing chemicals, are today classified as hazardous and registered in official lists (EU Commission 2000b). These lists are often confused with other "priority" lists, such as the list of about 2,800 high-production-volume chemicals (those produced in more than 1,000 tons per year; see Bro-Rasmussen et al. 1996), or with the much lower numbers of chemicals considered hazards in the environmental domain (EU Commission 1982). For comparison, consider the more carefully regulated domain of new chemicals introduced in the past twenty years. The EU Commission recently reported that 70 percent of all the substances that had been marketed during this period as new chemicals had been registered as hazardous (EU Commission 2001a). This is some indication of the inadequately precautionary practice that prevails in the regulation of existing chemicals.

Approval Schemes: Positive Lists

It was a change and a major precautionary improvement when, in schemes for approval, *positive lists* for pharmaceuticals, pesticides, and food additives were introduced in most industrialized countries during 1955 to 1970. As a new regulatory principle such lists implied that producers or users were only permitted to market and use chemicals that had been tested, evaluated (for human health and environmental effects), and registered for safety and quality standards prior to use. Further, these chemicals would generally only gain approval and be listed if their claimed applications were specifically justified. Under this policy, any chemical not on a list is not permitted. Positive lists have been developed for only a few groups of substances, such as food additives and pesticides.

Positive lists are a developed and well-described strategy for chemical regulation. The lists are efficient, forceful tools for the promotion of more and better data, and they have ensured comprehensive documentation for many hazardous chemicals. Despite many discussions and signs of inadequate application of precaution in the development of these lists, the strategy is widely considered to be an important tool for bringing hazardous chemicals and their practical applications under restrictive and scrutinizing controls.

Since their introduction, therefore, positive lists have reduced both numbers and volumes of many hazardous and controversial chemicals and products. For example, industrial producers have claimed that thousands of chemical substances possess plant-protective properties, whereas only about 1,250 ever gained acceptance as pesticides by crop protection organizations (British Crop Protection Council and Royal Society of Chemistry 1994). And even out of these, only about 100 to 200 active substances have been able to hold their positions as acceptable on official national lists of pesticides in various European countries, for example, in Scandinavian countries (Danish National EPA 2000).

Thus, over the years, positive list regulations have been a tool for reducing the numbers of pesticides as well as for phasing out many of the most potent and hazardous, which has to a certain extent lessened the human and environmental impact that would otherwise have resulted from unregulated and uncontrolled use of these chemicals. A wider application of positive lists for chemicals, therefore, may well be significant and useful as contributions to precautionary chemical legislation. They may be applied to other chemicals of related types or uses as has already been recognized in the EU, where approval schemes have been introduced in directives on biocides (EU Commission 1998) and (partially) on cosmetics (EU 1976).

The rapid increase in allergic suffering frequently tied to extensive use of cosmetics is one indication that many more compounds used in these products should be restricted (WHO and EEA 2002), which could be facilitated by extended application of the positive list as a precautionary tool. This is also the case for many industrial chemicals, such as solvents, paints, varnishes, and flame retardants, to which many people have direct, often continuous exposure. Several of these chemicals are today being linked to increasing concerns about known, uncertain, or ill-described hazards such as carcinogenicity and endocrine disruption (Andersson et al. 2001). Such an "authorization" scheme has been proposed for certain hazardous industrial chemicals in the EU (EU Commission 2001a).

Effects/Hazard Assessment: Processes and Schemes

The development of negative lists is primarily the result of hazard evaluations of individual substances. These are classified from existing data of mostly experimental origin, either in a screening process or resulting from full assessments focusing on intrinsic hazard potential. This eventually leads to establishment of limit values that may function as legal thresholds. In principle such values are the basis for regulation of emissions, intending to balance the levels of exposure against estimated not-harmful levels—but at the same time also to give legal acceptance for such exposures.

Precautionary Examples

Phaseouts and bans of individual chemicals as precautionary measures are relatively rare, but some examples include the banning of asbestos in the late 1990s and the gradual withdrawal of chlorofluorocarbon gases. These might be called "late lessons from early warnings" rather than examples of precautionary actions (see chapter 13) (EEA 2002). More illustrative, therefore, may be the cases that are tied to the withdrawal of individual food additives or pesticide chemicals on positive lists. Perhaps the best-known precautionary policy in this regard is the Delaney Clause of the U.S. Food and Drug Act of 1958, whereby animal carcinogens were banned from processed foods.

This is paralleled by the EU adoption in 1980 of a so-called practical zero tolerance, which is a legal threshold set at 0.1 μg per liter for residues of pesticides in drinking water, corresponding to a detection limit that was estimated to cover all active pesticides (EU 1980). This decision was reiterated as a restrictive (precautionary) protection of drinking water quality by the 1996 Water Quality Directive (EU 1998). By setting this as a "group value," the EU thereby disregarded "safe limits" set by the World Health Organization for individual pesticides (WHO 1992) at levels that were 10 to 1,000 times higher than 0.1 μg per liter.

More recently, the so-called European generation target establishes a goal for the elimination of discharges of all hazardous chemicals into the North Sea by 2020 (Esbjerg Declaration 1995). The interpretation of this ministerial decision is still being negotiated among the countries involved (OSPAR 2001). It remains to be seen, therefore, whether the cause for a precautionary elimination of chemical impacts will be pursued in earnest, or if the process will result only in reduction of individual risks that are already recognized, being connected to prioritized chemicals (see EU Commission 1982).

A thorough understanding of these and similar examples will be necessary for the further definition and structuring of the precautionary principle. Critical, of course, is the question whether actual decisions are made on the basis only of existing data, which are usually interpreted as adequate, or whether uncertainty is considered. Consequently, attention must be given to the question of whether the "gray zone," which unavoidably exists immediately beyond the "scientifically based risk zone," will be included or excluded from those exposure levels that are ultimately considered "safe" (see figure 7.1). It is necessary, then, to understand scientific uncertainties and their impacts on the scientific and policy process. Some of these uncertainties include variability and susceptible subpopulations, extrapolation to real-life settings, lack of knowledge, and indeterminacy. They are both scientific and social in nature, demanding a shared science-policy response.

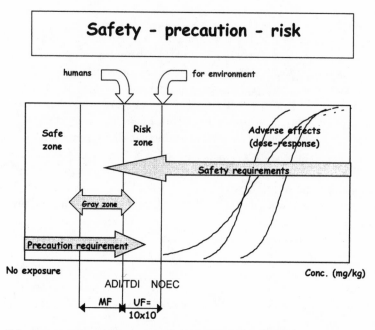

Figure 7.1. Establishment of "safe limits" for humans at acceptable or tolerable levels (normally called ADI/TDI) is tied directly to the experimental no-effect level (NOEC) by the application of standard uncertainty factors (UF), while protection levels for environment refer to the no-effect levels only. Precautionary measures beyond this may be achieved either by supplementary modifying factors (MF) to cover gray zones or by direct and restrictive exposure reductions (referred to as risk reductions).

Dealing with Uncertainties in Chemical Assessment

Toxicological/ecotoxicological practices have to deal with both trivial variability and scientifically complex uncertainties, both of which have received increasing attention over the years. However, this does not mean that full scientific justification has been developed for these methods, nor have they been harmonized and standardized.

In human as well as environmental toxicology there is a continuous search for the "most sensitive" endpoints—the effects that appear at lowest exposures, the most vulnerable species, and so forth—that are both significant and relevant. Frequently, however, there is lack of evidence for an adequate selection of the particular types of endpoints to examine, such as structural or functional effects.

One can also question whether decision criteria and guidance are sufficient for the selection of measurement endpoints, such as the identification of concentrations at which there are no effects or low effects. The establishment of no-effect levels is dependent on both experimental (e.g., selection of functional or structural endpoint) and statistical design (e.g., choice of distribution models). These points are critical issues, especially in ecotoxicology, because routine testing to a great extent relies only on acute effects and usually refers to experimental results obtained in only a few, easily reared laboratory species. Neither testing of chemical mixtures nor multispecies testing is part of fundamental hazard testing procedures. Similarly, there is a complete lack of criteria that would take into account broader effects on the structure, biodiversity, and functions of ecosystems.

Guidelines exist for regular statistical variance analyses of experimental data, although these are more valid in human toxicology than in ecotoxicology. In routine cases, the choice between statistical models (e.g., normal versus logarithmic distribution models) does not seem to be critical. But considerable difficulties often remain in cases of estimating and evaluating effects in the low-concentration ranges, which is of special significance when dealing with measurement of sensitive endpoints or vulnerable species (Hauschild and Pennington 2002).

Frequently, little or no attention is given to the precision of a so-called no-effect level for a given endpoint. Measurements of low-effect concentrations are often accepted (e.g., at the 5 percent level) as a substitute for actual no-effect concentrations. In so doing, the preferable but more stringent concept of "lower safe limits" is neglected (the so-called benchmark method), which may give a better basis for judging the need for uncertainty margins. Such uncertainties may be difficult to quantify.

Often the dilemma can be described as a choice between selection of results that are scientifically valid but of low relevance and approaches that are experimentally of lower validity but, because they focus on marginal situations, may be highly relevant. At other times, unfortunately, the validity of available data may be low simply as a result of insufficient requirements and lack of relevant criteria for a precautionary decision process—a situation that is critical especially in the field of ecotoxicology.

The extrapolation of experimentally obtained hazard results—that is, the transfer of animal laboratory findings to human life—is an important and often most problematic stage for the uncertainty evaluations. In the domain of human toxicology this process is dominated by the standardized use of safety—or rather, uncertainty—factors following the Lehman–Fitzhugh rationale (Lehman and Fitzhugh 1954) of a two-orders-of-magnitude margin to allow for differences between experimental animals and humans (interspecies), and for variability among individuals within the human population (intraspecies).

This application of the two-orders-of-magnitude margin of safety has gained general acceptance in the field of human toxicology, whereas in ecotoxicology only pragmatic factors (called AF, application factors) are practiced, tied to statistical variability studies. This is a difficult situation. On the one hand, there is a smoldering debate about the scientific basis for the overall use of the two-orders-of-magnitude margin in human toxicology. Industrial stakeholders (ECETOC 1995) have requested that the factors should be significantly reduced, while Renwick (1993) has advocated that the 10×10 margin be maintained in order to account separately for variability (uncertainty) in both physiological and metabolic processing (toxicokinetic) and the actual toxicodynamic effect of chemicals in the organisms.

On the other hand, researchers increasingly recognize from environmental impact studies that uncertainties (variability) tied to evaluation of ecotoxicological (and possibly also human toxicological) effects may be grossly underestimated in our present practices (Pennington et al. 2001). This leaves us with the continued use of safety (or uncertainty) factors as the obvious choice. The extrapolation factors that are presently applied in the EU (EU Commission 1994b) and recommended by OECD (OECD 1992) are merely expressing trivial, statistical variability of test results obtained from standardized laboratory testing and are referring only to the number and the credibility of the available data. Beyond this, scientifically or otherwise argued needs for precautionary margins—for example, those concerned with relevance and sufficiency of endpoints, impact over time and space, influence of complex mixtures, sensitive subpopulations, or chronic exposures—are not considered. Officially, the goal of a 95 percent protection level represents the pragmatic attempt to achieve what is called an environmental no-effect exposure level for vulnerable species. However, this recommendation is not based on an underlying problem description or considerations of uncertainty built into the assessment procedure.

In the field of human toxicology, it seems that the extrapolation approach introduced as a safety measure provides a rational answer to uncertainty when scientific experience is transferred and interpreted across species. However, the safety margin factors, which are arbitrary order-of-magnitude figures, are only quasi-scientific. There are still open questions that could justify further safety measures, eventually in the form of supplementary "modifying factors," as occasionally introduced in U.S. and international practice. In addition to uncertainty connected to general lack of knowledge and ignorance, some unsolved problems are attached to observations of low dosage/concentration effects. Uncertainty connected to our dealing with effects of chemical mixtures is still a subject that gives rise to both debate and concerns.

In ecotoxicology, the additional—but most important—concerns are tied to the meager data requirements that spring from the standard hazard assessment scheme, which is often characterized as less than a handful of single species tests. Major shortcomings are connected to unsolved problems in the evaluation of chronic exposures, and especially to the universal lack of endpoints with significance for the broader structures and functions of the ecosystem.

In general, therefore, a critical review of precautionary measures and extrapolation practices is well advised. It should deal with both trivial and complex matters of minimum data requirements, data appraisal, and exploitation. More systematic analyses of the uncertainty elements in present extrapolation practice are called for.

In ecotoxicology, there is a special need for safety or, rather, supplementary uncertainty factors in the estimation of tolerable concentrations in order to cover extrapolation from laboratory findings to the interaction among species in the ecosystem, including effects on functions other than those observed in single-species toxicity tests, and effects on ecosystems that are already stressed.

Lack of knowledge, insufficient understanding, or even ignorance of biological functions and structures is frequently given as the basis for factoring uncertainty into hazard (and risk) evaluations. This gives rise to concern in cases when suspicion of cause-and-effect relationships arises or when the nature and mechanisms of toxic lesions are not known. Obviously, this was the background for the Delaney Clause, when it was introduced by the U.S. Food and Drug Act of 1958. At a time of rising threats of chemical hazards, this statutory measure was important as a precautionary action. It accommodated a significant public concern and established a principle, which came to prevail in most countries, although it often did not gain the same legal status as in the United States.

It is unfortunate, therefore, and it can only be interpreted as a defeat for the precautionary principle that the Delaney Clause (for pesticides in food) was abandoned in the United States by the 1996 Food Quality Protection Act. The zero tolerance for pesticide residues in food as a driver for phasing-out or for possible banning of individual pesticide chemicals was replaced by the legal but nonscientific term, "*de minimis* risk," which gives room for judging chemicals as safe in cases when alternative test options or lack of knowledge has not been given full consideration. Scientifically, the Delaney Clause with its restrictive exclusion of potential carcinogens from food and drinking water became a challenge for toxicological researchers, who attempted to give definite answers and to develop adequate test systems for the identification of human carcinogens, and thereby to establish safe levels for these and other chemicals. Much more detailed knowledge has been gained since then, including an increased possibility for establishing pragmatic dose-effect thresholds for some carcinogens.

However, doubts still exist concerning actual thresholds for many carcinogens, and further unanswered questions have arisen. This is presently demonstrated also in the debate on low-concentration effects of synthetic, endocrine-disrupting chemicals. New categories of disturbing low-dose/long-term effects are being identified in the form of reproduction failures and hormonal disorders, which are increasingly being reported and can be linked to chemical exposures. Concerns are being raised with reference to both humans and the environment, and there is an increasing awareness among epidemiologists, endocrinologists, and basic researchers about the need for preventive actions despite difficulties in giving precise guidance for specific measures (Andersson et al. 2001).

The testing of chemicals for endocrine disrupting properties is presently under development. In the EU a priority selection procedure has resulted in a candidate list of about 550 synthetic substances. However, methodologies are still inadequate, and science will not, for the foreseeable future, be able to ensure the predictive power of possible test schemes (EU Commission 2001b). The inadequacy is evident as regards the identification of appropriate endpoints and choice of critical test windows such as mechanisms of action, exposure at various periods of sexual differentiation, and reproductive malformations. Experimentally, the need for identifying low-dose and long-term effects, including significance of multigenerational exposures, seems insurmountable (see chapter 8).

This situation is clearly pointing to the need for precautionary measures, including the following:

- Precautionary steps for protecting humans and the environment against chemicals with potential for irreversible, low-dose damages, to account in these cases for insufficient knowledge, inconclusive data, and inadequacy of present tools of protection; and
- Improved screening methods to be developed (e.g., based on group classification concepts [Bro-Rasmussen et al. 1996] or QSAR, quantitative structure activity relationship), making it conceivable that otherwise nonassessed chemicals are selected for temporary restrictions or that insufficiently tested chemicals can be "overclassified" (i.e., be considered as the most toxic in their class) until further data are made available.

Risk Assessment Rationale

The key position that risk assessment holds today as the basis for regulatory decisions has grown over the most recent decades (EU Commission 1994a). In its rationale it balances the hazard/effect assessments with release, distribution, and exposure evaluations. These, in turn, will usually be based on a combination of data from environmental or human monitoring programs and predictions based on theoretical, generic models (EU Commission 1994b) (figure 7.2). The risk concept is formulated as a mathematical relationship between potential hazards and predicted exposure levels. This comparison can be made at different levels of sophistication, depending on the volume of data available (OECD 1984), although this is not general practice, due to the mostly

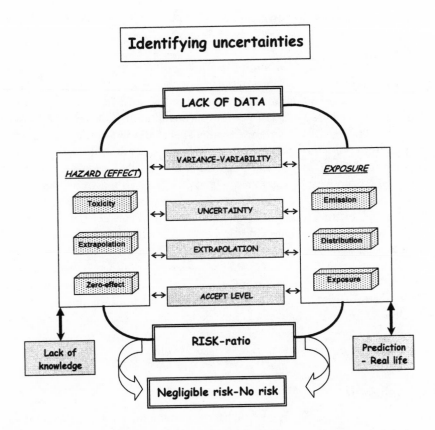

Figure 7.2. Technical guidelines have been developed in recent years as a reference for risk assessments (see EU Commission 1994b). The so-called risk ratio is hereby defined by comparing the estimated potential no-effect concentration (PNEC) with predicted exposure concentrations (PEC). Evaluation of uncertainties in this process is of paramount importance as a basis for precautionary measures. It will be essential also to take into account uncertainties as well as indeterminacies of both scientific and socioeconomic origin—and at all levels of the risk assessment system.

unspoken expectation that a full and flawless assessment must be made of the chemical in question before any action can be taken.

Generally, the use of statistical variance analysis, deterministic modeling of worst case scenarios, and the use of conventional extrapolation factors are the tools applied to account for uncertainty and to endow assessments with credibility. Thus, the development of the risk analysis system and the definition of its role are clearly interpreted in terms of cause-and-effect relations and processes. This reflects its origin in technological sciences. In the present guidelines, no direct uncertainty analyses are generally included. For observers as well as administrators and managers, the risk assessment schemes are seemingly based on assumptions that processes leading to risks are well defined and deterministic.

In real life, however, this is not true. Biological processes in an open world are increasingly being described statistically in stochastic models. Similarly, the fate and effects of chemicals will have to be described in nondeterministic terms in a dynamic, evolutionary environment and human society. Risk assessments are developed and performed under the impression of finality, and neither political nor administrative reasoning seems presently to be ready to give room for precautions beyond the pragmatic and officially established risk ratio, comparing exposure to hazard on the basis of a predicted environmental concentration (PEC) and an estimated potential no-effect concentration (PNEC).

Without any theoretical or practical guidance, this concept is in fact defining an acceptance of certain risks, and at the same time it "rules out" other options, such as negligible risk (achieved by applying a $100\times$ risk reduction factor beyond the PEC/PNEC value as suggested from the Netherlands (see van der Wielen 2000) or—ultimately—the concept of a no-risk terminology. These seem to be far-off goals, unless further guidelines for precaution are developed. Such guidelines should include options for a banning of individual chemicals or possibly introduction of closed system restrictions resulting in emissions and/or exposures at zero or near zero levels.

Conclusion

Some analysts (e.g., Wynne 1992) have recommended that socioeconomic uncertainty and indeterminacy should be taken into account when the fate and effect of chemicals are assessed. From an operational point of view, it seems hard to expect that such intervention may be accepted as an immediate background for precautionary measures. This is demonstrated already by the rulings and limitations given by the EU Commission (EU Commission 2000a) for the application of a precautionary principle. But, it is also clear from the discussion above concerning lack of knowledge and insufficiencies already accepted as part of scientific evaluations when dealing with exposures and processes, which may lead to adverse effects and damages. Ignorance becomes an inescapable part of scientific knowledge, and uncertainty becomes difficult to handle or even quantify.

It is true, however, that indeterminacy becomes an integrated part of the risk that results from the open scientific as well as socioeconomic systems within which we operate, deal with the uncertainty, and make decisions on precautions. Obviously, scientific uncertainty can never be fully ascertained and finalized and, accordingly, form the sole basis for precautionary decisions. It is a logical consequence that it will not be possible to leave the responsibility for setting criteria for "good practices" solely to chemical or biological scientists, risk assessors, or, for that matter, to any technical discipline at all. It will, however, be important or even indispensable that accords be established between science and the socioeconomic world before conclusions can be drawn.

The requirements to achieve this are multiple. They will include future strategies for controlling hazards, as well as scientific willingness to open up, review, and revise present practices regarding uncertainty and precaution. Also needed is a stakeholder acceptance that precautions are installed in a transparent way as parts of both hazard and risk assessment/management schemes in order to eliminate or reduce risks.

Future strategies in the field of chemical regulations should support basic requirements for prevention and precaution in a more comprehensive way. The need and responsibility for full evaluation of uncertainty and indeterminacy and for processing the results of assessment and management schemes must be a shared responsibility of scientists, managers, and politicians. This will include a need for guidelines to be established in order to reach commonly agreed goals (respecting the target of sustainability) and common metrics. This will be enabled by the following:

• Restrictions and controls at the political level to master indeterminacy, obviously based on improved strategies for the use of all chemicals;
• Principles for the possible alleviation of the pressure of uncertainties, which springs from ignorance or still missing knowledge, for example, by more open and liberal application of uncertainty factors, either by more restrictive market policy or by extended use of precautionary uncertainty factors when the impact of uncertainty and indeterminacy becomes unacceptable, for instance, regarding carcinogenic and endocrine-disrupting chemicals; and
• Opening and development of scientifically established "safety concepts" in order to include, in a transparent way, the evaluation of uncertainty as an integral part of all hazard and exposure assessments, covering both human and environmental issues and including not only risk characterization, but also aiming for an extension of the risk concept in order to embrace such terms as negligible risks and even zero risks.

References Cited

Andersson, A.-M., K. Grigor, E. Rajper-de Meyts, H. Leffers, and N. E. Skakke- bæk, eds. 2001. *Hormones and Endocrine Disrupters in Food and Water: Possible Impact on Human Health.* International Workshop at University Hospital, Copenhagen, 27–30 May 2001. Munksgaard: Copenhagen.
British Crop Protection Council and Royal Society of Chemistry. 1994. *The Pesticide Manual. A World Compendium,* 10th ed. Bath, U.K.: The Bath Press.

Bro-Rasmussen, F. 1999. "Precautionary Principle and/or Risk Assessment." *Environmental Science and Pollution Research* 6, no. 4.

Bro-Rasmussen, F., H. B. Boyd, C. E. Jorgenen, P. Kristensen, E. Laursen, H. Lokke, K. M. Nielsen, and J. Grundahl. 1996. *The Non-Assessed Chemicals in EU.* Report no. 1996/5 from the Danish Board of Technology, DK-1106 Copenhagen. www.tekno.dk.

Copenhagen Chemicals Charter. 2000. *Chemical Awareness,* no. 12 (December). Adopted by European Consumer and Environment Organisation at the Chemicals Under the Spotlight conference, Copenhagen, 27–28 October.

Danish National EPA. 2000. *List of Formulations Recognized and Registered as Pesticides.* In Danish. Miljøstyrelsen: Copenhagen.

ECETOC. 1995. *European Centre for Ecotoxicology and Toxicology of Chemicals. Assessment Factors in Human Health Risk Assessment.* Technical Report no. 68, August.

Esbjerg Declaration. 1995. Ministerial Declaration of the 4th International Conference on the Protection of the North Sea. Esbjerg, Denmark, 8–9 June.

EU. 1976. Council Directive 76/768 of 27 July 1976 on rules governing cosmetic products in the EU.

———. 1980. Council Directive 80/779 of 15 July 1980 on quality of drinking water for human consumption.

———. 1998. Council Directive 98/83 of 3 November 1998 relating to the quality of drinking water.

EU Commission. 1982. "List of 129 Chemicals Hazardous to the Aquatic Environment." *Official Journal of the European Communities,* 14 July, no. C176, p. 3. Communication from the Commission to the Council on dangerous substances that might be included in List I of Council Directive 76/464.

———. 1994a. "Risk Assessment for Existing Chemicals in Accordance with EU Council Regulation no. 793/93." Regulation no. 1488/94.

———. 1994b. "Environmental Risk Assessment for Existing Substances in Support of Commission Regulation no. 1488/94." Technical Guidance Document, p. 46, Report XI/919/94/EN. Ispra, Italy: European Chemical Bureau.

———. 1998. "Directive 98/8 of 16 February 1998 Concerning the Placing of Biocidal Products on the Market." *Official Journal of the European Communities,* no. L123, April 24, pp. 1–63.

———. 2000a. "Communication from the Commission on the Precautionary Principle." Com(2000) 1. Brussels: Europe Commission. Available at http://europa.eu.int/prelex/detail_dossier_real.cfm?CL=en&DosId=154698.

———. 2000b. "List of Hazardous Substances." Commission Directive 2000/32/EC of 19 May 2001, Published in the European Union Official Journal vol. L 136, 08/06/2000 P. 0001-0089. Brussels: European Union.

———. 2001a. "White Paper on Strategy for a Future Chemicals Policy." COM(2001)88, 8 February 2001. Brussels: European Commission. Available at http://europa.eu.int/prelex/detail_dossier_real.cfm?CL=en&DosId=162599. Accessed 8/12/02.

————. 2001b. European Workshop on Endocrine Disrupters Workshop Report. Brussels: European Commission. Available at www.europa.int./comm/environment/chemicals.pdf/workshop_report.pdf.

European Environmental Agency (EEA). 2002. *Late Lessons from Early Warnings: Environmental Issue Report no. 22 on the Precautionary Principle 1986–2000.* Copenhagen.

Hauschild, M., and D. Pennington 2001. "Indicators for Ecotoxicity in Life Cycle Impact Assessment." In Udo de Haes, et. al. (eds.), *Life Cycle Impact Assessment: Striving Toward Best Practice.* Pensacola: Society for Ecological Toxicity and Chemistry (SETAC).

Lehman, E. J., and O. G. Fitzhugh. 1954. "100-Fold Margin of Safety." *Association of Food Drug Officials of the U.S. Quarterly Bulletin* 18, no. 1: 33–35.

OECD. 1992. "Report from the Arlington Workshop on Hazard Assessment of Chemicals." Environment Monograph no. 59, 10–12 December 1990.

OECD, Environment Committee. 1984. "Chemicals on Which Data Are Currently Inadequate. Selection Criteria for Health and Environmental Purposes." Paris: OECD.

OSPAR. 2001. Convention for the Protection of the Marine Environment of the North-East Atlantic. Working Group on Priority Substances, Document SPS(2)01/8/1-E, October.

Renwick, A. G. 1993. "Data-Derived Safety Factors for the Evaluation of Food Additives and Environmental Contaminants." *Food Additives and Contaminants* 10, no. 3: 275–305.

van der Wielen, A. 2000. "Risk Management in the Netherlands: Acceptability Levels." *Chemaware* 9, no. 5.

World Health Organization (WHO). 1992. "Guidelines for Drinking Water Quality." Revision of the Guidelines in Report from a Task Group Meeting, Geneva, September. PEP/IPCS/EURO/92.1.

WHO and EEA. 2002. "Children's Health and Environment: A Review of Evidence." Report by WHO, Regional Office for Europe, Copenhagen, jointly with European Environment Agency, Copenhagen.

Wynne, B. 1992. "Uncertainty and Environmental Learning: Reconceiving Science and Policy in the Preventive Paradigm." *Journal of Global Environmental Change,* 2, no. 2:111–27.

CHAPTER 8

Chemicals Policy and the Precautionary Principle: The Case of Endocrine Disruption

Joe Thornton

How can scientific knowledge be used effectively to build a sustainable society? Most current environmental policies seek to manage hazardous activities within "acceptable" limits, using science-based predictions of safe levels of disruption. I call this framework the *risk paradigm,* because it views health hazards as manageable, quantifiable "risks" and uses risk assessment and risk management as its primary scientific and policy-making tools (Thornton 2000). An alternative model, the *ecological paradigm,* based in part on the precautionary principle (Raffensperger and Tickner 1999), begins from the view that scientific knowledge of the links between complex economic activities and complex natural systems is imprecise and incomplete. In the ecological paradigm, policies focus not on permitting potentially severe or irreversible risks but on avoiding them through technological change. The role of science in this framework is to provide early warnings of hazards and to evaluate alternatives so that society can choose the most sustainable technologies to fulfill its needs (Thornton 2000).

The conflict between these two models is starkly illustrated by the issue of chemical pollutants that can disrupt the endocrine system—hormones, the glands that produce them, and the response of target cells and tissues (see McLachlan 2001; Guillette and Crain 2000). Hormones are the body's natural signaling molecules: they circulate in very small quantities in the bloodstream and bind specifically to receptor proteins in or on target cells. In recent years it has been shown that a significant number of synthetic chemicals can disrupt

103

the endocrine system by augmenting or interfering with the action of natural hormones, leading to severe impacts on reproduction, development, behavior, and immunity at very low doses (NRC 2000; Royal Society 2000). Ecoepidemiological and experimental evidence link reproductive and developmental dysfunction in wildlife populations in some large ecosystems to exposure to endocrine disrupters (Fry 1995; Olsson et al. 1994; Reijnders 1986). Moreover, the doses and body burdens of some chemicals to which the general human population is exposed have reached the range that is known to cause endocrine disruption and/or its consequent health impacts (Ulrich et al. 2000; Devito et al. 1995; Koopman-Esseboom et al. 1994).

In the current regulatory system, industrial chemicals are generally permitted to be produced and discharged, so long as releases of individual substances from permitted facilities remain within limits predicted by risk assessment to be safe or acceptable (products incorporating industrial chemicals are for the most part unregulated). This approach makes the de facto assumption that our knowledge of chemical hazards is more or less complete. Although risk assessments partially address information gaps with safety factors and assumptions designed to be conservative, health impacts that have not been studied or for which dose-response relationships have not been quantified—as well as all substances for which no toxicological or exposure data are available—make no contribution to the risk estimate. A lack of evidence is thus misconstrued as evidence of safety, and untested or poorly understood chemicals are largely unregulated. For new chemicals, some premarket testing data must be submitted to the U.S. Environmental Protection Agency (EPA), but existing chemicals—about 99 percent of substances by volume on the market today—have never been subject to these requirements. Little or no toxicological data are available for the vast majority of the approximately 87,000 chemicals now in commerce (EPA 1998; Roe 1997; NRC 1984).

The limits of scientific knowledge are particularly pressing with regard to endocrine disruption. According to the EPA, very few of the thousands of substances in commerce have been tested for endocrine disruption, because standard toxicological protocols are inadequate to detect the subtle impacts and mechanisms involved (EPA 1998). Moreover, endocrine disruption occurs in some cases at doses orders of magnitude lower than those required to produce more obvious forms of toxicity or high cancer risks, so discharge and exposure limits based on other health effects do not necessarily protect against the impacts of endocrine disruption (e.g., Hayes et al. 2002).

In the risk paradigm, the primary constructive response to uncertainty is to gather new data. The EPA, for example, has recognized that "there currently is not enough scientific data available on most of the estimated 87,000 chemicals in commerce to allow us to evaluate all potential risks" (EPA 2002). The agency has therefore established the Endocrine Disrupter Screening and Testing Program (EDSTP) "to detect and characterize endocrine activity of pesti-cides, commercial chemicals, and environmental contaminants" and character-ize their dose-response relationships (EPA 2002). According to the agency, "The EDSP will enable EPA to gather the information necessary to identify endocrine disrupters and take appropriate regulatory action" (EPA 2002).

In this chapter I examine the complexity of endocrine disruption as a case study in the adequacy of the chemical-specific, test-and-permit approach to chemical regulation. I show that available scientific techniques and back-ground knowledge, such as those used in the EPA's EDSTP, are intrinsically inadequate to provide the information required to protect health and the environment in the current framework. This argument is not a critique of the EPA's strategy as a scientific program but a critical examination of the way science is used in the current policy system and the way decisions are made in the presence of limited knowledge. The problems associated with studying and regulating endocrine disrupters are equally applicable to many other types of toxicological endpoints: neurotoxicants, reproductive toxicants, and so forth.

To rectify the unrealistic demands that the risk makes upon science, I pro-pose a precautionary framework that takes better account of the remarkable complexity and diversity of endocrine systems and the limited knowledge and tools that are now available to understand and manage them.

Identifying Endocrine Disrupters

The EPA's EDSTP, adopted by the agency after consultation with a diverse group of scientists and stakeholders and still in final formulation, is a large-scale, resource-intensive effort that seeks a reasonable compromise between comprehensiveness and efficiency (EDSTAC 1998). Subjecting each chemical in commerce to a detailed investigation that could rule endocrine disruption definitively in or out would take centuries and billions of dollars, so the EPA has adopted a hierarchical approach to screening and testing (see figure 8.1). The strategy begins by using existing data on chemical production volume, exposure, and health impacts, along with the results of computational methods that predict binding to hormone receptors, to prioritize chemicals that may be endocrine disrupters. Those substances that are given high priority are then screened using a small battery of short-term in vivo (whole animal) and in vitro (test tube) assays; chemicals that are given low priority are held and are not likely to be evaluated in the near future. Of those chemicals that are sub-ject to rapid screening, those that do not generate a weight of evidence in favor of endocrine disruption are judged nonendocrine disrupters, and those that are positive continue on for comprehensive, multigenerational testing for a variety of endocrine-mediated impacts.

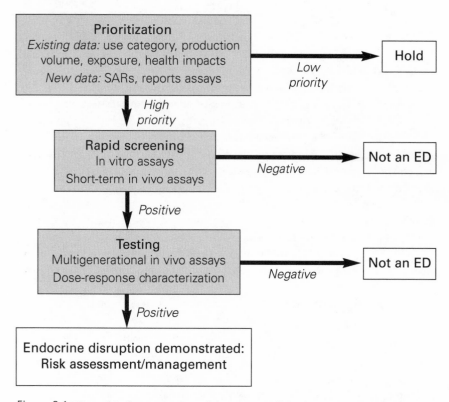

Figure 8.1. Hierarchical organization of the U.S. EPA's Endocrine Disrupter Screening and Testing Program. Only substances for which there is evidence of endocrine-disrupting potential in early stages are forwarded to full-scale testing in a timely fashion. ED, endocrine disrupter; SAR, structure-activity relationship.

For a chemical-by-chemical testing strategy to provide an adequate basis for regulations in the risk paradigm, it should avoid false-negative judgments; indeed, a failure to recognize even a small proportion of the large number of endocrine disrupters—or just a few high exposure or very potent ones—will result in laissez-faire treatment of chemicals that may cause significant health damage. The program should also provide reasonably accurate dose-response relationships to ensure that permitted exposure levels are truly safe. Can a realistic testing strategy be relied on to truly minimize false negatives and potency underestimates?

Diversity of Hormone Systems

The first difficulty arises from the sheer diversity of hormones and receptors. Like much current research in the field, the EDSTP will seek to identify only those substances that disrupt the estrogen, androgen, and thyroid (EAT) hormones and their nuclear receptors. According to the EPA, "These three hormone systems are presently among the most studied of the approximately 50 known vertebrate hormones. In vitro (test tube) and in vivo (whole animal) test systems to examine EAT hormone-related effects exist and are currently the most amenable for regulatory testing. For now, the EAT effects test systems represent a scientifically reasonable focus for the Agency's EDSP" (EPA 1998). As a description of the state of the science, and as an attempt to focus what would otherwise be an extraordinarily broad research program, the EPA's statement is undoubtedly correct. But focusing regulatory assessment solely on well-studied hormones ignores the well-established possibility that less-studied hormones and receptors may also be disrupted, posing potentially significant health and environmental hazards.

Despite very limited research, it is already known that synthetic chemicals can disrupt many of the dozens of nuclear receptors other than the EAT receptors which are of major biological importance (Giguere 1999). In addition to the nuclear receptors and their fat-soluble hormones, there are dozens of other hormones, neurotransmitters, and their receptors, and these too can be subject to chemical disruption (see Okino and Whitlock 2000; Seegal et al. 1994; Shafer et al. 1999; and Cooper et al. 2000 for examples).

The molecular diversity of the endocrine system clearly outstrips a testing strategy focused on three hormone/receptor systems. In principle, assays for the dozens of other hormones, signaling molecules, and receptors might be added to a strategy, but doing so would expand the scope of a testing program and the time and resources it demands, not a practical option when the goal is to provide a foundation for a rapid response to a public health risk. Further, assays for disruption of most non-EAT hormones mechanisms do not currently exist, and in many cases even background knowledge of the receptors' physiological effects is inadequate to allow the design of effective in vivo assays for their disruption.

Diversity of Endocrine-Active Substances

Because it cannot realistically test the thousands of chemicals now in production (or the thousands of chemicals generated as by-products of industrial processes or breakdown products of degradation that not included in the EPA's program), the EPA has proposed to prioritize a subset of commercial chemicals for screening. Chemicals will be prioritized if they are agricultural chemicals, and they have high production volumes, and there is evidence of large-scale human exposure, and there is evidence of likely endocrine disruption from previous toxicological studies or computational analysis of their structures. The first batch of prioritized chemicals fulfilling these criteria is expected to include several hundred substances, and it will take a number of years to evaluate them, at which time the program will be reviewed and another batch can then be assessed (Timm 2002).

Diversity of Mechanisms

The most rapid methods for identifying endocrine disrupters do not require testing of whole animals but use in vitro or computational methods. Many of these methods are intended to detect only compounds that bind to receptors and thereby activate or inhibit their ability to regulate gene expression. These include structure-activity relationship analysis, ligand-binding assays—in which the affinity of a receptor for a chemical is assessed—and reporter gene assays, a molecular method that the EPA plans to use for screening and, if a high-throughput method can be developed, in the effort to prioritize chemicals, as well (Timm 2002).

In addition to direct interactions with receptors, however, there are myriad other mechanisms by which synthetic chemicals can disrupt the endocrine system. For example:

• The pesticides lindane, methoxychlor, and Roundup (glyphosate) reduce the body's natural hormone levels by inhibiting the expression of regulatory proteins and enzymes that are required for the synthesis of steroids (Walsh and Stocco 2000, Walsh et al. 2000, 2001; Akingbemi et al. 2000).
• Dioxins down-regulate expression of the estrogen receptor in many tissues, reducing these tissues' sensitivity to the body's own hormone (Romkes et al. 1987).
• Some pollutants bind to hormone carrier proteins in the blood. For example, pentachlorophenol, atrazine, and several other organochlorines have high affinity for steroid hormone carriers (Danzo 1997); by displacing natural hormones from carrier proteins, synthetic chemicals may increase the rate at which the hormone is excreted, causing an antihormone-like effect, and also disrupt signaling through membrane receptors that specifically recognize hormone-bound carriers (Rosner et al. 1999).
• Chlordecone and a DDT metabolite block the ability of progesterone to trigger the maturation of oocytes via a receptor in the cell membrane (Das and Thomas 1999).

Because none of these mechanisms involves binding to the classic intracellular hormone receptors, they may not be effectively addressed. A few relatively rapid assays have been developed (and are included in the screening phase of EPA's program) for specific nonreceptor-mediated impacts, such as the "minced testis" assay for disruption of steroid hormone synthesis; for many other mechanisms, however, no such methods are available.

Complexity of Receptor Activity

Rapid molecular assays have played an important role in identifying endocrine disrupters to date and will continue to do so in the EPA's testing program, serving as important elements of screening efforts to select potential endocrine disrupters for full-scale testing. These assays are extremely useful for clarifying mechanisms of action, but their reductionist design limits their ability to detect some endocrine disrupters for several reasons (see also Rooney and Guillette 2000). First, the results of reporter assays for some ligand often depend upon the cell type and its stage of development. In fibrosarcoma cancer cells, for example, glucocorticoid and retinoic acid receptors activate gene expression in response to their natural ligands only during later developmental stages of the cancer (Vivanco et al. 1995). The drugs known as selective estrogen response modulators, including raloxifene and tamoxifen, are estrogen antagonists in some tissues but agonists in others (Shang et al. 2000). A receptor's response to a ligand also depends on the specific DNA response element to which it binds, and there are large numbers of response elements throughout the genome that vary in subtle or extreme ways from each other.

Finally, most receptors have multiple isoforms—variant forms encoded by duplicated genes or produced by differential splicing of RNA transcripts from a single gene. The EPA's reporter assays, for example, are slated to be carried out with a single form of the estrogen, androgen, and testosterone receptors, despite the fact that, for example, there are two estrogen receptor genes in mammals (Kuiper et al. 1996) and three in teleost fish (Hawkins et al. 2000).

In summary, the effect that a chemical may have on a receptor is fundamentally dependent on context. Molecular assays that are restricted to a single isoform, one or a few cell types, and a single type of response element can detect some agonists and antagonists, but they cannot be relied on to detect them all.

Complexity and Diversity of In Vivo Impacts

Toxicological testing in whole animals can circumvent many of the problems associated with simplified in vitro and computational assays, because the organism has all of its receptors, isoforms, and response elements intact. But the manifestations of endocrine disruption at the level of the whole organism can also be diverse, subtle, and challenging to detect. Rapid in vivo assays, like those used for screening in the EPA's program, are designed to detect the kinds of classic short-term effects that are caused by strong EAT mimics and blockers, when exposure occurs during adult or juvenile stages. Comprehensive multigenerational tests for a broad variety of subtle and severe endocrine impacts will detect many, perhaps even the majority, of endocrine disrupters that produce EAT-like impacts. Even these assays, however, are subject to a number of limitations that can produce additional false-negative judgments or underestimates of potency.

Timing

By definition, short-term assays must exclude at least some of the periods in which organisms are most sensitive to endocrine disruption. The EPA's proposed screening battery, for example, involves dose periods of three days to three weeks in length, in adult or juvenile animals, followed by immediate examination for designated endpoints. But it is well established that the most sensitive periods for endocrine disruption are during early development; in utero exposure to hormones and endocrine disrupters at doses much lower than required to cause irreversible effects in adults can produce permanent disruption of reproductive development (e.g., Bigsby et al. 1999; Hayes et al. 2002). Further, developmental and some other effects of hormones and their disrupters are often not manifest until months or years later (Thayer et al. 2001; Howdeshell et al. 1999).

Nonclassic Impacts

The manifestations of endocrine disruption are diverse and not easily represented in an efficient testing program. Most toxicological work, like the EPA's planned screening battery, focuses on a limited number of classic effects, such as increases in the weight of the uterus or the height of the vaginal lining after exposure to estrogen, or an acceleration of metamorphosis in frogs exposed to thyroid hormone. More comprehensive assays are planned for the full-scale testing stage, and these are focused primarily on reproductive and developmental outcomes. Hormones, however, cause a wide range of physiological and developmental impacts in addition to the classic effects. For example, EAT hormones and disrupters may also cause behavioral changes, reduce cognitive capacity, alter the function of the cardiovascular system, alter immune response, and disrupt lipid metabolism and bone homeostasis, in some cases at doses far lower than required to produce classic effects (Porterfield 2000; McEwen 2001; Dubey and Jackson 2001; McMurray 2001). Other nuclear receptors may also cause a wide variety of impacts (Giguere 1999). There are gaps in knowledge of the functions and potential impacts on even the most studied hormone/receptor systems.

At least some of these nonclassic functions appear to be mediated by mechanisms distinct from those that produce classic EAT effects (e.g., Hisamoto et al. 2001), raising the possibility that some disrupters cause only nonclassic impacts. The extent to which this will be a major problem is unknown, but existing data suggest it is likely to occur to at least some extent. For example, bisphenol-A, a component of certain common plastics, disrupts estrous cycling, alters male reproductive development, and accelerates puberty in the offspring of exposed rats at very low doses; only at very high doses does it increase uterine growth, and in some studies it does not do so at all (Rubin et al. 2001; Ashby et al. 2000; Markey et al. 2001).

Studying effects of endocrine disruption on biological systems is compounded by gaps in our knowledge of hormone functions. For example, even the functions of the well-studied estrogen and progesterone receptors are relatively unknown.

Subtlety and Statistical Power

Another limitation of full-scale testing is that some effects of endocrine disrupters can be subtle and therefore difficult to establish with statistical confidence. The EPA's protocols for reproductive toxicity testing require only twenty animals per group (Toppari et al. 1996). But consider a weak endocrine disrupter that reduces by 5 percent some functional capacity that varies naturally in a population, such as sperm density, cognitive ability, or immune capacity; detecting such an effect with statistical significance would require many hundreds of animals per dose group. Practical considerations therefore restrict statistically significant findings to impacts that are moderate to strong. No-observed effect levels—and the "acceptable" exposure standards derived from them—thus often reflect the point at which methods are no longer powerful enough to identify statistically significant impacts, not a truly safe dose. Subtler and more difficult to document impacts, however, may be very important from a public health point of view, considering the universal nature of exposure to many endocrine disrupters.

Dose-Response Relationships

Another potential source of difficulty is the complexity of the relationship between dose and response. Many impacts of hormones have nonlinear dose-response curves, particularly U- or J-shaped relationships (e.g., vom Saal et al. 1997; Melo and Ramsdell 2001). Standard toxicological tests use only moderate to high doses, based on the assumption that high doses will produce the same kinds of effects as low doses, but with greater severity. An endocrine disrupter that produces a U- or J-shaped dose response, however, may have little or no impact at the doses tested but cause significant effects at the lower levels that are environmentally relevant. The temporal organization of doses also matters: the pesticides thiram and chlordimeform have no effect on reproduction when the dose is continuous—as it is usually is—but they disrupt ovarian cycling when exposure occurs in a brief, appropriately timed burst (Stoker et al. 2001).

When testing is done according to standard protocols, complex dose-response relationships may therefore be missed or inaccurately estimated.

Mixture Effects

Testing of compounds in isolation, as is standard practice, will underestimate the effects of substances that are more potent endocrine disrupters in complex mixtures. Synergistic toxicological interactions can be common and quite powerful (Krishnan and Brodeur 1994; Mehendale 1994); they usually occur when two or more compounds contribute to the same endpoint through different mechanisms. Greater-than-additive interactions have been demonstrated for a variety of hormones and synthetic endocrine disrupters (Abbott et al. 1994; Bergeron et al. 1994; Edwards et al. 1999; Bergeron et al. 1999; Suzuki et al. 2001; but see Payne et al. 2000 for a lack of synergy when endocrine disrupters act through a common mechanism). Mixture effects are a particular problem for risk assessments and discharge permits, which are generally based on estimates of a chemical's potency derived in isolation. A mixture of weak estrogenic pesticides, for example, all at levels below their thresholds in isolation, has a strong effect on the proliferation of cultured breast cancer cells (Payne et al. 2001). Levels licensed as safe based on single-chemical testing may therefore be hazardous in the context of real-world exposures.

It is possible, of course, to determine whether endocrine disruption and other impacts are caused by the complex mixtures found in the environment by exposing animals to these mixtures, and a number of studies have demonstrated such a link (de Swart et al. 1996; Daly 1991). This kind of experiment does not specify which substances are responsible for the impacts, however, and this is precisely what is required for action in the current policy regime in which chemicals are regulated individually.

Variability Across Species

Most toxicological studies are conducted on one or a few common model organisms, but there are significant differences among species that undermine predictions made on the basis of a few model species, usually rats or mice. There is considerable variation among animal taxa in the affinity of receptors for natural hormones. The potency of synthetic endocrine disrupters also varies considerably among species: endosulfan, dieldrin, and methoxychlor bind with significant affinity to the ER-alpha of trout but not to those of human, mouse, chicken, or lizard (Matthews et al. 2000). Bisphenol-A activates the steroid X receptor of humans, which is involved in metabolism of steroid hormones and xenobiotic substances, but it has no effect on the same receptor in mice (Takeshita et al. 2001).

It is possible to expand the range of species tested beyond the usual model rodents. For example, the EPA plans to assay substances for endocrine activity in model species of rodents, birds, amphibians, and fish, an unusually broad group for a toxicological assessment program. But certain substances may be endocrine disrupters only in nonmodel organisms. The paint additive tributyltin, for example, disrupts the endocrine system of mollusks, causing females to grow male reproductive structures, but there is no evidence of an endocrine-disrupting effect on vertebrates (Smith 1981). Similarly, synthetic chemicals that activate or block the ecdysone receptor, which regulates metamorphosis and/or molting in insects, crustaceans, and many other invertebrates, could trigger premature maturation or block it entirely (Oberdorster et al. 2001); these substances would not be detected in a program focused on vertebrates, which lack this receptor. Steroids and related hormones even play critical roles in nitrogen-fixation in some bacteria (Fox et al. 2001), growth regulation in plants (Wang et al. 2001), and sexual development in some fungi (Brunt and Silver 1986). The first of these processes is known to be disrupted by bisphenol-A and the pesticides pentachlorophenol and methyl parathion (Fox et al. 2001). Results from the most thorough tests on vertebrates cannot always be extended to these distantly related but ecologically critical kinds of organisms.

Policy Implications of Endocrine Complexity and Scientific Limits

An ambitious testing program like the EPA's EDSTP will undoubtedly detect many new endocrine disrupters and provide useful information on mechanisms, dose-response relationships, and physiological impacts. For the purpose of policy, however, the critical question is whether such a program can provide information of the type, scope, and accuracy that is required for effective regulations to be formulated in the current regime. If testing does not detect or predict the effects of some chemicals, then substances that are hazardous but falsely thought safe will be legally produced and discharged into the environment without limit, and adverse impacts on public health may result. Similarly, substances whose toxicity is underestimated will be licensed for release at potentially unsafe doses. Unfortunately, current law in the United States places a very high burden of proof on the public before a substance's production or use can be restricted.

One can conclude that the EPA's program, like any single-chemical strategy, will result in many false-negative conclusions in prioritization, screening, and full-scale testing. While errors could be reduced through extensive additional testing, this would demand a prohibitive amount of time and resources, making impossible timely and comprehensive action on a problem of pressing need. Even with an unlimited amount of time and money, it would be difficult to design a fully comprehensive program, because we lack the required background knowledge of what chemicals are produced, what substances we are exposed to, and how the endocrine system works.

The risk paradigm of individual chemical testing and regulation fails to adequately protect health and the environment because it makes unrealistic and unreasonable demands on science. Current scientific tools can provide important knowledge of the effects of chemicals on endocrine and other systems, but because these systems are so complex and diverse, they cannot provide the comprehensive, detailed, and specific knowledge required for effective chemical-by-chemical permit-based regulations. *We need a new approach that takes account of both scientific knowledge and its limits, a policy framework that does not impose a scientifically unrealistic knowledge burden before action can be taken to protect health and the environment.*

It is not only the endocrine system that is complex and diverse; most biological and ecological systems have these qualities, too. A testing program to make a comprehensive list of neurotoxicants, or immunotoxicants, or developmental toxicants would be limited for the same reasons the EDSTP process is. Thus, the changes necessary to address the shortcoming of the current policy regime for endocrine disrupters are similar to those required to deal with many other health impacts of chemical pollution.

A Precautionary Chemicals Policy

We need a broad chemicals management policy that does not begin by trying to make an exhaustive list of all endocrine disrupters. Solutions and strategies do not have to be conceived according to the same categories that we use to describe problems; endocrine disruption is a category of biological effects, or, more precisely, of mechanisms for the production of those effects, but this category does not map clearly onto any classification of pollution sources, chemical applications, problematic chemical classes, or kinds of superior technology. A focus on the mechanism and endpoints per se therefore does not lead directly to any practical solutions.

The ecological paradigm represents a broad effort to move society toward safer and more sustainable materials. This approach begins with the precautionary principle: where there is reason to believe that a practice may cause severe, widespread, and irreversible damage to health and the environment, that practice should be replaced by the safest available alternatives. Some critics have argued that the precautionary principle would stop all productive activities, because every activity poses some risk. However, the principle applies only when a prima facie case can be made that some practice poses hazards with particular problematic qualities that make advance prevention essential, and even then it does not require the activity to cease but rather that the safest available methods to accomplish the same purpose be used.

From Risk Assessment to Alternatives Assessment

The licensing of "acceptable" discharges cannot prevent global contamination and health impacts, for three reasons. First, chemicals that persist in the environment or in the food web build up over time to higher and higher levels, and even dilute discharges eventually reach unacceptable concentrations. Second, permitting of industrial discharges of toxic chemicals ignores releases that may occur at unregulated points in the substance's life cycle. Facility-based regulations do nothing to prevent health impacts from hazardous chemicals that are deliberately dispersed into the economy through products. Third, when discharges of individual chemicals from individual sources are each assessed and permitted separately, even at "acceptable" levels that do not pose local health hazards, they contribute to a global pollution burden that may pose significant health risks.

To cope with these failings, the ecological paradigm shifts the focus of policy from permitting the production, use, and discharge of hazardous chemicals to preventing them whenever possible. Chemicals reasonably judged to have the potential to cause widespread, long-term, and severe forms of damage should be replaced, whenever feasible, with safer alternatives, rather than permitted in "acceptable" amounts. To move toward this end, the ecological paradigm emphasizes the technological principle of *clean production:* preventing the generation of hazardous substances by changing the inputs, products, and/or processes involved in materials production and use (Jackson 1993; Geiser 2001).

To implement this technological strategy, the nature of regulatory decision making must change. In the risk paradigm, safer substitutes, even if they are available, need not be used, so long as polluting processes can comply with discharge or exposure limits. In the ecological paradigm, assessment of alternatives for potentially hazardous chemicals shifts to the center of environmental decision making (O'Brien 2000). For each application in which synthetic chemicals that have the potential to cause widespread, long-term, and severe forms of damage are used or produced, the central question of the policy process becomes, "Is there a safer, feasible method for fulfilling society's needs that does not require the use of potentially dangerous chemicals?" In this framework, quantifying the exposure level predicted to be hazardous or safe for each chemical is not necessary; instead, the focus is on identifying opportunities to continually reduce the generations of hazardous substances, whether or not quantitative predictions or proof of harm has been established.

Managing Chemical Classes

How can these goals be realized? Two obvious questions arise: (1) how do we judge which chemicals are potentially severe hazards, and (2) how do we determine which is the safest available alternative? In both cases, science must play a central role, but the way science is used changes fundamentally from the current system.

Chemicals can be provisionally judged to pose potentially severe hazards (and therefore subjected to alternatives assessment) if they are members of chemical classes for which a prima facie case of harm can be established. (Individual chemicals for which available evidence constitutes a specific prima facie case of hazard should also be subject to alternatives assessment even if they are members of nonprioritized classes.) A prima facie case that a class is environmentally problematic would require a demonstration that its members often are persistent, bioaccumulative, mutagenic, carcinogenic, and/or toxic at low doses to development, reproduction, and function of the nervous, immune, or endocrine systems. Whether a class tends to exhibit these properties can be evaluated inductively—by examining the frequency of their occurrence among class members that have been evaluated—and based on fundamental chemical and biological principles, since effects on ecosystems and organisms arise because of basic chemical properties such as chemical stability, oil solubility, and molecular size and shape.

Based on considerations of this type, it has been argued (Thornton 2000; APHA 1994; International Joint Commission 1992) that organochlorine substances represent one class of chemicals that should be prioritized for substitution. This is based on the inductive finding that most organochlorines that have been tested have been found hazardous (APHA 1994; Henschler 1994). Further, the chlorination of organic molecules, because of the chemistry of the chlorine atom, often increases persistence and almost invariably increases bioaccumulation, toxicity, carcinogenicity, and endocrine-disrupting potential (Thornton 2000; Henschler 1994). Other classes that are good candidates for prioritization are other organohalogens (to which the same logic used for organochlorines applies); heavy metals, which are by definition persistent, and whose members include the extremely toxic lead, cadmium, arsenic, and mercury; and synthetic aromatic organics, the ring structures of which can lead to an increase in persistence, bioaccumulation, and affinity for hormone receptors compared to nonaromatics.

Framing environmental decision making in terms of chemical classes does not prevent individual substances and processes from being rigorously evaluated; in fact, it requires it. In the ecological paradigm, once a class of chemicals has been prioritized, its members are presumed to be hazardous, but they are specifically evaluated for the availability of safer alternatives, and the hazards of the original process and its putative replacements are weighed against each other. If proponents of some member of a hazardous class can demonstrate to a reasonable degree of certainty that no safer alternative is available, then that substance or process should continue to be used. For example, certain chlorine-based pharmaceuticals have no effective alternatives at present. In this framework, comprehensive testing would no longer be necessary to guarantee the safety of every chemical in commerce but only those for which its proponents are arguing there is no safer alternative available. This approach is similar to the concept behind pharmaceutical regulation, where medicines are supposed to be demonstrated safe and effective before being licensed.

This strategy would not eliminate false-negative judgments of safety, because there is always the possibility of some type of health hazard we have not considered in the design of our assays. Replacing untested or partially tested members of problematic classes with safer alternatives, however, could drastically reduce both the number and the consequences of false-negative judgments.

The ecological paradigm would not eliminate false-positive judgments of hazard, either; some members of prioritized classes might be treated as harmful when in fact further testing might have shown them to be more or less safe. But the ecological paradigm's emphasis on substitution of safe alternatives means that the consequence of false-positive judgments is not complete cessation of useful activities but continual conversion to and innovation in greener processes, a positive outcome. For each hazardous application, the goal of alternatives assessment is to identify a suite of superior alternative materials and/or processes. Society can then use its legal, regulatory, and economic tools to encourage and require implementation of these substitutes and to spur innovation in even better methods.

Evaluating Alternatives

This class-based, alternatives-focused approach would shift environmental regulation from micromanagement focused on each chemical to a kind of macromanagement, in which hazardous sectors of the economy are moved in a planned conversion process toward *sustainability*—compatibility with natural materials cycles and biological processes. The concept of choosing more sustainable alternatives brings us to the second question: by what criteria should processes and materials and their alternatives be judged, and how do we decide which technologies are acceptable for a given application? This effort is particularly important so that one hazardous substance or process is not replaced with another. The evaluation of materials and technologies will rely in large part on science-derived data, but it also requires judgment and open debate about the most desirable way for society to accomplish its material and economic goals. The process should therefore be both scientifically rigorous and politically transparent.

The relative merits of materials and processes can be interpreted based on the principles of green chemistry (Anastas and Warner 1998), with highest priority given to avoiding characteristics that can lead to widespread, severe, and irreversible hazards. Ultimately, all synthetic chemicals that remain in the economy should be products of green chemistry—chemical engineering with the express purpose of designing sustainable materials and techniques that are compatible with ecological processes. Green chemistry seeks to develop synthetic organic chemicals composed of the elements employed in natural mainstream biochemistry (carbon, nitrogen, oxygen, hydrogen, sulfur, and phosphorous, plus ionic or metallic forms of iron, zinc, potassium, calcium, and chloride); its goal is to produce materials that degrade in the environment and are reintegrated into natural materials cycles without causing harm to living organisms or the systems that sustain them (see chapter 19).

In evaluating alternatives, relevant knowledge will include fundamental chemical characteristics, specific test results, and predictions and supportable generalizations based on similar or related chemicals. Attention should be paid not only to the potential hazards of a substance itself but also the by-products and breakdown products produced during its life cycle. The degree of uncertainty related to the substance or process should also be carefully considered; if a substance is a member of a hazardous class, it can be presumed hazardous until evidence is available to indicate that it is not likely to have the kinds of problematic qualities discussed above. There is therefore an important benefit in continued toxicological testing, as comprehensive as is feasible, so that the information available on both new and existing processes is as complete as possible. The major difference from the current system is that partial information and data gaps do not serve as a barrier to environmentally sound technological change.

The point of a precautionary policy focused on broad classes of hazardous substances is to drive a planned, gradual, and society-wide technological conversion toward sustainable production technologies. The potential priority classes I have suggested are used in a wide variety of industries, so a critical approach to their use presents a major technological, economic, and social challenge. For this reason, the implementation of the transition should be democratically determined and transparent, with broad public and stakeholder guidance over the identification of priority classes, identification and selection of alternative technologies, conversion timelines and the granting of exemptions for socially necessary uses, and programs to minimize and compensate for economic dislocation.

Precaution and Science

As I have shown, science plays a major role in the ecological paradigm. It fills many of the same roles it does in the current system—gathering data on specific chemicals, for example. But science has additional important roles in the ecological paradigm that are absent in the risk paradigm, such as evaluating prima facie cases of hazard for classes of substances and processes and evaluating alternative technologies. Precautionary policies also require an expanded use of science to monitor the integrity of health and the environment, so that we can evaluate whether our actions are yielding environmental improvements, adjust them accordingly, and identify new hazards that are not presently known. The ecological paradigm requires continued investigations into the structure and dynamics of living systems, since better understanding leads to better foresight. For example, ambitious research on the endocrine system and its disruption by chemicals should continue in order to improve background knowledge, identify new classes of potential disrupters, develop new methods for diagnosis and treatment, and provide early warnings of new forms of health damage and its causes.

The ecological paradigm is intended to provide not only a better and more complete use of science than the risk paradigm but better ethics and politics as well. In the current system, decisions about production technologies are made by private parties and are limited only by the ability of pollution control devices to meet discharge limits. Even when safer alternatives are available, they often languish, because society does not have the legal means or political will to require them to be used. Indeed, the current regulatory system, by restricting debate to how much of a given pollutant is acceptable, prevents society from considering broad questions about the technologies of production and the availability of alternatives (Thornton 2001). In the twenty-first century, one company's decisions about how materials should be manufactured, food produced, or people transported, for example, have the potential to affect the health and well-being of people all over the world for many generations. Choices about production technology are therefore a politically and ethically proper matter to be determined democratically. The ecological paradigm is intended to provide a framework in which a broad base of citizens and stakeholders, including but not limited to industry, can participate in selecting the technologies that society uses to meet its material needs. The goal of the ecological paradigm is to create a policy regime that uses science in a way that effectively informs, rather than restricts, careful democratic deliberation and decision making about health, the environment, and technology.

Acknowledgments

I am grateful to Joel Tickner, David Crews, and two anonymous reviewers for very helpful comments on the manuscript.

References Cited

Abbott, B. D., G. H. Perdew, A. R. Buckalew, and L. S. Birnbaum. 1994. "Interactive Regulation of Ah and Glucocorticoid Receptors in the Synergistic Induction of Cleft Palate by 2,3,7,8-Tetrachlorodibenzo-p-dioxin and Hydrocortisone." *Toxicology and Applied Pharmacology* 128: 138–50.

Akingbemi, B., R. Ge, G. Klinefelter, G. Gunsalus, and M. Hardy. 2000. "A Metabolite of Methoxychlor, 2, 2-Bis (P-Hydroxyphenyl)-1, 1, 1-Trichloroethane, Reduces Testosterone Biosynthesis in Rat Leydig Cells through Suppression of Steady-State Messenger Ribonucleic Acid Levels of the Cholesterol Side Chain Cleavage Enzyme." *Biology of Reproduction* 62, no. 3: 571–78.

American Public Health Association (APHA). 1994. "Resolution 9304: Recognizing and Addressing the Environmental and Occupational Health Problems Posed by Chlorinated Organic Chemicals." *American Journal of Public Health* 84: 514–15.

Anastas, P., and J. Warner. 1998. *Green Chemistry: Theory and Practice.* Cambridge: Oxford University Press.

Ashby, J., J. Odum, D. Paton, P. A. Lefevre, N. Beresford, and J. P. Sumpter. 2000. "Re-evaluation of the First Synthetic Estrogen, 1-keto-1,2,3, 4-tetrahydrophenanthrene, and Bisphenol A, Using Both the Ovariectomised Rat Model Used in 1933 and Additional Assays." *Toxicology Letter* 115: 231–38.

Bergeron, J. M., D. Crews, and J. A. McLachlan. 1994. "PCBs as Environmental Estrogens: Turtle Sex Determination as a Biomarker of Environmental Contamination." *Environmental Health Perspectives* 102: 780–81.

Bergeron, J. M., E. Willingham, C. T. Osborn III, T. Rhen, and D. Crews. 1999. "Developmental Synergism of Steroidal Estrogens in Sex Determination." *Environmental Health Perspectives* 107: 93–97.

Bigsby, R., R. Chapin, G. Daston, B. Davis, J. Gorski, L. Gray, K. Howdeshell, R. Zoeller, and F. vom Saal. 1999. "Evaluating the Effects of Endocrine Disruptors on Endocrine Function During Development." *Environmental Health Perspectives* 107, no. 4: 613–18.

Brunt, S. A., and J. C. Silver. 1986. "Steroid Hormone-Induced Changes in Secreted Proteins in the Filamentous Fungus Achlya." *Experimental Cell Research* 63: 22–34.

Cooper, R. L., T. E. Stoker, L. Tyney, J. M. Goldman, and W. K. McElroy. 2000. "Atrazine Disrupts the Hypothalamic Control of Pituitary Ovarian Function." *Toxicological Sciences* 53: 297–307.

Daly, H. B. 1991. "Reward Reductions Found More Aversive by Rats Fed Environmentally Contaminated Salmon." *Neurotoxicology Teratology* 13: 449–53.

Danzo, B. J. 1997. "Environmental Xenobiotics May Disrupt Normal Endocrine Function by Interfering with the Binding of Physiological Ligands to Steroid Receptors and Binding Proteins." *Environmental Health Perspectives* 105: 294–301.

Das, S., and P. Thomas. 1999. "Pesticides Interfere with the Nongenomic Action of a Progestogen on Meiotic Maturation by Binding to Its Plasma Membrane Receptor on Fish Oocytes." *Endocrinology* 140: 1953–56.

de Swart, R. L., P. S. Ross, J. G. Vos, and A. D. Osterhaus. 1996. "Impaired Immunity in Harbour Seals (Phoca vitulina) Exposed to Bioaccumulated Environmental Contaminants: Review of a Long-term Feeding Study." *Environmental Health Perspectives* 104, suppl. 4: 823–28.

DeVito, M. J., L. S. Birnbaum, W. H. Farland, and T. A. Gasiewicz. 1995. "Comparisons of Estimated Human Body Burdens of Dioxinlike Chemicals and TCDD Body Burdens in Experimentally Exposed Animals." *Environmental Health Perspectives* 103: 820–31.

Dubey, R. K., and E. K. Jackson. 2001. "Cardiovascular Protective Effects of 17-beta-Estradiol Metabolites." *Journal of Applied Physiology* 91: 1868–83.

EDSTAC. 1998. *Endocrine Disruptor Screening and Testing Advisory Committee (EDSTAC) Final Report.* Washington, D.C.: U.S. Environmental Protection Agency.

Edwards, C. J., K. Yamamoto, S. Kikuyama, and D. B. Kelley. 1999. "Prolactin Opens the Sensitive Period for Androgen Regulation of a Larynx-specific Myosin Heavy Chain Gene." *Journal of Neurobiology* 41: 443–51.

Environmental Protection Agency (EPA). 1998. Endocrine Disrupter Screening Program. *Federal Register* 63, no. 248 (28 December): 71542–68.

———. 2002. Endocrine Disruptor Screening Program Web Site. http://www.epa.gov/scipoly/oscpendo/#currentstatus.

Fox, J. E., M. Starcevic, K. Y. Kow, M. E. Burow, and J. A. McLachlan. 2001. "Nitrogen Fixation. Endocrine Disrupters and Flavonoid Signaling." *Nature* 413: 128–29.

Fry, D. M. 1995. "Reproductive Effects in Birds Exposed to Pesticides and Industrial Chemicals." *Environmental Health Perspectives* 103, suppl. 7: 165–71.

Geiser, K. 2001. *Materials Matter.* Cambridge, Mass.: MIT Press.

Giguere, V. 1999. "Orphan Nuclear Receptors: From Gene to Function." *Endocrine Reviews* 20: 689–725.

Guillette, L. J., and A. D. Crain, eds. 2000. *Environmental Endocrine Disrupters: An Evolutionary Perspective.* New York: Taylor and Francis.

Hawkins, M. B., J. W. Thornton, D. Crews, J. K. Skipper, A. Dotte, and P. Thomas. 2000. "Identification of a Third Distinct Estrogen Receptor and Reclassification of Estrogen Receptors in Teleosts." *Proceedings of the National Academy Sciences of the USA* 97: 10751–56.

Hayes, T. B., A. Collins, M. Lee, M. Mendoza, N. Noriega, A. A. Stuart, and A. Vonk. 2002. "Hermaphroditic, Demasculinized Frogs after Exposure to the Herbicide, Atrazine, at Low Ecologically Relevant Doses." *Proceedings of the National Academy of Sciences of the USA* 99: 5476–80.

Henschler, D. 1994. "Toxicity of Chlorinated Organic Compounds: Effects of the Introduction of Chlorine in Organic Molecules." *Angewandte Chemie International Edition* 33: 1920–35.

Hisamoto, K., M. Ohmichi, H. Kurachi, J. Hayakawa, Y. Kanda, Y. Nishio, K. Adachi, K. Tasaka, E. Miyoshi, N. Fujiwara, N. Taniguchi, and Y. Murata. 2001. "Estrogen

Induces the Akt-dependent Activation of Endothelial Nitric-Oxide Synthase in Vascular Endothelial Cells." *Journal of Biological Chemistry* 276: 3459–66.

Howdeshell, K. L., A. K. Hotchkiss, K. A. Thayer, J. G. Vandenbergh, and F. S. vom Saal. 1999. "Exposure to Bisphenol A Advances Puberty." *Nature* 401: 763–64.

International Joint Commission. 1992. *Sixth Biennial Report on Great Lakes Water Quality.* Windsor, Ont.: International Joint Commission.

Jackson, T. 1993. "Principles of Clean Production: An Operational Approach to the Preventive Paradigm." In *Clean Production Strategies: Developing Preventive Environmental Management in the Industrial Economy,* edited by T. Jackson, 143–64. Boca Raton: Lewis.

Koopman-Esseboom, C., D. C. Morse, N. Weisglas-Kuperus, I. J. Lutkeschipholt, C. G. Van der Paauw, L. G. Tuinstra, A. Brouwer, and P. J. Sauer. 1994. "Effects of Dioxins and Polychlorinated Biphenyls on Thyroid Hormone Status of Pregnant Women and Their Infants." *Pediatric Research* 36: 468–73.

Krishnan, K., and J. Brodeur. 1994. "Toxic Interactions Among Environmental Pollutants: Corroborating Laboratory Observations with Human Experience: Mechanism-Based Predictions of Interactions." *Environmental Health Perspectives* 102, suppl. 9: 11–17.

Kuiper, G. G., E. Enmark, M. Pelto-Huikko, S. Nilsson, and J. A. Gustafsson. 1996. "Cloning of a Novel Receptor Expressed in Rat Prostate and Ovary." *Proceedings of the National Academy of Sciences of the USA* 93: 5925–30.

La Baer, J., and K. R. Yamamoto. 1994. "Analysis of the DNA-binding Affinity, Sequence Specificity and Context Dependence of the Glucocorticoid Receptor Zinc Finger Region." *Journal of Molecular Biology* 239: 664–88.

Markey, C. M., C. L. Michaelson, E. C. Veson, C. Sonnenschein, and A. M. Soto. 2001. "The Mouse Uterotrophic Assay: A Reevaluation of Its Validity in Assessing the Estrogenicity of Biosphenol A." *Environmental Health Perspectives* 109: 55–60.

Matthews, J., T. Celius, R. Halgren, and T. Zacharewski. 2000. "Differential Estrogen Receptor Binding of Estrogenic Substances: A Species Comparison." *Journal of Steroid Biochemistry and Molecular Biology* 74: 223–34.

McEwen, B. S. 2001. "Estrogen's Effects on the Brain: Multiple Sites and Molecular Mechanisms." *Journal of Applied Physiology* 91: 2785–801.

McLachlan, J. A. 2001. "Environmental Signaling: What Embryos and Evolution Teach Us about Endocrine Disrupting Chemicals." *Endocrine Reviews* 22: 319–41.

McMurray, R. W. 2001. "Estrogen, Prolactin, and Autoimmunity: Actions and Interactions." *International Immunopharmacology* 1: 995–1008.

Mehendale, H. M. 1994. "Amplified Interactive Toxicity of Chemicals at Nontoxic Levels: Mechanistic Considerations and Implications to Public Health: Mechanism-Based Predictions of Interactions." *Environmental Health Perspectives* 102, suppl. 9: 139–50.

Melo, A. C., and J. S. Ramsdell. 2001. "Sexual Dimorphism of Brain Aromatase Activity in Medaka: Induction of a Female Phenotype by Estradiol." *Environmental Health Perspectives* 109: 257–64.

National Research Council (NRC). 1984. *Toxicity Testing: Strategies to Determine Needs and Priorities.* Washington, D.C.: National Academy Press.

————. 2000. *Hormonally Active Agents in the Environment.* Washington, D.C.: National Academy Press.

Oberdorster, E., M. A. Clay, D. M. Cottam, F. A. Wilmot, J. A. McLachlan, and M. J. Milner. 2001. "Common Phytochemicals Are Ecdysteroid Agonists and Antagonists: A Possible Evolutionary Link between Vertebrate and Invertebrate Steroid Hormones." *Journal of Steroid Biochemistry and Molecular Biology* 77: 229–38.

O'Brien, M. 2000. *Making Better Environmental Decisions.* Cambridge, Mass.: MIT Press.

Okino, S. T., and J. P. Whitlock Jr. 2001. "The Aromatic Hydrocarbon Receptor, Transcription, and Endocrine Aspects of Dioxin Action." *Vitamins and Hormones* 59: 241–64.

Olsson, M., B. Karlsson, and E. Ahnland. 1994. "Diseases and Environmental Contaminants in Seals from the Baltic and the Swedish West Coast." *Science of the Total Environment* 154: 217–27.

Paech, K., P. Webb, G. C. Kuiper, S. Nilsson, J. Gustafsson, P. J. Kushner, and T. S. Scanlan. 1997. "Differential Ligand Activation of Estrogen Receptors ER-alpha and ER-beta at AP1 Sites." *Science* 277: 1508–10.

Payne, J., N. Rajapakse, M. Wilkins, and A. Kortenkamp. 2000. "Prediction and Assessment of the Effects of Mixtures of Four Xenoestrogens." *Environmental Health Perspectives* 108: 983–87.

Payne, J., M. Scholze, and A. Kortenkamp. 2001. "Mixtures of Four Organochlorines Enhance Human Breast Cancer Cell Proliferation." *Environmental Health Perspectives* 109: 391–97.

Porterfield, S. P. 2000. "Thyroidal Dysfunction and Environmental Chemicals: Potential Impact on Brain Development." *Environmental Health Perspectives* 108, suppl. 3: 433–38.

Raffensperger, C., and J. Tickner, eds. 1999. *Protecting Public Health and the Environment: Implementing the Precautionary Principle.* Washington, D.C.: Island Press.

Reijnders, P. J. 1986. "Reproductive Failure in Common Seals Feeding on Fish from Polluted Coastal Waters." *Nature* 324: 456–57.

Roe, D., W. Pease, K. Florini, and E. Silbergeld. 1997. *Toxic Ignorance: The Continuing Absence of Basic Health Testing for Top-Selling Chemicals in the United States.* Washington, D.C.: Environmental Defense Fund. Available at http://environmentaldefense.org.

Rogatsky, I., K. A. Zarember, and K. R. Yamamoto. 2001. "Factor Recruitment and TIF2/GRIP1 Corepressor Activity at a Collagenase-3 Response Element That

Mediates Regulation by Phorbol Esters and Hormones." *EMBO Journal* 20: 6071–83.

Romkes, M., J. Piskorska-Pliszczynska, and S. Safe. 1987. "Effects of 2,3,7,8-Tetrachlorodibenzo-p-Dioxin on Hepatic and Uterine Estrogen Receptor Levels in Rats." *Toxicology and Applied Pharmacology* 87: 306–14.

Rooney, A. A., and L. J. Guillette. 2000. "Contaminant Interactions with Steroid Receptors: Evidence for Receptor Binding." In *Environmental Endocrine Disrupters: An Evolutionary Perspective,* edited by L. J. Guillette and A. D. Crain, 82–125. New York: Taylor and Francis.

Rosner, W., D. J. Hryb, M. S. Khan, A. M. Nakhla, and N. A. Romas. 1999. "Sex Hormone-Binding Globulin Mediates Steroid Hormone Signal Transduction at the Plasma Membrane." *Journal of Steroid Biochemistry and Molecular Biology* 69: 481–85.

Royal Society. 2000. *Endocrine Disrupting Chemicals (EDCs).* London: The Royal Society.

Rubin, B. S., M. K. Murray, D. A. Damassa, J. C. King, and A. M. Soto. 2001. "Perinatal Exposure to Low Doses of Bisphenol A Affects Body Weight, Patterns of Estrous Cyclicity, and Plasma LH Levels." *Environmental Health Perspectives* 109: 675–80.

Seegal, R. F., and S. L. Schantz. 1994. "Neurochemical and Behavioral Sequelae of Exposure to Dioxins and PCBs." In *Dioxins and Health,* edited by A. Schecter, 409–48. New York: Plenum.

Shafer, T. J., T. R. Ward, C. A. Meacham, and R. L. Cooper. 1999. "Effects of the Chlorotriazine Herbicide Cyanaine on GABA-A Receptors in Cortical Tissue from Rat Brain." *Toxicology* 142: 57–68.

Shang, Y., X. Hu, J. DiRenzo, M. A. Lazar, and M. Brown. 2000. "Cofactor Dynamics and Sufficiency in Estrogen Receptor-Regulated Transcription." *Cell* 103: 843–52.

Smith, B. S. 1981. "Male Characteristics on Female Mud Snails Caused by Antifouling Bottom Paints." *Journal of Applied Toxicology* 1: 22–25.

Stoica, A., B. S. Katzenellenbogen, and M. B. Martin. 2000. "Activation of Estrogen Receptor-alpha by the Heavy Metal Cadmium." *Molecular Endocrinology* 14: 545–53.

Stoker, T. E., J. M. Goldman, and R. L. Cooper. 2001. "Delayed Ovulation and Pregnancy Outcome: Effect of Environmental Toxicants on the Neuroendocrine Control of the Ovary." *Environmental and Toxicology Pharmacology* 9: 117–29.

Suzuki, T., K. Ide, and M. Ishida. 2001. "Response of MCF-7 Human Breast Cancer Cells to Some Binary Mixtures of Oestrogenic Compounds In-Vitro." *Journal of Pharmacy and Pharmacology* 53: 1549–54.

Takeshita, A., N. Koibuchi, J. Oka, M. Taguchi, Y. Shishiba, and Y. Ozawa. 2001. "Bisphenol-A, an Environmental Estrogen, Activates the Human Orphan Nuclear Receptor, Steroid and Xenobiotic Receptor-mediated Transcription." *European Journal of Endocrinology* 145: 513–17.

Thayer, K. A., R. L. Ruhlen, K. L. Howdeshell, D. L. Buchanan, P. S. Cooke, D. Preziosi, W. V. Welshons, J. Haseman, and F. S. vom Saal. 2001. "Altered Prostate Growth and Daily Sperm Production in Male Mice Exposed Prenatally to Subclinical Doses of 17alpha-ethinyl Oestradiol." *Human Reproduction* 16: 988–96.
Thornton, J. 2000. *Pandora's Poison: Chlorine, Health, and a New Environmental Strategy.* Cambridge, Mass.: MIT Press.
———. 2001. "Implementing Green Chemistry: An Environmental Policy for Sustainability." *Pure and Applied Chemistry* 73: 1231–36.
Timm, G. 2002. U.S. Environmental Protection Agency Office of Pesticide Programs. 26 April. Personal communication, telephone interview.
Toppari, J., J. C. Larsen, P. Christiansen, A. Giwercman, P. Grandjean, L. Guillette, B. Jegou, T. Jensen, P. Jouannet, N. Keiding, H. Leffers, J. McLachlan, O. Meyer, J. Muller, E. Rajpert-De Meyts, T. Scheike, R. Sharpe, J. Sumpter, and N. Skakkebaek. 1996. "Male Reproductive Health and Environmental Xenoestrogens." *Environmental Health Perspectives* 104, suppl. 4: 741–803.
Ulrich, E. M., A. Caperell-Grant, S. H. Jung, R. A. Hites, and R. M. Bigsby. 2000. "Environmentally Relevant Xenoestrogen Tissue Concentrations Correlated to Biological Responses in Mice." *Environmental Health Perspectives* 108: 973–77.
Vivanco, M. D., R. Johnson, P. E. Galante, D. Hanahan, and K. R. Yamamoto. 1995. "A Transition in Transcriptional Activation by the Glucocorticoid and Retinoic Acid Receptors at the Tumor Stage of Dermal Fibrosarcoma Development." *EMBO Journal* 14: 2217–28.
vom Saal, F. S., B. G. Timms, M. M. Montano, P. Palanza, K. A. Thayer, S. C. Nagel, M. D. Dhar, V. K. Ganjam, S. Parmigiani, and W. V. Welshons. 1997. "Prostate Enlargement in Mice Due to Fetal Exposure to Low Doses of Estradiol or Diethylstilbestrol and Opposite Effects at High Doses." *Proceedings of the National Academy of Sciences of the USA* 94: 2056–61.
Walsh, L. P., C. McCormick, C. Martin, and D. M. Stocco. 2000. "Roundup Inhibits Steroidogenesis by Disrupting Steroidogenic Acute Regulatory (StAR) Protein Expression." *Environmental Health Perspectives* 108: 769–76.
Walsh, L. P., and D. M. Stocco. 2000. "Effects of Lindane on Steroidogenesis and Steroidogenic Acute Regulatory Protein Expression." *Biology Reproduction* 63: 1024–33.
Wang, Z. Y., H. Seto, S. Fujioka, S. Yoshida, and J. Chory. 2001. "BRI1 Is a Critical Component of a Plasma-Membrane Receptor for Plant Steroids." *Nature* 410: 380–83.

CHAPTER 9

Uncertainty and Global Climate Change: The Case of Mosquitoes and Mosquito-Borne Disease

Alistair Woodward

My aim is to shed some light on the variety of uncertainties that accompany science and to prompt reflection on how uncertainty does and should affect public policy. To do this, I have chosen a modern environmental problem that is unusually challenging because of its scale. Climate change is one of the first truly global environmental issues, and because of the longevity of greenhouse gases and the inertia of the world's climate system, the effects span centuries. A problem so widespread and associated with such a long timescale brings with it enormous uncertainties. To a large extent, policy decisions depend on how these uncertainties are handled. Meaningful cuts in greenhouse emissions will require substantial social changes, but the consequences of taking no action may also be very expensive. Furthermore, choices have to be made before all the relevant information is available. It would be convenient if a decision on resetting the global thermostat could be delayed until it was absolutely clear that a warmer world was a harmful one. However, climate systems cannot be turned on and off like an air-conditioning unit. The considerable lag period between greenhouse emissions and climate impacts means that policy makers must make decisions many years before the full effects are apparent. This article deals with uncertainties about how global warming will affect us, rather than with uncertainties about whether and to what extent global warming is occurring. It should be noted that not long ago, the uncertainties policy makers had to deal with were predominantly about whether it would occur at all.

This paper focuses on one category of potential climate change impacts: mosquito-borne diseases. It suits my purpose because the problem is complex—the frequency of these diseases depends on many factors, demographic, economic and ecological—and views differ on what advice scientists should be providing to policy makers. Disagreements may be fruitful—sometimes they provide a window for a clearer view of judgment, inference, and assessment of uncertainty. For example, Conrad Brunk investigated reasons for the widely differing results of three Canadian risk assessments of the pesticide alachlor and showed the major effects on the outcome of various risk-tolerant or risk-sensitive judgments in the scientific process (Brunk et al. 1991).

Background

There is no doubt that human activity has altered the composition of the earth's atmosphere. What has happened to the world's climate as a result is less certain, but most of the warming that has occurred in the past fifty years is probably human induced. It is more difficult to forecast what will happen in the future—complicating factors include future trends in greenhouse emissions, biophysical feedback forces, and threshold phenomena in the climate system that by their very nature are difficult to anticipate. The effects of climate change on human health depend on a sequence of events that produces a "cascade of uncertainty" (Schneider et al. 1998). Health effects will depend not only on the nature and rate of climate change at the local level, but also on human capacities for adaptation and the ability of ecological systems to buffer climate variability.

Mosquito-borne diseases are a major cause of human ill health worldwide. Each year there are hundreds of millions of cases of malaria and dengue, the two most common mosquito-transmitted infections (Kovats et al. 2001). Mosquitoes cannot regulate their internal temperatures and are therefore exquisitely sensitive to external temperatures and moisture levels. Specifically, temperature influences the size of mosquito populations (via rates of growth and reproduction) and infectivity (resulting from the effects on mosquito longevity and pathogen incubation). As a result, no assessment of the impacts of climate change can overlook possible effects on the global pattern of malaria, dengue, and other mosquito-borne diseases.

Climate change will be associated with changes in temperature, precipitation, and possibly soil moisture, and these factors may all affect disease prevalence. But mosquitoes are not only temperature-dependent; they are also very adaptable. For example, these insects can adjust to adverse temperature and moisture by exploiting microenvironments such as containers and drains (and as a result can overwinter in places where the temperatures are theoretically too cold). Mosquitoes can exploit climate changes that work in their favor, yet also have a good capacity to adapt to changes that might be expected to work against them.

There are few instances in which effects of long-term climate change on human health have been directly observed, and where there have been substantial changes in rates of mosquito-borne disease in recent times, there is little evidence that climate has played a major part (Kovats et al. 2001). This means that assessments of the impact of future climate change rely on expert judgment, informed by "analogue" studies, deduction from basic principles, and modeling of health outcomes related to climate inputs.

The Third Assessment Report of the Intergovernmental Panel on Climate Change

I have chosen to look in detail at the Third Assessment Report (TAR) of the Intergovernmental Panel on Climate Change (IPCC) because it is the most wide-ranging and detailed study of the impacts of climate change (McCarthy et al. 2001). The IPCC is a large, international body of scientists that has been convened on three occasions in the past ten years to advise governments on the magnitude and causes of climate change, the likely effects, and measures that may be taken to reduce impacts. The scientists who contribute to the work are divided into working groups along these lines. As a means of exploring the uncertainties associated with such assessments, I will examine the differences between the conclusions reached by Working Group II of the IPCC and one individual scientist who disagrees strongly with the position of the IPCC.

To estimate the likely effects of climate change on malaria incidence, the IPCC considered the following lines of evidence:

• Observed variations in disease incidence and range in association with short-term climate variability
• Long-term disease trends and the factors that may be responsible
• Projections based on first-principle relations between temperature and vector growth and activity
• Empirical disease models based on the current geographical distribution of malaria

Recognizing the complexity of the task and the gaps in current knowledge, the authors of the TAR strove to achieve consistent expressions of certainty and to be explicit about the ways in which judgments were made. Two approaches were applied: assessment of the "degree of belief" among the authors that the event would occur given current knowledge and an assess-

Table 9.1. Climate change and mosquito-borne diseases: the conclusions of the Third Assessment Report

Health impact	Confidence
Mosquito-borne infections likely to move to higher altitudes	Medium–high
Move to higher latitudes	Medium–low
Transmission seasons extended	Medium–high
Increased proportion of world population in areas of potential transmission for malaria and dengue	Medium–high
Decreased transmission where temperatures already high	Low–medium
Other vector-borne infections affected at margins of current distributions	Medium–high

ment of the quality of the science presently available. The degree of belief assessment was spread over five points ranging from 95 percent or greater (very high confidence) to 5 percent or less (very low confidence). The quality of the science was judged against a four-point standard (well established, established but incomplete, competing explanations, and speculative).

Overall, the TAR judged that rising temperatures and changing rainfall would have mixed effects on the potential for infections such as malaria and dengue worldwide (table 9.1). It reported that in areas with limited public health resources, warming, in conjunction with adequate rainfall, will cause certain mosquito-borne infections to move to higher altitudes (medium to high confidence) and higher latitudes (medium to low confidence). Transmission seasons will be extended in some endemic locations (medium to high confidence). Where temperatures are already close to the upper tolerable threshold for the disease vector (a relatively uncommon limiting factor), it is expected that transmission of disease will decrease (low to medium confidence). Models suggest the proportion of the world's population living in areas in which malaria or dengue could potentially be transmitted would increase by a "modest" amount (medium to high confidence). (This forecast is accompanied by the qualifier that such forecasts do not take account of other important factors besides climate, such as socioeconomic and demographic factors.)

Other Views

Not all scientists agree with the IPCC's conclusions on mosquito-borne disease. On one side, some scientists believe that simple climate change models provide no useful information about future disease rates because factors other than climate are bound to be more important (Reiter 2000). On the other side, some scientists argue that the IPCC is overcautious, and global warming is already extending the geographic range of significant mosquito-borne diseases (Epstein 2000). A close look at the reasons for these differences provides insights into the ambiguities of science for policy. I will examine in particular the contrast between the Reiter and IPCC positions.

Paul Reiter is a senior scientist in the U.S. Centers for Disease Control, with many years' experience of vector-borne infectious diseases. He has written widely on the influence of climate and other factors on the history of mosquito-borne infections (Reiter 2000; Dye and Reiter 2000). His papers are frequently cited when questions are raised about the accuracy of climate change forecasts (Masood 2000). In these papers, Reiter points to factors other than climate that have caused substantial shifts in the pattern of malaria and other infections. Examples include forest clearance, urbanization, changing agricultural practices, and improved health care. As an illustration, malaria was common in most of northern Europe until the middle of the nineteenth century, but in the late 1800s the incidence of locally transmitted disease dropped sharply in many countries, including England, France, Germany, and Denmark (Reiter 2001). The reasons for this decline had little to do with climate, but were more likely to be found in the clearing of swamps, new farming methods, better housing, more widespread use of quinine as a treatment for malaria, and lower densities of human populations in rural areas (Bruce-Chwatt and de Zulueta 1980). In parts of Europe where these social and demographic changes did not occur (such as Russia and Poland), malaria persisted into the twentieth century. In the 1920s, outbreaks of malaria in the Soviet Union occurred as far north as Archangel, demonstrating the ability of disease-carrying mosquitoes to survive in locations that are climatically adverse.

On the basis of this evidence, Reiter argues that the causation of mosquito-borne disease is complex and multifactorial, and climate is unlikely to be any more important in the future than it has been in the past as an explanation the rise and fall of malaria, dengue, and yellow fever. Therefore, "It is inappropriate to use climate-based models to predict future [disease] prevalence" (Reiter 2001).

Similarities and Differences

When comparing Reiter's papers with the TAR, it is striking how much common ground there is. Both emphasize the complexity of mosquito-borne disease and the importance of factors such as population susceptibility, insecticide resistance, primary health care, land use patterns, and subtle alterations in microenvironments. The IPCC shares Reiter's view that climate has not been a major determinant in the historical ebb and flow of mosquito-borne disease. Reiter agrees with the IPCC that unfavorable social and environmental changes could lead to a resurgence of malaria in parts of the world that are currently disease-free (Dye and Reiter 2000). The critiques of disease modeling are similar also; there seems to be no disagreement about the limitations of the available methods for forecasting future changes.

But the conclusions are different. Reiter argues that because malaria rates are strongly affected by nonclimate factors, one cannot conclude that there will be warming-induced increase in disease. The IPCC says that for a given configuration of co-factors (e.g., population susceptibility, adequate rainfall, deficient disease control activities), climate change would cause the disease to become more common (increasing its range or transmission season or both). So should policy makers take account of malaria, dengue, and other mosquito-borne diseases in weighing up the costs and benefits of reducing greenhouse emissions? One conclusion inclines to a negative (we can't tell whether there will be impacts and in any case, other factors than climate are more important) and the other to a positive position (in certain settings climate change could make things substantially worse).

Possible Reasons for Different Views

Why should scientific assessments that agree on the evidence and the methods end up pointing in different directions? First, the IPCC conclusion is couched in terms of disease potential (what would happen *if* certain conditions were to apply). In contrast, Reiter focuses on predictions of actual disease incidence (not what *might* happen, but what *will* happen). Underlying both positions are presumptions about what scientists can usefully contribute to the policy debate. The focus on what might happen is consistent with a view that scientists have important insights to offer policy makers, even when they cannot predict the future with a high degree of certainty. The focus on what *will* happen suggests that the role of science is to establish with a high degree of confidence whether causal relationships exist. From this perspective, science cannot be used as the basis for precautionary action where such confidence cannot be achieved.

Some have argued that scientists may actually cause more harm than good when they comment on complex and ambiguous problems that lie outside the limits of empirical research, since they invest what is really no more than lay opinion with what appears to be the authority of science (Taubes 1995). My view is that science has a legitimate function in tackling "what if?" type questions. The prediction criterion is too severe: it presumes a deterministic universe that does not exist. In fact no scientific study (not even the most tightly controlled experiment) can predict the future with complete certainty. Levels of confidence are much higher in some areas than others, but in many fields, science necessarily deals largely with questions about "what might happen."

Extrapolations from the research environment to the real-world setting are always hedged by conditions. For example, it may not be explicit, but the conclusion of every laboratory investigation takes the form "*if* the setting in the laboratory were to be replicated, *then* the following outcomes would be expected . . ."

Second, the IPCC and Reiter make different assumptions about the state of a future world. Any summary statement about the likely effects of a changed climate assumes a certain level of susceptibility, whether it is the status quo, what might be predicted from current trends, or a worst case scenario. Reiter's position assumes a configuration of high disease control capacity and low population susceptibility (as a result of socioeconomic change, for example). "If the present warming trend continues, human strategies to avoid these temperatures—particularly indoor living and air conditioning—are likely to become more prevalent" (Reiter 2001). The IPCC, in contrast, considers possible futures in which disease susceptibility is greater, such as settings in which public health services deteriorate, economic productivity declines, and social order unravels. Reiter concludes that "reestablishment of the disease [in Italy] is unlikely unless living standards deteriorate drastically"(Reiter 2001); the IPCC holds a similar view but frames it differently— "malaria could become established again [in Europe] under the prolonged pressures of climatic and other environmental-demographic changes if a strong public health infrastructure is not maintained."

Third, scale is important. The factors that produce the most noticeable changes over a short time period may not be the same as those that cause long-term changes in disease rates. Similar considerations apply to spatial distributions. Malaria has been pantropical for centuries, and this is essentially a function of climate (Sachs and Malaney 2002). The recent retreats and advances cited by Reiter, such as the retreat of malaria from the eastern and southern United States, are not due to climate and have had important consequences for public health, but on a global scale, they are movements on the margins. McMichael has applied this idea to the metaphoric "pool of disease," emphasizing the difference between factors that agitate the surface (short-term, localized variations) and the conditions that determine the depth of the pool (McMichael 1995). Causes in the latter category have also been described as the "driving forces" of disease incidence (Corvalan et al. 1999). The relative importance of different causes will depend on the scale that is chosen, and the choice of what time- and space-defining windows to apply depends on the problem in hand. An investigation of an outbreak of disease will naturally focus on the causes of short-term and localized variations in incidence, but an assessment of long-term future disease trends might give greater weight to large-scale influences that occur upstream in the sequence of causes. In general, the Reiter interpretation tends toward the "outbreak" frame of reference, whereas the IPCC interpretation is more attuned to larger scale, "driving forces" explanations.

The differences in the conceptual frameworks applied to the problem of mosquito-borne disease and climate change might be illustrated in this way. Imagine a children's playground. City planners propose to turn the street alongside the playground into a major arterial road. How would this decision affect children's health? We overlook, for the sake of the example, problems caused by air pollution and noise and concentrate on the risk of injury. In the past few years, as part of urban renewal projects supported by substantial economic growth in the city, the town authorities have built around the playground fences and gates that are considered by experts in the safety field to be "childproof."

Some might argue that with the fence and gates in place, traffic considerations are irrelevant. If children can't get out of the playground they are not exposed to the risk of injury, so it doesn't matter how many cars there are, nor how quickly they are traveling. Others might argue that if the street becomes an arterial road the *potential* for injury will increase. According to this view, traffic-based models of injury are not "irrelevant" to public health. It may be unwise to assume that the playground is permanently child proofed. *If* gaps in the fence were not fixed promptly or the gates were not used as intended, *then* children would be at much increased risk of serious injury in a high-traffic environment. There also may be children who are not protected by fences and gates because they play outside the playground (such as those who live in less wealthy city areas ten blocks down the street).

The likeness is imperfect, but one might compare the public health infrastructure required to deal with malaria with the fence around the playground, and global warming with increases in traffic. In both instances there are multiple causes of illness or injury, and an assessment of the possible effects of changes in one factor will depend on what assumptions are made about the levels of other, component causes.

Science and Uncertainty

The TAR raises the question (without giving an answer) of whether "science for policy" is different from "science itself." Answering particular questions that are important at the time for policy makers may not be not the same as pursuing science for its own sake, but does the standard of proof vary? Perhaps one could argue that there should be greater leeway for judgment when the emphasis lies on responding in a timely fashion to policy makers' questions, but the distinction does not appear to be a very useful one. If there were two kinds of science, how would one distinguish them from each other? Is any scientist practicing "science for policy" when he or she comments on the public implications of his or her work? These questions raise important additional questions about methods. Should all attempts at quantification be abandoned where the underlying science is not well understood and there is only a weak empirical base for estimating the likelihood of future events? Are there ways to retain quantitative estimates while tempering them with a full description of uncertainty bounds? Possible approaches included in the IPCC report are standardized graphical displays (see figure 7.2 for an example in the TAR) (McCarthy et al. 2001) and verbal summaries of confidence categories. The IPCC approach to uncertainty is clearly imperfect (e.g., there are substantial inconsistencies between the reports of Working Groups I and II), but it points the way for future, large-scale scientific assessments.

Studies of risk perception show that values and social position have an important bearing on how individuals view environmental hazards (Flynn et al. 1994). This is likely to be true for climate change scientists as well. Van Asselt and Rotmans (1996) liken risk assessment to a group of hikers crossing an unfamiliar landscape. Although the hikers start together and face the same terrain, it would not be surprising if they chose different routes and, as a result, arrived at different destinations. Their choices along the way (to cross a dangerous river or to take a lengthy diversion; to follow footpaths or forge new routes; to continue when the weather is threatening or take shelter) depend on past experiences, preferences, interests, and preconceptions about the nature of the land ahead. Similarly, no one begins an assessment of the effects of climate change with a completely open mind. Scientists bring to the task expectations, attitudes, and values that influence the questions that are asked, help to make sense of the data and, inevitably, shape the meaning given to the results. Important dimensions of difference include presumptions about nature (capricious, benign, forgiving, or fragile), the limits of human capacities, and priorities given to core values (such as equity and individual liberties). It is not clear whether a large group, such as the IPCC, is less prey to such problems than individual scientists might be.

Even between disciplines, there may be major differences. A survey of experts in the field found that natural scientists' estimates of the total damage caused by climate change tended to be much greater than those of economists. It was suggested that "economists know little about the intricate web of natural ecosystems, whereas scientists know equally little about the incredible adaptability of human economies" (Nordhaus 1994). At a most basic level, the manner in which results are presented may differ and affect the message that is taken from the data (Morgan and Henrion 1990). A simple example is the difference between absolute and relative measures of effect. Models by Martens et al. (1995), for instance, suggest that global warming will increase the numbers of people living in regions of potential malaria transmission by 2080 by about 260 to 320 million. This appears a large effect, but in relation to the number of people already living in malaria zones the change is small (about 4 percent). The level of carbon dioxide in the earth's atmosphere has increased by about 30 percent in the past one hundred years. Climatologists tend to emphasize the relative increase (and the rate of change) (McCarthy et al. 2001); a different impression is given when the change is expressed in absolute terms (from 0.029 percent to 0.037 percent) (Reiter 2001).

Do multidisciplinary assessments, such as those carried out by the IPCC, underestimate uncertainty? In theory, they might do so. Collaborations like the IPCC have important strengths: no single discipline holds all the knowledge required to deal with complex environmental problems like climate change. But an interdisciplinary group may have more difficulty providing a complete account of uncertainties than is possible when the authors all speak the same technical language. Uncertainties to do with issues of measurement and attribution tend to be smoothed over when scientific debates move from specialized forums to public settings (Godard 1996). A similar process has been observed with interdisciplinary scientific panels in the past. "Doubts and uncertainties of core specialists are diminished by the overlaps and interpenetrations with adjacent disciplines . . . the net result is a more secure collective belief in the policy knowledge, or the technology, than one might have obtained from any of the separate contributing disciplines" (Wynne, cited in Godard 1996). But in regard to mosquito-borne disease, at least, the tone of the IPCC assessment does not appear to have been greatly influenced by the mixing of disciplines. The discussion of uncertainties in the TAR differs little from that contained in papers written separately by the scientists involved (for instance, Githeko et al. 2000; Patz et al. 2000). In general, the challenge for interdisciplinary groups may be to develop a clear language for communicating about uncertainty and, at the same time, give due weight to disagreements between specialists.

Conclusion

In the case study I have examined here, major sources of uncertainty concerning links among climate change, mosquitoes, and human health become apparent when one examines reasons for the variations of views from the IPCC position. Statistical uncertainty is one component, but not a major one. On the other hand, semantic issues—such as distinctions between "forecasts" and "predictions" or between "future disease rates" and "potential future disease rates"—are important causes of disagreement. To a large extent, the difference between the IPCC's conclusions and those presented by Reiter can be explained by differences in basic assumptions and fundamental worldviews. The IPCC view might be characterized as a cautious approach that does not take historic social advances for granted and sees humans as relatively minor players on a very large stage. The Reiter analysis places greater weight on the human capacity to shape and control environments, combined with confidence that social and economic achievements of western countries in the past century will be sustained worldwide over the next hundred years.

Advice of any kind, scientific or otherwise, always has two components: a particular framing of the problem in hand and an account of what lies within the frame. Where there are differences in the advice, the cause may lie with the choice of windows and not the description of the view. Arguments about the quality of the science will not resolve differences if the explanation lies fundamentally in the way the problem has been conceived.

Scientists are trained primarily to describe and analyze data, and may be less inclined (or possibly less well prepared intellectually) to stand back and reflect on the values and assumptions that permeate scientific method. When policy makers weigh up scientific advice, it may be helpful to know where the advice is coming from. In a sense, the users of scientific advice must decide not only whether the science is "robust" but also whether they agree with the values that frame a particular position. For this to happen, scientists need to be explicit about the worldview they bring to their work.

A great deal has been written about the way models and theories shape what is seen (and what is not seen) (Ziman 1984; Krieger 1994). It is more difficult to find practical suggestions in the literature about how scientists can be more open about their starting positions and, in this way, reduce uncertainty by making clear where disagreements truly lie. Perhaps a step in the right direction is the requirement by some journals that authors supply a declaration of interests (Editors 1993). This policy has been introduced because editors know how powerful the framing effect can be, and they believe that the readers will be better placed to assess a paper if they know its origins (Smith 1998). The focus is mostly on conflicts that may arise due to funding and employment, but a similar approach could perhaps be applied to a broader definition of "interests."

The approach taken by the IPCC in its climate change assessment is, in many ways, an attractive one. Positive features include the wide range of disciplines involved, the iterative nature of the assessment, involvement of scientific peers and policy makers in the review process, and the attempt to make the process of collecting and weighing up scientific evidence as transparent as possible. This case study suggests a number of changes that might assist providers and consumers of scientific advice: more explicit discussion of the sources of uncertainty, sticking to the difficult task of building a common framework between disciplines, and treating disagreement between scientists as potentially illuminating rather than problematic.

Acknowledgments

I am grateful for detailed comments on an earlier draft by Simon Hales, Phil Weinstein, Margot Parkes, Kris Ebi, Tony McMichael, and Ralph Chapman.

References Cited

Bruce-Chwatt, L., and J. de Zulueta. 1980. *The Rise and Fall of Malaria in Europe.* Oxford: Oxford University Press.

Brunk, C. G., L. Haworth, and B. Lee. 1991. *Value Assumptions in Risk Assessment. A Case Study of the Alachlor Controversy.* Waterloo, Ont.: Wilfrid Laurier University Press.

Corvalan, C. F., T. Kjellstrom, and K. R. Smith. 1999. "Health, Environment and Sustainable Development: Identifying Links and Indicators to Promote Action." *Epidemiology* 10, no. 5: 656–60.

Dye, C., and P. Reiter. 2000. "Climate Change and Malaria: Temperatures without Fevers?" *Science* 289, no. 5485: 1697–98.

Editors. 1993. "Conflict of Interest." *Lancet* 3417: 42–43.

Epstein, P. R. 2000. "Is Global Warming Harmful to Health?" *Scientific American* 283, no. 2: 50–57. See comments.

Flynn, J., P. Slovic, and C. K. Metz. 1994. "Gender, Race, and Perception of Environmental Health Risks." *Risk Analysis* 14: 1101–8.

Githeko, A. K., S. W. Lindsay, U. E. Confalonieri, and J. A. Patz. 2000. "Climate Change and Vector-Borne Diseases: A Regional Analysis." *Bulletin of the World Health Organization* 78, no. 9: 1136–47.

Godard, O. 1996. *Integrating Scientific Expertise into Regulatory Decision-Making.* Florence: European University Institute.

Kovats, R. S., D. Campbell-Lendrum, A. J. McMichael, A. Woodward, and J. S. Cox. 2001. "Early Effects of Climate Change: Do They Include Changes in Vector-Borne Disease?" *Philosophical Transactions of the Royal Society of London, Series B: Biological Sciences* 356: 1–12.

Krieger, N. 1994. "Epidemiology and the Web of Causation: Has Anyone Seen the Spider?" *Social Science and Medicine* 39: 887–903.

Martens, W. J. M., L. W. Niessen, J. Rotmans, T. H. Jetten, and A. J. McMichael. 1995. "Potential Risk of Global Climate Change on Malaria Risk." *Environmental Health Perspectives* 103, no. 5: 458–64.

Masood, E. 2000. "Biting Back." *New Scientist* 167, no. 2257: 41–42.

McCarthy, J. J., O. F. Canziani, N. A. Leary, D. J. Dokken. and K. Se. White. 2001. *Climate Change 2001: Impacts, Adaptation, and Vulnerability. Contribution of Working Group II to the Third Assessment Report of the Intergovernmental Panel on Climate Change.* Cambridge: Cambridge University Press.

McMichael, A. J. 1995. "The Health of Persons, Populations, and Planets: Epidemiology Comes Full Circle." *Epidemiology* 6: 633–36.

Morgan, G. M., and M. Henrion. 1990. *Uncertainty: A Guide to Dealing with Uncertainty in Quantitative Risk and Policy Analysis.* New York: Cambridge University Press.

Nordhaus, W. D. 1994. "Expert Opinion on Climatic Change." *American Scientist* 82: 45–51.

Patz, J. A., D. Engelberg, and J. Last. 2000. "The Effects of Changing Weather on Public Health." *Annual Review of Public Health* 21: 271–307.

Reiter, P. 2000. "From Ague to West Nile." *Scientific American* 283, no. 6: 10.

————. 2001. "Climate Change and Mosquito-Borne Disease." *Environmental Health Perspectives* 109, suppl. 1: 141–61.

Sachs, J., and P. Malaney. 2002. "The Economic and Social Burden of Malaria." *Nature* 415: 680–85.

Schneider, S. H., B. L. Turner, and H. M. Garriga. 1998. "Imaginable Surprise in Global Change Science." *Journal of Risk Research* 1, no. 2: 165–85.

Smith, R. 1998. "Beyond Conflict of Interest." *British Medical Journal* 317: 291–92.

Taubes, G. 1995. "Epidemiology Faces Its Limits." *Science* 269, no. 5221: 164–69.

van Asselt, M. B. A., and J. Rotmans. 1996. "Uncertainty in Perspective." *Global Environmental Change* 6: 121–57.

Ziman, J. 1984. *An Introduction to Science Studies: The Philosophical and Social Aspects of Science and Technology.* Cambridge: Cambridge University Press.

CHAPTER 10

The Precautionary Principle as a Guide to Environmental Impact Analysis: Lessons Learned from Global Warming

*Donald A. Brown**

The precautionary principle does not simply require policy makers to take action in the face of identified hazards; it also requires scientists and other analysts to identify clearly all the plausible effects of a proposed course of action, even when there is considerable uncertainty about the magnitude or nature of those effects. Successful implementation of the precautionary principle as a guide for decision making depends on responsible management of uncertainty not only in the final stages of policy formation, but also at earlier stages of analysis.

Appropriateness of Applying the Precautionary Principle to Global Warming

The precautionary principle is an appropriate guide to policy making on global warming programs, for two reasons. First, the precautionary principle is based on the rather uncontroversial ethical norm that persons should not perform acts that could cause harm to others, even if there is some uncertainty about whether those consequences will occur (Brown 2002). Aspects of the

---

*The opinions expressed in this article are those of the author and do not necessarily reflect the positions of his current or former employers.

141

global warming problem that make it appropriate to apply the precautionary principle include:

- The enormous adverse potential impacts on human health and the environment from global warming
- The disproportionate effects of global warming on the poorest people of the world who have not consented to or benefited from the use of fossil fuels in the developed nations
- The fact that much of the science of the climate change problem is not in dispute even if one acknowledges uncertainty about timing or magnitude global warming impacts
- The fact that the longer nations wait to take action, the more difficult it will be to stabilize greenhouse gases in the atmosphere at levels that do not create serious damage to human health and the environment

Therefore, a very strong case can be made that ethical considerations alone compel the use of this principle in formulating global warming policy (Brown 2002). However, if there is any doubt about an ethical justification for the use of the precautionary principle in global warming policy making, the second reason this chapter assumes the appropriateness of this principle to global warming decision making is because this issue is a settled matter of international law under the United Nations Framework Convention on Climate Change (UNFCCC). The UNFCCC included a version of the precautionary principle as a mandatory guide to all nations that are parties to the UNFCCC. Article 3 of the UNFCCC states:

> The Parties should take precautionary measures to anticipate, prevent or minimize the causes of climate change and mitigate its adverse effects. Where there are threats of serious or irreversible damage, lack of full scientific certainty should not be used as a reason for postponing such measures, taking into account that policies and measures to deal with climate change should be cost-effective so as to ensure global benefits at the lowest possible cost. (UNFCCC 1992)

The precautionary principle must thus guide analyses of environmental and human health impacts caused by global warming.

In the sections that follow, I examine how three technical documents have dealt with uncertainty in estimates of the likelihood of ecological and other effects from global warming and how they have shaped policy. I first examine the Second Assessment of the Intergovernmental Panel on Climate Change (IPCC), published in 1995 (IPCC 1995a–d). Second, I compare two studies that examine possible future impacts of global warming on the

Mid-Atlantic region of the United States. These two studies are based on the same scientific information and models, but draw dramatically different conclusions. Two lessons can be drawn from these case studies. First, scientific reports should make clear the importance of points they are unable to quantify, yet recognize as plausible. Second, economists and other analysts working to draw policy conclusions from such reports must not ignore those plausible effects that happen not to be amenable to quantification.

Second Assessment of the Intergovernmental Panel on Climate Change

Recognizing the need to consolidate knowledge about global climate change, the World Meteorological Organization and the United Nations Environment Programme established the IPCC in 1988 (IPCC 1995a). The international community assigned IPCC the role of assessing the scientific, technical and socioeconomic information necessary to understand the likelihood and implications of human-induced climate change (IPCC 1995a).[1]

The IPCC completed its First Assessment Report in 1990. This report was very influential in establishing the need to create the Intergovernmental Negotiating Committee, the group established by the UN General Assembly to negotiate the UNFCCC. The UNFCCC, which entered into force in 1994, provides the overarching international legal policy framework for addressing climate change.

The IPCC has continued to provide scientific, technical and socioeconomic advice to the world community, and in particular to the 170-plus parties to the UNFCCC through its periodic assessment reports on the state of knowledge of causes of climate change, its potential effects, and options for response strategies. Published in 1995, the IPCC Second Assessment Report provided key input to the negotiations that led to the adoption of the Kyoto Protocol to the UNFCCC in 1997 (IPCC 1995a–d). The IPCC also prepares special reports and technical papers on topics where independent scientific information and advice is deemed necessary. A third IPCC report was issued in 2001, but is not examined in this chapter (IPCC 2001).[2]

Identifying Global Warming Effects: Inherent Uncertainties

The IPCC's Second Assessment summarized the state of scientific knowledge on human-induced climate change, made specific predictions about hazards associated with climate change, and discussed mitigation options. In order to make specific predictions about timing and magnitude of future climate change and its effects on human health and the environment, the IPCC had to rely on computer models that describe the entire climate system, including interactions among oceans, soils, vegetation, and the atmosphere; the effects of global ice and snow cover on the reflection or absorption of incoming solar radiation; sun variability; the type and extent of cloud formation; and the natural carbon cycle. The computer models that attempt to describe these complex interactions suffer from inherent limitations that are not likely to be resolved in the near future. The IPCC also acknowledges additional uncertainties in its predictions of the response of the climate system because positive and negative feedbacks within the climate system are not well understood (Brown 2002).

In order to predict future climate events associated with human release of greenhouse gases, it is necessary to run models making varying assumptions about future levels of human population, technology, and fuel use. The IPCC developed a range of future greenhouse gas emissions scenarios based on several stated assumptions concerning population and economic growth, land use, technological changes, energy availability, and fuel mix during the period 1990 to 2100 (IPCC 1995b: 3). The IPCC used these emissions scenarios to develop projections both of atmospheric greenhouse gas concentrations and of future human-induced climate change and its effects on human health and the environment.

Although the IPCC Second Assessment expressed growing confidence in the computer models it used, it also recognized inherent limitations in the models' ability to predict human health and environmental impacts accurately. These limitations include computing limitations, assumptions about future events, and ecological impact uncertainties.

Computing Limitations

A climate model that explicitly included all our current knowledge about the climate system would be too complex to run on any existing computer (IPCC 1997). Therefore, for practical purposes, compromises must be made in attempts to model climate. The "art" of modeling includes selecting models of varying complexity for different aspects of the climate system. Differences among models include the following.

• *Number and size of spatial dimensions of the model.* Because some physical quantities such as temperature, humidity, and wind speed must be understood as they vary in space, it is necessary to represent these quantities at different locations in space in the computer model. The most general models for climate change employed by the IPCC are "general circulation models," which describe interactions among the atmosphere, oceans, and terrestrial ecosystems (IPCC 1997). Many of the more complex climate models array variables in three dimensions, with typical horizontal grids of several hundred kilometers and vertical layers of one kilometer. The greater the number of grids or blocks, the greater the complexity of the computing challenge and the higher the cost of running various scenarios. Because of the large number of variables entailed by this scheme, computer models are limited in their ability to model phenomena at smaller scales (IPCC 1997).

- *The need to parameterize some variables.* Computing limitations do not allow models of high enough resolution to resolve important subgrid processes. To account for subgrid phenomena, models average, rather than compute, subgrid effects such as cloud formation and cloud interactions with atmospheric radiation. This process of substituting empirical averages for subgrid phenomena is known as "parameterization" of subgrid phenomena (IPCC 1997).
- *The number of physical processes represented.* In order to make calculations manageable, models make simplifying assumptions about the nature of reality. Not all variables of concern can be included in models, and modelers must make choices among variables.

Assumptions about Future Events

To make predictions about the future effects of human activities on the climate system, scientists must make assumptions about the quantity of greenhouse gases that humans will put into the atmosphere. Factors that will help determine this quantity include population levels, the mix in use of different fuels, the efficiency of future combustion techniques and energy use, and level of economic activity. To accommodate these unknowns, computer models are usually run for several scenarios. In 1995, the IPCC reviews adopted six scenarios. Each of these scenarios made different assumptions about population growth, cost of energy supplies, deforestation rates, and economic growth rates (IPCC 1995b: 3).

Ecological Impact Uncertainties

Even if predictions about future temperature increases could be made with acceptable levels of certainty, predicting the consequences of specific temperature increases on ecological systems requires confrontation with many additional issues that are plagued by scientific uncertainty. The science of ecology is much too soft to predict ecosystem-wide responses to stress with certainty (Shrader-Frechette and McCoy 1993).

The IPCC's Predictions

For all these reasons, attempts to predict effects of human activities on the climate system and of climate change on ecological systems are frustrated by considerable scientific uncertainties. Particularly difficult to predict are the unexpected, large, and rapid climate system changes (as have occurred in the past) that are referred to as *climate surprises*. Opponents of emissions reduction programs in the United States argue that because of these uncertainties, the United States should take no action to reduce its emissions (see, for example, Mobil Corporation 1997).

Despite uncertainties, in its 1995 report the IPCC made a number of specific predictions. These predictions identified adverse climate impacts caused by rising temperatures that would include rising seas, increases in vector-borne disease, increases in intensity of storms, threats to food supplies, changes in the carbon cycle, adverse impacts on ecosystems, threats to forests, increases in deserts, and increases in droughts and floods. The severity of these impacts will depend on how much human-caused temperature change is experienced in this century, which the IPCC predicted will range from a high value of 3.5°C to a low value of 1°C (IPCC 1995a: para. 5).

These predictions were derived from complex computer models of the global climate system that predict mostly smooth and gradual responses of the climate, carbon, and ecological systems to increases in greenhouse gases (Chang 2001). Yet, the IPCC acknowledged that there are a number of potential nonlinear and sudden responses of the climate system. These nonlinear responses of the climate system—climate surprises (Streets and Glantz 2000)—of concern include the following.

- *Abrupt changes in ocean circulation patterns.* In the historical record, very rapid climate changes are believed to have been caused by sudden shifts in the ocean circulation patterns of the North Atlantic. If this were to occur again, there could be dramatic changes in the flow of the Gulf Stream, which would quickly cause sudden and major changes in regional climates (Stocker 2000: 302).
- *Large increases in methane release from melting permafrost.* If temperature increases rapidly enough, large amounts of methane could be released from the arctic permafrosts into the atmosphere, causing even larger amounts of global heating (Stocker 2000: 302).
- *Large sudden increases in sea level caused by breakup of polar ice cap.* A large part of the West Antarctic Ice Shelf could break off and instantaneously increase sea levels by as much as 10 meters (Stocker 2000: 312).
- *Nonlinear responses of the earth's carbon cycle.* Warming could have a nonlinear effect on the ability of natural processes to absorb CO_2 from the atmosphere. Since the solubility of CO_2 in sea water decreases as the water temperature increases, as oceans increase in temperature they may be less able to remove CO_2 from the air (Stocker 2000: 312). Higher temperatures could also increase plant respiration, thereby increasing CO_2 emissions (Stocker 2000: 312).

These rapid nonlinear climate changes are not included in the model-derived changes. Although the IPCC acknowledged that these climate surprises could create more severe impacts from global warming, the most frequently quoted IPCC predictions have been those derived from the models, such as the prediction that the earth will warm by a best estimate of 2°C and that the sea level will rise 50 cm. The IPCC could do a better job of making policy makers aware that climate surprises are plausible responses that could ultimately prove that the IPCC's more quantitative predictions have underestimated global warming impacts.

One can only speculate why policy makers have tended to ignore the IPCC's acknowledged concerns about climate surprises. One possible reason is that most of the text of the IPCC report is devoted to the impacts generated by the computer models, whereas very little of the analysis is concerned with climate surprises. The sheer volume of material on the computer-generated models may create the impression that they are more important than the other predictions. Another explanation is that the IPCC quantitative predictions of impacts derived from the models are more likely to be used than the narrative descriptions of the surprises precisely because they are quantitative, whereas the discussion of climate surprises includes little quantitative analysis.

Policy makers may prefer to work with quantitative predictions where they exist, simply because it is easier to incorporate these quantitative predictions into quantitative methods for policy analysis. For instance, when identifying the costs and benefits of various global warming programs to reduce the threat of global warming, many economists have used the impacts derived from the models, not those entailed by the surprises, as the basis for cost-benefit analyses (Shogren and Toman 2000). These analyses do not mention that much more dangerous effects, with much higher costs, are both possible and plausible, potentially underestimating the costs of nonaction. As a result, governments have not considered the full potential costs of global warming in figures used to compare cost and benefits of global warming programs.

Decision makers who have relied on cost-benefit analyses to make decisions about global warming programs lacked critical information. Even if the probability of the catastrophic surprises were small, the possibility of irreversible catastrophic effects might be a decisive factor for a decision maker.

This demonstrates that if impact analyses do not identify all plausible impacts, decision makers may be without important information that they may need to make an informed decision. One can conclude that information about all plausible impacts should be included at all stages of impact analyses. It is critically important that if cost-benefit analyses of emissions reduction programs omit consideration of some impacts, this information must be prominently displayed in the cost-benefit analysis.

Problems with IPCC's Reliance on Peer-Reviewed Research

The IPCC does not conduct independent research to determine impacts, but rather makes judgments based on published and peer-reviewed material. The reliance on published and peer-reviewed material may actually limit the IPCC's ability to consider scientifically plausible global warming impacts. This is so because only scientific conclusions that have reached conventional scientific standards of proof are generally published in the scientific publications that the IPCC relies on. Thus, by relying on published and peer-reviewed material to make its quantitative judgments, the IPCC may be excluding scientifically plausible but uncertain impacts of great concern.

Scientists normally follow procedures designed to prevent findings of causation that are not supported by high levels of proof. This approach is clearly not "value neutral," but rather is tilted on the side of protecting against making errors in prediction that would overstate possible serious consequences of global warming. The standard scientific approach of achieving 95 percent certainty of a causal relationship often makes sense in the laboratory, where research questions can be narrowly defined and complicating factors minimized. Yet, when predicting ecological outcomes it is often impossible to reach high levels of confidence about relationships between human action and ecological effects. As a result, the application of science to public policy concerned with identifying plausible but unproven impacts requires the use of procedures that do not insist on 95 percent confidence levels of correlation between cause and effect.

There is some evidence that in a few cases the IPCC's reliance on peer-reviewed and published studies has understated likely impacts from global warming. For instance, a very recent assessment using new tools to predict ecological impacts has found that climate change impacts might be much worse than those predicted by the 1995 IPCC report. This assessment found, among other things, that net ecosystem productivity may decrease significantly after the 2030s, causing terrestrial carbon sink to decrease significantly, with the possibility of it becoming negative by 2100, thus increasing the release of carbon to the atmosphere. By 2025 the number of people living in countries with water stress may increase by 50 to 100 million. Food prices and the risk of hunger will increase, as will the number of people at risk of coastal flooding (Parry et al. 1999).

This new assessment has been able to describe more serious impacts than those predicted by the IPCC in 1995 because there are now more sophisticated ways of modeling ecological impacts. These more sophisticated models indicate that much more carbon will be released from soils and vegetation than that formerly assumed by the IPCC in its 1995 report (Parry et al. 1999). In another example, a recent study has concluded that the nonlinear and rapid responses of the climate system entailed by the climate surprises may be likely rather than of low probability as assumed by many (Chang 2001).

The failure to identify all plausible impacts of climate change and the practice of limiting discussion to predictions supported by evidence that satisfies conventional scientific standards of proof raise many ethical issues. These include:

- how to achieve informed consent from those who may be adversely affected by a decision to take no action to reduce global warming;
- who should have the burden of proof in decisions relating to action on global warming; and
- whether the present generation has the right to gamble on issues that may affect future generations' quality of life.

Because there is considerable value in identifying the impacts that are being derived from the computer models that rely on peer-reviewed research to form the basis of equations used, the IPCC should not completely abandon this practice. It should, however, give greater emphasis to other plausible impacts from global warming that are yet not understood well enough to be included in the equations that constitute the computer models used to identify impacts. Perhaps the IPCC could continue to identify impacts using current methods but closely follow these descriptions of impacts with a listing of other plausible impacts of concern to the scientific community. The IPCC should clearly indicate both that these other impacts are speculative, yet that current scientific knowledge suggests they are plausible. In addition, research institutions that produce global warming science on which the IPCC and others will rely—to describe global warming impacts or consider the costs and benefits of global warming reduction strategies—must be allowed to identify scientifically plausible impacts even if the conclusions about these impacts have not been proven with high levels of statistical correlation between hypothesized cause and effect.

Two U.S. Climate Change Impact Analyses

Two recent reports prepared under the guidance of the U.S. Global Change Research Program (USGCRP) predict global warming effects on the Mid-Atlantic states in the United States. These two reports are based the same two models and many of the same assumptions, but produced very different descriptions of likely impacts. The first study is entitled "Preparing for a Changing Climate: The Potential Consequences of Climate Variability and Change, Mid-Atlantic Overview" (Mid-Atlantic Regional Assessment Team 2000) and is usually referred to as the MARA report. MARA was one of sixteen regional assessment reports on global warming impacts coordinated by USGCRP. The second report, also prepared by the USGCRP, is a synthesis report of global warming impacts for all of the United States. This national report includes a separate section on the Northeastern United States, a region that includes most of the territory covered by the MARA report. This national report is entitled "Climate Change Impacts on the United States: The Potential Consequences of Climate Variability and Change" and was prepared by the National Synthesis Team (National Synthesis Team, 2000).

Both reports were issued in 2000 under the auspices of the USGCRP. The USGCRP not only prepared the National Synthesis Team Report but also provided MARA with the population, income, and growth projections to be used in the assessment and specified the models that should be used so that all the regional reports could be compared (Mid-Atlantic Regional Assessment Team 2000). Both reports used the same two models: one produced by the Hadley Center for Climate Prediction and Research in Great Britain and one developed by the Canadian Center for Climate Modeling and Analysis (Mid-Atlantic Regional Assessment Team 2000: 12; National Synthesis Team 2000: 14). Despite using the same models and considering the same geographic area, MARA and the National Synthesis Team Report convey very different impressions of how global warming may affect the region.

MARA

MARA's executive summary placed impacts into categories of uncertainty: "most certain," "moderately uncertain," and "most uncertain" (Mid-Atlantic Regional Assessment Team 2000). Impacts identified in these categories included the following. Under "most certain" impacts, MARA lists increased growth in soybeans, corn, and fruit trees, whereas tobacco becomes less competitive; increased likelihood of urban heat stress; and increased sea-level rise, creating potential damage to the coastal zone's structure, wetlands, and estuaries. Under "moderately certain" impacts, MARA predicts that forests will grow a little faster, although extreme events could disrupt the pattern of revenues from forestry; and people could experience fewer health problems from cold-related stress. Under "most uncertain" impacts, MARA lists biodiversity losses; changes in precipitation and stream flow; water quantity and quality impacts; changes in ecological function, including less diverse mix of forest species; decline in cold water fisheries; and increases in disease carried by insects and animals.

After classifying these global warming impacts, the MARA executive summary concludes that the region's economy is resilient to these changes. But, a close reading of the individual chapters reveals several potential serious impacts that are not mentioned in the executive summary. For example, the report includes a picture showing likely changes in tree species. The picture indicates that the dominant hardwoods in Pennsylvania (maple, beech, birch, elm, ash, and cottonwood) will be replaced by an oak and hickory forest (Mid-Atlantic Regional Assessment Team 2000: 20). Yet it is reasonable to question why an impact as significant as the region's loss of major hardwood trees would not be mentioned in the executive summary.

The National Synthesis Report

The National Synthesis Team Report contains a short, six-page description of likely impacts on the Northeastern United States (National Synthesis Team 2000: 40–45). Unlike the MARA report, the National Synthesis report does not classify effects according to levels of certainty. Instead it groups effects in the following categories: (1) extremes in weather extremes; (2) stresses on estuaries, bays, and wetlands; (3) multiple stresses on urban areas; (4) recreation shifts; (5) human health; and (6) species change (National Synthesis Team 2000: 40–45). Within these categories, the National Synthesis Team predicts the following effects in the Northeastern United States:

- Likely decreases in extreme cold and snowfall, with likely increases in precipitation events over frozen ground and rapid snow melting events that will increase flooding; increases in heavy precipitation events at all times of year; and potential increases in intensity and frequency of hurricanes
- Likely increases in pollution stresses on estuaries and bays because of climate-caused reductions in oxygen in water
- Very likely rising sea levels that increase losses to wetlands and marshes
- Multiple increased stresses on urban areas caused by rising sea levels, elevated storm surges, damage to transportation systems, heat-related illness and death, and adverse impacts on water supply systems
- Likely impacts on human health from increases in certain vector-borne diseases, although the frequency and distribution of these diseases are difficult to predict
- Potential changes in predator-prey relationships, invasive species, habitat for migratory birds, trout populations, ecosystems, and tree types
- Recreation shifts, including loss of beachfront, reduction in fall foliage–related recreation, reduced winter skiing, and increased seasons for warm weather recreation

Comparison of the Two Reports

Both government personnel and employees of nongovernmental organizations have stated to the author that based on a reading of MARA, they saw little reason for concern about global warming impacts in the Mid-Atlantic region. Some of the same individuals have reported a greater understanding of why global warming should be of great concern in the Mid-Atlantic region after reading the National Synthesis Team Report (personal communication, 2001). Differences between the MARA executive summary and the National Assessment Team Report probably explain the different reactions to these reports even though later chapters of the MARA report reduce some of the inconsistencies. Three key differences between these reports are the following:

- The MARA report placed into the "most uncertain" category some impacts that, in fact, are very likely to happen, but where the exact nature and magnitude of impact cannot be predicted with high levels of certainty. As a result, many readers may have misinterpreted the "most uncertain" category, for instance, to designate "low probability" impacts. Differences in the types of questions asked may have changed the "most uncertain" categorizations.
- The MARA report placed conclusions about increased growth of soybeans, corn, and fruit trees in the "most certain" category. If this is interpreted as "highest probability," then one could be led to the conclusion that the most likely impacts from global warming are positive in nature.
- The MARA report concludes that the economy in the region is resilient to global warming impacts. Yet this conclusion might be interpreted as resilience to only those identified impacts characterized by high levels of certainty, or those that would be experienced from the lower or middle end of the expected temperature range. Also, the economic analysis aggregates impacts across the entire MARA region, without mentioning the possibility of devastating subregional or distributive effects of climate change, such as to the winter recreation or tourism industries or agriculture.

One cannot read the MARA report without concluding that its authors were very concerned about not going further in making predictions than the science would allow. This is a laudable goal; yet identification of plausible but unproven impacts is not "going beyond the science" if there is a reasonable scientific basis for identification of plausible impacts. There is nothing inherently inconsistent in avoiding scientifically unfounded speculation while clearly identifying all scientifically plausible impacts. Because many uncertainties will not be resolved until it is too late to prevent the impact, demanding high levels of certainty before identifying an issue as an object of concern is a dubious value judgment and inconsistent with the precautionary principle.

Conclusions and Recommendations

Both the IPCC experience and the experience of the MARA and the National Synthesis Team reports demonstrate that scientists need to identify scientifically plausible impacts that they believe to be important, yet are unable to quantify. Furthermore, economists drawing conclusions from studies in the natural sciences should not ignore potentially serious, irreversible, or catastrophic events simply because their probability is not amenable to quantification or their magnitude and timing cannot be predicted with a high level of certainty.

The IPCC should do a better job of making policy makers aware that the climate surprises are plausible responses of the climate system and the carbon cycle that may very well ultimately prove that the IPCC's more quantitative predictions have underestimated global warming impacts. To do this, the IPCC should throughout its reports include a narrative description of scientifically plausible but unquantifiable impacts. This description should put policy makers on notice that the quantified impacts are not the only impacts that should be of concern. These descriptions of plausible impacts must appear in the same parts of impact analyses that contain quantified impacts.

As we have seen in the examples discussed in this chapter, if impact analyses do not identify all plausible impacts, decision makers may be without important information that they may need to make an informed decision. For example, the MARA report encourages confusion about "certainty" about magnitude, timing, and other aspects of climate change effects with "likelihood" of those effects. As a result, MARA is a flawed decision-making tool for policy makers. Information about all plausible impacts must be included at all stages of impact analyses. In addition, cost-benefit analyses of emissions reduction programs (or other proposed measures to mitigate environmental impacts of a human activity) should not omit consideration of plausible, significant impacts whose probability or likely magnitude has not been quantified.

If environmental impact analyses identify and describe only those effects that can be characterized with high levels of certainty, there is no way to apply precautionary decision-making processes successfully. Those responsible for preparing impact analyses should instruct scientists conducting the analysis that they are expected to identify all scientifically plausible impacts.

To give a decision maker full information on human activities that could cause serious or irreversible harm, the precautionary principle requires that impact analyses must identify all scientifically plausible impacts—not only impacts that have been identified through the use of traditional scientific procedures requiring high levels of statistical correlation between hypothetical cause and effect.

The appropriate question to ask scientists in cases where the precautionary principle is applicable is whether there is a reasonable scientific basis to identify a potential impact, not whether the impact has been proven to be a consequence of human behavior. Where potential but unproven impacts are of concern, those impacts should be clearly identified as only plausible, that is, not proven impacts. For these impacts, scientists should describe the chain of events and mechanisms that could create the impacts of concern so that the reasoning that leads to the identification of these unproven impacts can be subjected to peer review.

Scientific uncertainty is not the same as low probability. Impact analyses that emphasize only highly certain impacts may be very misleading about the state of the scientific knowledge. Even when timing and magnitude of an event or effect is uncertain, there may be a strong scientific basis for concern about the impact. For this reason, reports that identify potential impacts on human health and the environment that may be caused by human activities should make distinctions among highly likely, likely, unlikely but potentially catastrophic, and scientifically plausible impacts for which probabilities are not known.

Endnotes

1. The IPCC has three working groups. Working Group I assesses the scientific aspects of the climate system and climate change. Working Group II addresses the vulnerability of socioeconomic and natural systems to climate change, negative and positive consequences of climate change, and options for adapting to it. Working Group III assesses options for mitigating climate change. This chapter considers only the work of Working Groups I and II.
2. This third IPCC report concluded that the earth's average surface temperature could rise 2.5 to 10.4°F from 1990 to 2100—much higher than the IPCC's estimate five years before when it predicted a rise of 1.8 to 6.3°F in its second report.

References Cited

Brown, D. 2002. *American Heat: Ethical Problems with the United States Response to Global Warming*. Latham, Md.: Rowman and Littlefield.

Chang, K. 2001. "Drastic Shifts in Climate Are Likely, Experts Warn." *New York Times*, 12 December, A12.

Intergovernmental Panel on Climate Change (IPCC). 1995a. *The Science of Climate Change: Contribution of Working Group I to the Second Assessment of the Intergovernmental Panel on Climate Change. Summary for Policy Makers*. http://www.ipcc.ch/about/about.htm.

———. 1995b. *Climate Change, Impacts, Adaptations and Mitigation of Climate Change: Scientific-Technical Analyses, Summary for Policy Makers*. http://www.ipcc.ch/pub/sarsyn.htm.

———. 1995c. *Climate Change 1995: The Science of Climate Change: Contribution of Working Group I to the Second Assessment of the Intergovernmental Panel on Climate Change*. Cambridge: Cambridge University Press.

————. 1995d. *IPCC Second Assessment Synthesis of Scientific-Technical Information Relevant to Interpreting Article 2 of the UNFCCC.* http://www.ipcc.ch/pub/sarsyn.htm.

————. 1997. *An Introduction to Simple Climate Models Used in the Second Assessment Report.* http://www.ipcc.ch/pub.

————. 2001. *Climate Change, 2001: Synthesis Report. Summary for Policy Makers.* http://www.grida.no/climate/ipcc_tar/syr/pdf/fourthvolume.pdf.

Mid-Atlantic Regional Assessment Team. 2000. *Preparing for Changing Climate: The Potential Consequences of Climate Variability and Change.* University Park: Pennsylvania State University Press.

Mobil Corporation. 1997. "Climate Change: Where We Come Out." *New York Times*, Op-ed, 15 November.

National Synthesis Team. United States Global Research Program. 2000. *Climate Change Impacts on the United States: The Potential Consequences of Climate Variability and Change.* Cambridge: Cambridge University Press.

Parry, M., N. Arnell, M. Hulme, P. Martins, R. Nichols, and A. White. 1999. "Viewpoint, the Global Impact of Climate Change: A New Assessment." *Global Environmental Assessment* 9: S1.

Shogren, J., and M. Toman. 2000. "How Much Climate Change Is Too Much? An Economics Perspective." *Resources for the Future*, 14 September. http://www.rff.org/disc_papers/PDF_files/0022.pdf.

Shrader-Frechette, K., and E. McCoy. 1993. *Methods in Ecology.* Cambridge: Cambridge University Press.

Stocker, T. 2000. "Past and Future Reorganizations in the Climate System." *Quaternary Science Reviews* 19: 301–19.

Streets, D., and M. Glantz. 2000. "Exploring the Concept of Climate Surprises." *Global Climate Change* 10: 2.

United Nations Framework Convention on Climate Change (UNFCCC). 1992. UN doc. A:AC.237/18, Rio de Janeiro.

CHAPTER 11

Certainty Claimed: Science and Politics in Canadian GMO Regulations

Katherine Barrett and Stuart Lee

For those seeking to promulgate the precautionary principle, claims of scientific "certainty" regarding the safety of technologies are often encountered as major obstacles. Politicians, industry representatives, and regulatory scientists, for example, may defend the development of new technologies by proclaiming that they are adequately safe, that hazards are manageable, or that new products are "substantially equivalent" to existing ones. Proponents of more precautionary measures must then contest such claims by demonstrating sufficient plausibility of harm and uncertainty of impacts. Indeed, failure to critically analyze claims of scientific certainty may weaken grounds for implementing the precautionary principle. If specific classes or types of uncertainty are not acknowledged, application of the precautionary principle may appear unwarranted or, more significantly, a barrier to trade.

In this chapter, we examine more closely the dynamic and creative tension between claims of certainty and uncertainty. Specifically, we examine four constructions of certainty that occur at different sites of decision making within the Canadian regulatory approval process for genetically modified (GM) organisms. We argue that the claims of certainty that are used to justify policy decisions are not simply a function of scientific knowledge that exists independently of the political process. Rather, recognition of un/certainty depends on institutional, political, *and* scientific factors and is therefore shaped significantly by the context in which knowledge claims are generated, contested, and accepted.

This analysis illustrates the need to investigate not only the level of scientific uncertainty surrounding GM organisms (or other technologies) but how claims of certainty are put forward and defended within a particular political situation and, consequently, how uncertainty is downplayed or erased entirely.

The Global Context

With the adoption of the Cartagena Protocol on Biosafety in January 2000, biotechnology, uncertainty, and the precautionary principle have moved to the forefront of international negotiations on trade and the environment. The objective of the Cartagena Protocol, established under the 1992 Convention on Biological Diversity, is to prevent adverse effects of "living modified" organisms and food[1] on the conservation and sustainable use of biological diversity, particularly those effects that may result from the "transboundary" (international) movement of such organisms.

Decisions under the Cartagena Protocol must be based primarily on risk assessment as outlined in Annex III of the agreement.[2] Significantly, however, the protocol also invokes the "precautionary approach" in the preamble and includes precautionary language in the binding text. Articles 10 and 11 on decision procedures state that lack of scientific certainty regarding the extent of potential adverse effects of GM organisms shall not prevent parties from taking preventative action to avoid such harms. In other words, the protocol grants parties the right to refuse import of GM organisms even where there is scientific uncertainty about the nature or extent of harms that may occur.

Negotiations toward implementing the protocol began in late 2000, and it is not yet clear how the agreement will operate in practice. Ultimately, the strength and scope of the protocol will be determined in large part by its relationship to other international agreements, particularly the Agreement on the Application of Sanitary and Phytosanitary Measures (SPS Agreement) under the World Trade Organization. This agreement, signed in 1994, establishes rules for food safety and plant and animal health standards. These rules allow parties to adopt measures to protect human and animal health while ensuring that such measures have minimal negative effects on trade and "are not being used as an excuse for protecting domestic products" (WTO 1998). Like the Cartagena Protocol, decisions under the SPS Agreement must be based on risk assessment. However, Article 5.7 of the agreement also stipulates that "in cases where scientific evidence is insufficient" parties may provisionally adopt regulatory measures to prevent harm, while seeking information "for a more objective risk assessment." This relatively weak invocation of the precautionary principle has been interpreted to mean that parties must be *actively seeking* information to address identified uncertainties and must review "provisional measures" within a reasonable period of time (Campbell 2001). This reading implies that uncertainties are merely temporary gaps in scientific data and can eventually be filled, and thereby mitigated, with further research.

The relationship between these agreements remains ambiguous. Implementing these agreements will therefore raise key questions regarding application of the precautionary principle: what types of uncertainty should be recognized under a precautionary approach, and, thus, under which circumstances are precautionary measures warranted?

Social Perspectives on Scientific Un/certainty

These are not trivial questions because, to some extent, all activities are uncertain in their consequences. A rapidly growing body of research on complex systems underscores the difficulty of predicting events outside the reductionist confines of the laboratory (Gallagher and Appenzeller 1999; Gallopin et al. 2001; Holling 2001). Complexity takes us far from conceptions of uncertainty as temporary data gaps and places us firmly in the realm of ignorance and indeterminacy, characterized by open causal chain and networks (see Scoones 1999; Wynne 1992b, 1996).

In this chapter, we are interested in how claims of certainty about the safety of GM organisms are generated and defended against a background of pervasive uncertainty and provisional knowledge. It is our contention that certainty is achieved through social and political processes as much as it is through the generation of new data or evaluation of the "best available knowledge."

The social and political nature of knowledge production has been aptly demonstrated through case studies of laboratory experimentation and general scientific practice (Latour and Woolgar 1979; Latour 1987; Jasanoff 1990; Pickering 1995; Epstein 1996). These studies argue against the existence of a well-defined scientific method that allows scientists to definitively know the "truth" about a natural object or process. Instead, they show that doing science involves ongoing struggles with the materials studied, the instruments used to study them, and the political and social context in which studies are conducted. We argue that the explicitly political arena of regulatory science (Jasanoff 1990), where scientific claims are generated and defended in an often confrontational regulatory context, provides little incentive to highlight uncertainties because doing so may undermine political negotiation and action. This erasure of uncertainty weakens the ability to justify and undertake legitimate precautionary actions.

Our goal is to move discussion of uncertainty and precaution beyond analyses of science as knowledge disassociated from social and political circumstance. This allows for a more robust and indeed more authentic discussion of science, uncertainty, and precaution (Woodhouse and Nieusma 2001).

Case Study: Regulating the Environmental Impacts of GM Organisms In Canada

The Canadian Regulatory Process: Politics in Tension

Canada first approved GM crops for large-scale commercial planting in 1996. Since that time, the Canadian agricultural biotechnology industry has grown rapidly. Canada was the world's third largest producer of GM crops in 2000, with over three million hectares of farmland planted to transgenic varieties, mainly canola, corn, and soy (James 2000).

Rapid growth of Canadian agricultural biotechnology is due in part to steady support from the Canadian federal government. Indeed, the first National Biotechnology Strategy, announced in 1983, was mandated to "provide federal policy guidance and programme support to encourage the concerted action necessary to make commercial progress" in biotechnology (NBAC 1984). Regulations for assessing the health and environmental impacts of GM crops, food, and other organisms were developed approximately ten years later. An overarching "Regulatory Framework" was announced in 1993 and was touted as "both economically and scientifically sound" in that it could build on existing scientific expertise while "fostering competitiveness through timely introduction of biotechnology products to the marketplace" (Government of Canada 1993). This dual mandate—to both regulate and promote GM technology—reflects a long-standing tension in the Canadian government's stance on agricultural biotechnology, as we discuss later in this chapter.

The 1993 Regulatory Framework outlined four principles upon which subsequent, more specific guidelines were based (box 11.1) (AAFC 1993). While emphasizing protection of health and the environment, the framework also specified that no new legislation would be developed for GM organisms and that regulation would focus primarily on the final characteristics of GM "products" rather than the process of genetic engineering itself. The framework further stressed that regulatory decisions would be grounded in "science-based risk assessment" (AAFC 1993).

Since the introduction of these principles, the strength of scientific knowledge underpinning Canadian biotechnology regulations has been emphasized in policy and public relations documents, often as a means of affirming that regulatory practices are sufficient to ensure safety. For example, the Minister of Agriculture has stated that Canadian biotechnology regulations are "based on the latest and best scientific knowledge we have" (Vanclief 1999). Other government documents describe the regulatory process as "extensive" and "stringent," using "comprehensive," "full" and "science-based" assessments and "scrupulous examination of all data" to ensure that "every possible precaution

---

Box 11.1. Principles of the 1993 Federal Regulatory Framework for Biotechnology

Regulation will:

• Be based on the characteristics of the final product (including review of the process by which it is made)
• Use science-based risk assessment
• Aim to protect health and the environment
• Build on existing legislation and areas of responsibility

Adapted from Agriculture and Agri-Food Canada 1993.

---

is taken" (CFIA 1999a, 2001b–e). In fact, during negotiations leading to adoption of the Cartagena Protocol, the Canadian government and other major exporters of GM crops argued that existing methods of risk assessment were adequate to safeguard against the hazards of GM organisms and that including the precautionary principle in the protocol was therefore unnecessary (UNEP 1999; Falkner 2000).

Yet, there is no consensus within the scientific literature about the safety of GM organisms (Wolfenbarger and Phifer 2000; Letourneau and Burrows 2001; RSC 2001; Commoner 2002). Given this pervasive controversy and the complexity of biological and ecological systems under investigation, how can government regulators maintain such confidence in the decision-making process and in their conclusions regarding the environmental impacts of GM crops?

Forging Certainty

As we have discussed, answers to the above question reflect political goals, pressures, and commitments as much as current scientific knowledge.

In this section, we present four potential sources of uncertainty or controversy around the release of GM crops. We then outline the kind of work that is done to exclude this uncertainty from the regulatory process. In each case, we find that scientific knowledge contributes to decision making, but by no means determines the outcome of a risk assessment. Rather, policy outcomes might be more accurately viewed as a necessary balance between conflicting political pressures to regulate environmental impacts while nonetheless ensuring timely commercialization of new GM products.

Definitions

The terms used to define an entity or frame a problem can have a major impact on how the problem is approached or what aspects of the problem are acknowledged as salient. Terms used in the current biotechnology debate remain remarkably inconsistent and controversial. According to one view, "genetically modified" or "genetically engineered" organisms are those developed through recombinant DNA techniques. This view stresses that GM organisms are inherently novel and therefore pose new potential risks. An alternative view holds that recombinant DNA techniques are essentially the same as conventional hybridization and selection processes and, therefore, that GM organisms are merely an extension or refinement of practices generally recognized as safe. Such definitions are important; they accentuate particular characteristics of GM organisms, draw attention to a particular range of potential impacts, and therefore bear significantly on the level of regulatory oversight that is considered appropriate.

The Canadian government adopts the latter position, at least for regulatory purposes.[3] Under the framework outlined above, GM crops are regulated within a broader category called "plants with novel traits." This category includes all plants that have been genetically altered such that they are not "familiar" or "substantially equivalent" to plants already grown in Canada. Defining GM crops in this way focuses regulatory attention on expressed characteristics of the final *product* (e.g., biological and ecological traits) rather than on the process of genetic alteration used to develop the novel crop; new varieties created through hybridization, mutagenesis, and recombinant DNA techniques are evaluated through the same regulatory procedures (Barrett and Abergel 2000).

The product-based regulatory style of the Canadian government reflects an underlying "philosophy . . . that genetically engineered organisms are not fundamentally different from traditionally derived organisms" (CFIA 2001b). While this assumption does not necessarily exempt GM crops from regulatory oversight, it does explicitly restrict the scope of assessment. For example, claiming that GM organisms are not different from non–GM organisms allows regulatory agencies to respond with certainty to concerns about long-term testing of GM foods: "As the techniques of modern biotechnology do not introduce risks which are different from those already associated with the food supply, the potential for long-term effects from novel foods is no different than that from traditional foods which have been safely part of the Canadian diet for a long time" (CFIA 1999b). By this rationale, the potential hazards of GM organisms can be adequately predicted using "well-defined and understood principles of risk assessment" (CFIA 2001b) and measured relative to (ostensibly) well-known and acceptable impacts of non-GM varieties and conventional agricultural practices.

Through such judicious choice of definitions, the Canadian Regulatory Framework minimizes the scope of uncertainty associated with releasing GM organisms into the environment. The Canadian definition provides little incentive or rationale to anticipate unprecedented hazards and uncertainties that might result from the process of recombinant DNA technology. Such hazards might include unforeseen ecological interactions (Bergelson et al. 1998) as well as social and economic impacts resulting from disproportionate harms of GM crops on organic farmers or the scale-dependent impacts arising from the exceptionally rapid and extensive commercialization of GM crops. Significantly, Canada's "product-based" regulatory system also provides little incentive to investigate ecological impacts of the conventional production of non-GM crops, or the additive effects of GM and non-GM crops; conventional agriculture is fixed as the baseline of acceptable practice and acceptable risk. Through the deliberate and strategic choice definitions, certainty is reinforced and in-depth analysis of uncertainty is effectively constrained.

METHODS

Analysts in science and technology studies have repeatedly demonstrated the central importance of instruments and methods in determining the characteristics of the entities that scientists study (Callon 1986; Latour 1987, 1999; Mol and Law 1994; Pickering 1995; Shapin 1996). The selection of methods of sample preparation, measurement, and recording of results all strongly constrain what can be known about an object of analysis (Lynch 1990; Latour 1999). Some qualities are recorded and validated; other qualities remain unknown because they are not recorded or are deemed unimportant (Star 1991). This form of "methodological uncertainty" limits the degree to which research results can be extrapolated beyond the experimental conditions chosen by the researcher (Funtowicz and Ravetz 1992).

Methodological uncertainty is particularly acute when analyzing complex, large-scale problems that cannot be reduced easily to controlled laboratory investigation (such as climate change). How can small-scale experiments on such systems be appropriately extrapolated to represent real-world scales and contingencies? What test parameters, controls, and error margins are feasible and appropriate? Because these issues are central when according appropriate credibility or certainty to a particular knowledge claim, methodological uncertainty becomes problematic when it remains unacknowledged. For example, under the pressures of policy making, scientists and politicians alike may devise rhetorical strategies for reducing the perceived uncertainty of their knowledge claims (van der Sluijs et al. 1998).

In the Canadian regulatory system, criteria used to test the safety of GM organisms (i.e., familiarity and substantial equivalence) are outlined in regulatory documents and are summarized in box 11.2. Developers of "plants with

---

Box 11.2. Criteria Used to Assess Environmental Risks of "Plants with Novel Traits"

• Increased weediness in agricultural settings
• Increased invasiveness into nonagricultural settings
• Potential for becoming a "plant pest" (toxicity and/or allergenicity)
• Impacts on nontarget organisms
• Gene flow to related plants
• Impacts on biodiversity

Adapted from Agriculture and Agri-Food Canada 1994.

---

novel traits" are required to address all of these criteria and provide regulators with experimental test data or "valid scientific rationale" to support their claims (AAFC 1994). Based on this information, regulatory agencies approve or reject the application and publish the outcome in publicly available "Decision Documents." These documents outline the agencies' general conclusions about the environmental safety of GM crops and the factors considered in reaching these conclusions, but do not describe experimental methods, data, results, or information sources. Such details of the scientific process are not made public in Canada and, in fact, may be protected as confidential business information. Comprehensive public review of the science used in approving GM crops is therefore made impossible by a decision-making process that is closed to all but industry and government regulators.

In one case, however, scientific data used in the approval of a particular GM crop variety, an herbicide-tolerant canola, were obtained through the Canadian Access to Information Act (Barrett and Abergel 2002). This example may be illustrative of the methods used to assess and approve other GM crops and the ways in which methodological uncertainty is, or is not, addressed.

In the canola case, risk experiments were typically conducted over a period of one to two field seasons. Some of the risk assessment criteria required under government regulations (see box 11.2) were tested in a number of diverse geographic regions and replicated trial sites. Other criteria were tested through more limited methods. For example, growth rate, seed production, and maturity rate were examined as indicators of increased potential for weediness. These tests were conducted in a maximum of thirty-two replicated sites throughout western Canada. In contrast, tests for invasiveness of GM canola into natural settings were conducted in a single trial in a roadside ditch. No tests were conducted specifically to assess the impacts of GM canola on biodiversity. It therefore appears that different methodological standards were used to test each of the risk assessment criteria even though all criteria are important for evaluating environmental impacts.

Such differences are significant because, as discussed above, different methodological approaches will entail different sources and levels of methodological uncertainty. This is particularly important when conducting ecological field tests where contingencies such as fluctuating weather and soil conditions and the well-documented variability of canola populations (Kumar et al. 1998) will affect interpretation of results. Moreover, in the GM canola case, almost all experiments yielded negative conclusions; that is, no differences between GM and non-GM canola were detected. Based on these results, regulators stated conclusively that GM canola does not pose an increased risk in the required test parameters shown in box 11.2.

In such cases where negative conclusions (e.g., GM canola is not different from parent strain) are drawn from limited number of highly variable experiments, it is important to determine if the failure to detect a difference between GM and non-GM canola is representative of actual conditions (there really is no difference) or whether tests were simply not robust enough to a detect difference. It is significant in this respect that many of the experiments in the GM canola risk assessment included calculations of the chances of falsely stating that there are differences (type I error rate). In no case, however, were conclusions supported with calculation of the chances of a false stating that there are no differences (type II error rate) (see Peterman and M'Gonigle 1992; Barrett and Raffensperger 1999). Such precautionary measures are now standard requirements for many ecological studies, particularly those related to resource management and conservation (RSC 2001).

In fact, Decision Documents on herbicide tolerant canola and other approved GM crops published by the federal government contain no discussion of uncertainty whatsoever. No qualifications are placed on the capacity of experimental methods to predict complex ecological impacts. Rather, the uncertainty and heterogeneity of methodological approaches are effectively negated. For example, the Decision Document on GM canola states that this variety of GM canola "has no altered weed or invasiveness potential compared to currently commercialized canola varieties," that it will "not result in altered impacts on interacting organisms," and that "the potential impact on biodiversity . . . is equivalent to that of currently commercialized canola" (AAFC 1995).

How can we account for such discrepancies in methodological approaches and for failure to consider the effects of methodological uncertainty? One possible explanation is that some of the criteria examined in the GM canola risk assessment were also required to fulfill additional regulatory hurdles. Measures of yield and maturity rates in a number of geographic regions, for example, are required for "variety trials," which are used to evaluate the agronomic performance and uniformity of new crop varieties before they can be registered for commercialization. In the GM canola case, these trials were performed concurrently with risk assessment experiments. This not only suggests that the same tests and data were used to fulfill two very different regulatory hurdles, but also implies an urgency to market GM crops as expediently as possible. In contrast to criteria required for variety trials, tests for invasiveness into natural settings or impacts on biodiversity are not required under any other federal regulation. It therefore appears that in this case, regulatory pressures related more to marketing than ecological impacts influenced the scope of methods for determining the safety of GM canola and the degree to which methodological uncertainty was acknowledged.

CONSEQUENCES

In this section, we outline a related strategy to further reduce uncertainty: in cases where negative impacts are acknowledged, regulators assert that these impacts will nonetheless be minimal. Two general arguments are used to support this assertion. First, regulators claim that adverse events can be mitigated or controlled once they occur, and second, regulators claim that harmful consequences are, to some extent, offset by beneficial impacts of GM crops.

According to government-produced fact sheets and Decision Documents, there is significant probability that release of GM crops will result in some adverse ecological events. In discussing the transfer of GM traits to wild species or other agricultural crops, regulators state that for GM crops such as herbicide tolerant canola, these events are "indeed possible." However, the argument continues, because the resulting plants "would be controlled using other available chemical means," gene transfer "would not result in increased weediness or invasiveness" (AAFC 1995). Elsewhere, CFIA states:

> If a herbicide tolerant trait was passed onto a weed, it does not mean that the weed would become a bigger pest; it simply means that the weed would be tolerant to a specific herbicide. The weed could still be controlled using other management practices such as tillage or alternative herbicides. (CFIA 2001e)

In assuming that such events can be adequately controlled, the Canadian government downplays both scientific and sociopolitical uncertainties associated with herbicide tolerant weeds. For example, how will accumulation of multiple herbicide tolerance genes in crops and weeds be prevented or controlled (Hall et al. 2000)? How will changes in patterns of herbicide use affect populations of weeds and other organisms (Hartzler 1998)? Who "owns" and is liable for invasive weeds that have acquired patented herbicide tolerant genes?[4] These questions underscore the interplay between scientific and social factors but are not accounted for in regulators assertions that consequences can be sufficiently managed.

An equally powerful way to minimize the consequences of adverse ecological events is to assert that GM crops will provide compensating benefits. The biotechnology industry has received substantial support from the Canadian government since the first national biotechnology "strategy" was launched in the early 1980s. More recent government documents maintain that "there are many benefits to products derived from biotechnology" (AAFC 1996). For example:

> Biotechnology has the potential to increase Canada's international competitiveness and promote sustainable development in key economic sectors. (CBS 1998)

> With the advent of genetic engineering, it is possible that plants, animals, products and processes will be genetically engineered to be more productive, disease resistant or to meet consumer requirements more exactly. (AAFC 1996)

> By genetically engineering crops to be insect resistant or herbicide tolerant, farmers can manage their crops with less use of chemicals. (CFIA 2001a)

However, the environmental and economic benefits of GM crops have yet to be convincingly demonstrated. In some cases, GM crops have provided added convenience to farmers and reduced pesticide use. In other cases, GM crops have resulted in lower yields, added costs, and no change in pesticide use. Studies suggest that the relative costs and benefits of GM crops cannot be predicted easily, but are sensitive to a number of complex variables, including crop type, pest infestation levels, local ecological conditions, reporting methods, and agricultural practices (CEC 2000; Benbrook 2001; ERS 2001).

By portraying consequences as predictable and manageable, and by highlighting the purported benefits, Canadian regulators rhetorically minimize or dismiss the full range of scientifically valid uncertainties surrounding the large-scale commercialization and release of GM crops.

PARTICIPATION

As with many new technologies, there is a great range of opinions on the acceptability of, safety of, and need for GM crops. These diverse opinions, if allowed to inform the political process, would introduce greater complexity and uncertainty around social, political, and ethical as well as scientific issues. As sociologists have argued, dissension and resistance to new technologies arise not only from threats of physical harms to health and the environment, but equally—and reasonably—from lack of participation in and loss of control over decision making and a consequent loss of trust in decision makers (Wynne 1992a, 1996; GECP 1999).

The Canadian regulatory process effectively ignores these sources of uncertainty by maintaining a closed decision-making process and by limiting the range of knowledge, experience, and opinions that are taken into consideration. Two reports released in 2001 have stressed that current decision-making processes lack adequate mechanisms for independent peer review or open public participation. A report by an Expert Panel of the Royal Society of Canada, for example, raises "serious concerns" about the "barriers of confidentiality that compromise the transparency and openness" of the regulatory process and further warns that public interest in a regulatory system that is "science-based"—that meets scientific standards of objectivity, a major aspect of which is full openness to scientific peer review—is significantly compromised when that openness is negotiated away by regulators in exchange for cordial and supportive relationships with the industries being regulated (RSC 2001). Similar concerns have been expressed in a report by the Canadian Biotechnology Advisory Committee (CBAC), a multi-stakeholder group established by and reporting to the federal government (CBAC 2001).

How do these shortfalls—lack of transparency, closed decision-making processes, and "cordial" government-industry relations—affect recognition of uncertainty and application of the precautionary principle? First, technical uncertainties within scientific data are more likely to be overlooked in an exclusive decision-making process. There is always an element of judgment required in interpreting scientific information, and different conclusions can often be drawn from the same data set. These judgments are based on assumptions that in turn reflect the values, experiences, and commitments of those involved (Brunk et al. 1992). The tradition of peer review, while not infallible, does provide a greater diversity of interpretations and opinions on the plausibility of the research and is more likely to identify uncertainties or inconsistencies in the data that may be overlooked by more exclusive processes. This fundamental measure of openness and scientific credibility is currently absent from the Canadian regulatory system.

At a broader level of analysis, closed decision making among like interests not only restricts interpretation of the data and acknowledgment of "technical uncertainties," but also limits the range of questions that can be considered and, therefore, the type of data or knowledge that is considered relevant. For example, the current Canadian regulatory process for GM crops is explicitly limited to "scientific" issues—science that is largely defined by regulators and developers of GM crops. Underlying this focus is the assumption that public acceptance of GM food is ultimately a scientific question and that more and better science will eventually distinguish between acceptable and unacceptable GM products, thereby defusing further controversy.

This narrow approach, however, fails to acknowledge the depth of "normative uncertainty" (Levidow 1996) underlying the acceptance of GM food. It fails to recognize that public attitudes cannot be reduced to or resolved by science-based risk assessment alone. The Royal Society report stressed this point, stating that the "scientific" risks upon which the regulatory system is currently focused are inextricable both from socioeconomic risks and from risks to "fundamental philosophical, religious or 'metaphysical' values" (RSC 2001). Stakeholder consultations and a questionnaire survey conducted by CBAC in 2001 began to address these questions by asking stakeholders if and how social and ethical issues surrounding GM food should be addressed by regulatory agencies. The questionnaire and stakeholder consultations conducted by CBAC revealed marked differences in concerns about GM food between members of industry groups and members of Canadian consumer or nongovernment organizations. Differences were particularly striking on ethical issues related to conflict of interest within regulatory agencies, public access to scientific data used in decision making, and the right to mandatory versus voluntary labeling of GM foods. This diversity contrasts with the current regulatory process in which a limited range of interests determines what counts as appropriate questions, test methods, evidence, and conclusions about the safety and acceptability of GM crops. In the process of discounting differences of opinion and values, uncertainties, both technical and normative, are simultaneously ignored.

Conclusions: Ignoring Ignorance

Through analysis of Canadian regulations for GM crops, we have highlighted several types of uncertainty that are routinely omitted from the decision-making process. These uncertainties extend well beyond temporary data gaps to encompass definitions, methods, standards, interpretations, and norms as well as a fundamental *ignorance* about the workings of complex, interrelated ecological and social systems (Wynne 1992b; Dovers and Handmer 1999).

While our conclusions are specific to the Canadian political context, this type of analysis is important for and applicable to other regulatory systems. Our central point is that recognition of uncertainty—and acknowledgment of ignorance—are in large part a function of the context in which questions are posed and answers are generated. We argue that that both knowledge and ignorance are necessarily conditioned by social context. Thus, in the policy arena, recognition of un/certainty and application of the precautionary principle cannot be separated from the political goals, pressures, and commitments that structure decision making.

Therefore, it is important to examine critically the contextual factors that structure *how* we know what we claim to know—and how we account for, or discount, our ignorance. To this end, we suggest that implementing the precautionary principle be focused on designing a deliberative *process* that emphasizes and draws upon necessarily heterogeneous, conflicting, and partial perspectives. This process should foster public input and dialogue on the appropriate role of science as well as other factors in decision making, and it should strive to make explicit the values that frame, define, and fund public policy decisions.

Undoubtedly, this process will have limitations because of financial constraints, political pressures, and availability of data. However, we believe strongly that the diversity of people who live with the outcomes of policy decisions should have the opportunity to discuss and shape these limitations and thereby have the privilege of appreciating the necessarily provisional nature of "science-based" decisions.

Endnotes

1. The Convention on Biological Diversity defines living modified organisms in Article 3: " 'Living modified organism' means any living organism that possesses a novel combination of genetic material obtained through the use of modern biotechnology." We use the terms "genetically modified" and "living modified" interchangeably.
2. Article 26 of the protocol also allows for socioeconomic considerations to be taken into account.
3. Arguably, the Canadian government's support of biotechnology as an innovative and patentable technology contradicts this position.
4. This issue was highlighted in the recent court case between a Saskatchewan farmer and Monsanto (see Lyons 2000).

References Cited

Agriculture and Agri-Food Canada (AAFC). 1993. *Regulation of Agricultural Products of Biotechnology.* Ottawa: Biotechnology Strategies Coordination Office.

———. 1994. *Assessment Criteria for Determining Environmental Safety of Plants with Novel Traits.* Ottawa: Plant Industry Directorate.

———. 1995. *Determination of Environmental Safety of Monsanto Canada Inc.'s Roundup Herbicide: Tolerant Brassica napuscanola Line GT73.* Ottawa: Plant Products Division.

———. 1996. *BioInfo: Legislation and Agricultural Products.* Ottawa: Agriculture and Agri-Food Canada.

Barrett, K., and E. Abergel. 2000. "Breeding Familiarity: Environmental Risk

Assessment for Genetically Engineered Crops in Canada." *Science and Public Policy* 21, no.1: 2–12.

———. 2002. "Defining a Safe GM Organism: Boundaries of Scientific Risk Assessment." *Science and Public Policy* 29, no. 1: 47–58.

Barrett, K., and C. Raffensperger. 1999. "Precautionary Science." In *Protecting Public Health and the Environment: Implementing the Precautionary Principle,* edited by C. Raffensperger and J. Tickner, 106–22. Washington, D.C.: Island Press.

Benbrook, C. 2001. "Do GM Crops Mean Less Pesticide Use?" *Pesticide Outlook* 12, no. 5: 204–7.

Bergelson, J., C. Purrington, and G. Wichmann. 1998. "Promiscuity in Transgenic Plants." *Nature* 395: 25.

Brunk, C. G., L. Haworth, and B. Lee. 1992. *Value Assumptions in Risk Assessment. A Case Study on the Alachor Controversy.* Waterloo: Wilfrid Laurier Press.

Callon, M. 1986. "Some Elements of a Sociology of Translation: Domestication of the Scallops and the Fishermen of St. Brieuc Bay." In *Power, Action, and Belief: A New Sociology of Knowledge?* edited by J. Law, 196–233. London: Routledge.

Cameron, J. 1999. "The Precautionary Principle: Core Meaning, Constitutional Framework, and Procedures for Implementation." In *Perspectives on the Precautionary Principle,* edited by R. Harding and E. Fisher, 29–58. Annandale, NSW, Australia: Federation Press.

Campbell, F. T. 2001. "The Science of Risk Assessment for Phytosanitary Regulation and the Impact of Changing Trade Regulations." *BioScience* 51, no. 2: 148–53.

Canadian Biotechnology Advisory Committee (CBAC). 2001. *Improving the Regulation of Genetically Modified Food and Other Novel Foods in Canada: Interim Report.* Ottawa: Canadian Biotechnology Advisory Committee. Available at http://www.cbac.cccb.ca/documents/en/Improving_Regulation_GMFood.pdf.

Canadian Biotechnology Strategy (CBS). 1998. "Key Elements of the Renewed Canadian Biotechnology Strategy." strategis.ic.gc.ca/SSG/bh00229e.html.

Canadian Food Inspection Agency (CFIA). 1999a. "Food Safety Concerns." www.inspection.gc.ca/english/ppc/biotech/safsal/fooalie/shtml.

———. 1999b. "Questions about the Long Term Effects of Foods Derived from Biotechnology." www.inspection.gc.ca/english/ppc/biotech/safsal/longterme/shtml.

———. 2001a. "Agricultural Products of Biotechnology: A Brief Status Report." www.inspection.gc.ca/english/ppc/biotech/gen/statuse/shtml.

———. 2001b. "Concerns and Issues about Biotechnology." www.inspection.gc.ca/english/ppc/biotech/gen/issuse.shtml.

———. 2001c. "Frequently Asked Questions on Genetically Modified Foods." www.inspection.gc.ca/english/ppc/biotech/gen/faqe.shtml.

———. 2001d. "Outcrossing to Wild Species." www.inspection.gc.ca/english/ppc/biotech/enviro/transfe/shtml.

————. 2001e. "Weeds and Plants Produced through Biotechnology." www.inspection.gc.ca/english/ppc/biotech/enviro/weemaue.shtml.

Commission of the European Communities (CEC). 2000. "Economic Impacts of Genetically Modified Crops on the Agri-food Sector. A First Review." http://europa.eu.int/comm/agriculture/publi/gmo/fullrep/cover.htm.

Commoner, B. 2002. "Unraveling the DNA Myth." *Harper's Magazine,* February.

Dovers, S. R., and J. W. Handmer. 1999. "Ignorance, Sustainability, and the Precautionary Principle: Towards an Analytical Framework." In *Perspectives on the Precautionary Principle,* edited by R. Harding and E. Fisher, 167–89. Annandale, NSW, Australia: Federation Press.

Economic Research Service (ERS). 2001. "Impacts of Adopting Genetically Engineered Crops in the United States." www.ers.usda.gov/emphases/harmony/issues/genengcrops/genengcrops.htm.

Epstein, S. 1996. *Impure Science: AIDS, Activism, and the Politics of Knowledge.* Berkeley: University of California Press.

Falkner, R. 2000. "Regulating Biotech Trade: The Cartagena Protocol on Biosafety." *International Affairs* 76, no. 2: 299–313.

Funtowicz, S. O., and J. R. Ravetz. 1992. "Three Types of Risk Assessment." *Social Theories of Risk,* edited by S. Krimsky and D. Golding. Westport, Conn.: Praeger.

Gallagher, R., and T. Appenzeller, eds. 1999. "Beyond Reductionism." *Science* 284: 79–109.

Gallopin, G. C., S. Funtowicz, M. O'Connor, and J. Ravetz. 2001. "Science for the Twenty-first Century: From Social Contract to the Scientific Core." *International Social Science Journal* 53, no. 2: 219–29.

Global Environmental Change Programme (GECP). 1999. "Special Briefing No. 5. The Politics and GM Food: Risk, Science, and Public Trust. www.gecko.ac.uk/gm-briefing.html.

Government of Canada. 1993. *Federal Government Agrees on New Regulatory Framework for Biotechnology.* Press release, Ottawa.

Hall, L., K., Topinka, J. Hiffman, L. Davis, and A. Good. 2000. "Pollen Flow between Herbicide-resistant *Brassica napus* Is the Cause of Multiple-resistant *B. napus* Volunteers." *Weed Science* 48: 688–94.

Hartzler, B. 1998. "Are Round-Up Ready Weeds in Your Future?" *Weed Science Online* 3 November. www.weeds.iastate.edu/mgmt/qtr98-4/roundupfuture.htm.

Holling, C. S. 2001. "Understanding the Complexity and Economic, Ecological, and Social Systems." *Ecosystems* 4: 390–405.

James, C. 2000. *Global Status of Commercialized Transgenic Crops: 2000.* Ithaca, N.Y.: ISAAA.

Jasanoff, S. 1990. *The Fifth Branch. Science Advisors as Policy Makers.* Cambridge, Mass: Harvard University Press.

Kumar, A., G. Rakow, and K. Downey. 1998. "Isogenic Analysis of Glufosinate-ammonium Tolerant and Susceptible Summer Rape Lines." *Canadian Journal of Plant Science* 78, no. 3: 401–8.

Latour, B. 1987. *Science in Action.* Cambridge, Mass: Harvard University Press.

———. 1999. *Pandora's Hope: Essays on the Reality of Science Studies.* Cambridge, Mass: Harvard University Press.

Latour, B., and S. Woolgar. 1979. *Laboratory Life: The Social Construction of Scientific Facts.* Beverly Hills, Calif.: Sage Publications.

Letourneau, D. K., and B. E. Burrows, eds. 2001. *Genetically Engineered Organisms: Assessing Environmental and Human Health Effects.* New York: CRC Press.

Levidow, L. 1996. "Regulating GMO Release: Britain's Precautionary Dilemmas." *Natures-Sciences-Societes* 4, no. 2: 131–43.

Lynch, M. 1990. "The Externalized Retina: Selection and Mathematization in the Visual Documentation of Objects in the Life Sciences." In *Representation in Scientific Practice,* edited by M. Lynch and S. Woolgar. Cambridge, Mass.: MIT Press.

Lyons, M. 2000. "Farmer's Reaping No Fluke, Court Told: Schmeiser Planted Roundup Ready Canola Knowingly." *Saskatoon Star Phoenix* 6 June.

Mol, A., and J. Law. 1994. "Regions, Networks, and Fluids: Anaemia and Social Topology." *Social Studies of Science* 24: 641–71.

National Biotechnology Advisory Committee (NBAC). 1984. *Annual Report.* Ottawa.

Peterman, R. M., and M. M'Gonigle. 1992. "Statistical Power Analysis and the Precautionary Principle." *Marine Pollution Bulletin* 24, no. 5: 231–34.

Pickering, A. 1995. *The Mangle of Practice: Time, Agency, and Science.* Chicago: University of Chicago Press.

Royal Society of Canada (RSC). 2001. "Expert Panel on the Future of Food Biotechnology." www.rsc.ca/foodbiotechnology/indexEN.html.

Scoones, I. 1999. "New Ecology and the Social Sciences: What Prospects for a Fruitful Engagement?" *Annual Review of Anthropology* 28: 479–507.

Shapin, S. 1996. *The Scientific Revolution.* Chicago: University of Chicago Press.

Star, S. L. 1991. "Power, Technology, and the Phenomenology of Conventions: On Being Allergic to Onions." In *A Sociology of Monsters: Essays on Power, Technology, and Domination,* edited by J. Law, 26–55. London: Routledge.

United Nations Environment Programme (UNEP). 1999. "Draft Report of Extraordinary Meeting of the COP for the Adoption of the Protocol on Biosafety to the CBD." www.biodiv.org/biosafe/bswg6/bswg6.html#resses.

Vanclief, L. 1999. "Food Safety Will Not Be Compromised." *London Free Press,* June 25, A13.

van der Sluijs, J., J. van Eijndhoven, S. Shackley, and B. Wynne. 1998. "Anchoring Devices in Science for Policy: The Case of Consensus around Climate Sensitivity." *Social Studies of Science* 28, no. 2: 291–323.

Wolfenbarger, L. L., and P. R. Phifer. 2000. "The Ecological Risks and Benefits of

Genetically Engineered Plants: The Accumulating Evidence." *Science* 290, no. 5499: 2088–93.

Woodhouse, E. J., and D. Nieusma. 2001. "Democratic Expertise: Integrating Knowledge, Power, and Participation." In *Policy Studies Annual Review.*, edited by R. Hope et al., Policy Studies Annual Review. New Brunswick, N.J.: Transaction Books.

World Trade Organization (WTO). 1998. *WTO Agreements: Sanitary and Phytosanitary Measures.* Switzerland.

Wynne, B. 1992a. "Risk and Social Learning: Reification to Engagement." In *Social Theories of Risk,* edited by S. Krimsky and D. Golding, 275–97. Westport, Conn.: Praeger.

———. 1992b. "Uncertainty and Environmental Learning." *Global Environmental Change* 2: 111–27.

———. 1996. "May the Sheep Safely Graze? A Reflexive View of the Expert-Lay Knowledge Divide." In *Risk, Environment, and Modernity: Towards a New Ecology,* edited by S. Lash, B. Szerszynsk, and B. Wynne, 44–83. London: Sage Publications.

CHAPTER 12

Uncertainty and Biodiversity Conservation

Reinmar Seidler and Kamaljit Bawa

Is Conservation Biology Failing to Reach Its Goals?

In a 2000 issue of *Conservation Biology,* then-editor David Ehrenfeld pointed out that although the journal, and the discipline it represents, had flourished for over a decade, its contributors had rarely been able to report measurable progress toward conservation goals (Ehrenfeld 2000). Conservation biology, reports Ehrenfeld, has been largely failing to address core conservation issues in a practical way—failing to actually "do conservation." Despite extraordinary dedication to the task of conserving biodiversity, despite bringing a remarkable, even heroic, combination of skills, ingenuity, and hard work to bear on the problems, and despite having achieved many significant advances in understanding the nature and extent of threats to the natural world, Ehrenfeld says that conservation biologists have only occasionally succeeded in their central aim, which is to protect the earth's biodiversity. Published results have tended to consist more of "descriptions and recommendations than actual conservation achievements." Conservation biology has been described as a "crisis discipline." Evoking by analogy another crisis profession—medicine—Ehrenfeld suggests that conservation biologists must turn from diagnosis to action, indeed to "treatment."

Why has it proven so difficult to apply the lessons of the past and to adapt action to present needs? Undoubtedly, it is in part because conservation measures tend to be difficult to implement, slow to take effect, often expensive and therefore controversial. Outcomes are almost always uncertain. But Ehrenfeld suggests further that it is because ecological science simply cannot control, or

175

even predict, the course of global-scale events that the "huge forces out there that are sweeping biodiversity away," namely, economic globalization and industrial development, simply lie beyond the influence of the scientific community. In effect, Ehrenfeld is warning us against a potentially naïve faith in the efficacy of the scientific method divorced from its social context. The history of conservation biology over the past decades shows that establishing a rational basis for action or change will not necessarily motivate that action or change. "What began as an effort to provide the purely scientific foundation for conservation must now—to be effective—do much more," he says. Ehrenfeld's argument is partly about the scientific method itself and partly about the uses to which scientific knowledge is put. It is an argument not at all *against* science, but *for* the recognition that conservation science may be reaching its limits in certain respects.

Reductionist Science and Conservation Science

The growth of modern "reductionist" science since the seventeenth century has produced an extraordinary flowering of knowledge, techniques, and technologies. Scientists have seen their primary role as that of reducing uncertainty about some specific aspect of the world and secondarily as improving the lives of people. Social values and aims have proved variable and contradictory, and it has often been difficult to agree on goals and priorities.

One of the most eloquent advocates for an ambitious reductionist science has been ecologist E. O. Wilson. In *Consilience* (1998), Wilson describes reductionism as "the search strategy employed to find points of entry into otherwise impenetrably complex systems." Wilson assigns science the task not only of describing the natural world as it is, but of forming the underpinnings for a unified theory of brain, mind, and ultimately of all human culture. However, while the reductionist approach has historically been astonishingly successful when applied to systems with clearly defined scalar hierarchies, it has been less so in others. In physics, for instance, the particle, atomic, molecular, conventional, and astronomic scales are sufficiently discrete as to have individual regimes of activity, each of which may be analyzed to a high degree in isolation from the others. In the study of living systems, however, even the greatest familiarity with the dynamics of one level of organization will often produce only uncertain predictions about activity on another. This is true of work at the molecular, cellular, and organismal levels, and even more so at the community and ecosystem level. Living systems are simply the most complex systems known. In addition, most living systems are presently undergoing rapid and continuous change under pressure from human activities. Anyone dealing with the conservation or management of natural resources or ecosystems is therefore dealing with the interaction of multiple, complex, and imperfectly understood systems.

Thus, managers of natural resources or ecosystems are compelled by necessity to carry out interventions under conditions of great uncertainty. The process of deciding on actions is often hampered by lack of consensus about the ramifications of action, or even about the basic functioning of the systems to be acted upon. But practitioners of a "crisis discipline" must be prepared to take some actions even in the absence of certainty. The precautionary principle can provide a philosophical framework, if not operational guidelines, for taking action and defining policy in the face of uncertainty. Indeed, the biodiversity crisis is a systemic problem whose dimensions may offer a model of how the precautionary principle might be operationalized.

In this chapter, we focus on the intersection between the natural world and the activities of its human inhabitants. We treat the human economy and the natural world as distinct—though inextricably interwoven—systems. A short list of the principal problems faced by those interested in the conservation of biodiversity today will include high rates of species extinction, changing land use patterns resulting in massive habitat loss and the spread of invasive species, biogeochemical changes with probable consequences for ecosystem functioning, and climate change at local, regional, and global scales. These urgent problems simply cannot be adequately addressed within a meaningful time frame by a focus on reducing levels of uncertainty, except in locally defined circumstances. This applies especially to questions of the ecological roles played by biodiversity and the consequences of its loss.

We ask here whether conservation biology can make some changes in its tactics that might serve to strengthen the voice of science and "reason" in the social debate about the direction of economic development and the options for biodiversity conservation. Our main concern is uncertainty: what approaches we, as conservation biologists, can adopt to conserve biodiversity in the face of inherent uncertainty. We begin by looking at an example of a conservation problem that has generated a good deal of research and literature worldwide. We show that a practical decision about forest management implicates a series of basic but still unanswered questions about biodiversity and ecosystem function, and suggest that this problem belongs to a category of problems that have at their heart a level of scientific uncertainty that is effectively irreducible (at least within a reasonable time frame). We try to put this problem into context as a social phenomenon and suggest some ways in which science may be able to adapt to the necessity of guiding policy effectively within this context. We conclude with a comment about the implications of climate change for global biodiversity.

Tropical Forest Management and Biodiversity Conservation: An Example of Irreducible Uncertainty?

Loss and degradation of tropical forest habitats are two of the most acute problems facing the earth's biological diversity, because many tropical forests are so complex, so species-rich, and at the same time so understudied. Since most of the highly biologically diverse countries are also economically stressed, they rely heavily on international funding and direction to implement conservation measures. This fact highlights the importance of the following question: What proportion of available funding should go into the research and development of techniques of sustainable tropical forestry (sustainable forest management, SFM) on the one hand, and what proportion to the purchase or setting aside of natural tropical forest areas as conservation-oriented parks and to their protection against human exploitation on the other? These two positions represent poles of opinion on the potential utility of logged forests as repositories of biodiversity and on the likely future of such forests.

Those who advocate increased support for tropical silviculture point out that almost all forest areas will eventually come under some degree of exploitation and that silvicultural techniques must be devised to guide that exploitation into relatively nondestructive channels (Grieser Johns 1997; Szaro et al. 1999; Putz et al. 2001). Some of them argue that since most tropical forest parks and reserved areas are close to inhabited areas, projects on the model of Integrated Conservation and Development Projects (ICDPs) are needed. These are conservation projects designed to garner the support of local people by raising their incomes, bolstering economic security and often improving education, and reducing their dependence on destructive harvesting techniques (Kremen 2000). They also note that logging by itself has been responsible for almost no species extinctions or endangerments (Grieser Johns 1997) and that tropical forest parks are often only "paper parks" embedded in social conflict, impossible to police adequately, and vulnerable to incursions and illegal resource exploitation.

Those, on the other hand, who advocate greater support for preserved areas on the park model point out that attempts to legislate sustainable and nondestructive use patterns have often failed and have even backfired by becoming the pretext for opening previously inaccessible areas to exploitation through the building of roads and infrastructure (Brown et al. 2001). They observe that logging rarely happens in isolation, but is often followed by uncontrolled hunting, fishing, mining, wildfires, illegal or semilegal homesteading, even clearing, and ad hoc deforestation (Brandon et al. 1998). Further, they point out that "buffer zone" concepts, in which limited resource use is permitted within the zone surrounding an off-limits core conservation area, have rarely been satisfactory in practice because incomes to local people from limited resource use have been inadequate to lift them out of poverty (Wunder 2001). They point to the need for a more highly developed civil society and more effective regulatory environment before such detailed programs can be carried out successfully and suggest that until then, a return to funding strictly conservation-oriented projects may be the only way to protect most of the remaining unexploited areas (Brandon et al. 1998; Wunder 2001).

Perhaps the controversy is one of emphasis rather than clearly dichotomous viewpoints. Clearly, these options are not mutually exclusive on the global scale. They already coexist, even within nations and regions. But like many, perhaps most, controversies about biodiversity conservation, this one seems to take the general form of "use versus preserve." In a context of limited funding for conservation projects, and an equivocal track record of conservation success, there is a real question as to which general approach is most likely to succeed in a given area.

Forest Management: What Do We Need to Know?

In order to decide which of these apparently divergent options is preferable in a given situation, or what mix of funding and research emphasis is most likely to contribute to forest conservation over the next several decades of anticipated intense pressure, we will want to know what kinds of ecological effects to expect from logging. There have been several decades of careful research into the effects of logging on biodiversity in a selection of tropical forest environments (see references in, e.g., Grieser Johns 1997; Bawa and Seidler 1998; Frumhoff 1995). A large group of studies has examined the effects of conventional ("unsustainable") harvesting methods on groups of animals or plant species, and a few have looked at the effects of well-monitored and carefully managed silvicultural operations. However, the quantity and consistency of research directly geared toward ways of conserving biodiversity as an integral part of tropical forest management is still low. Few practical manuals exist for tropical forestry with conservation as an explicit goal. The fact that tropical forests vary widely in biogeographical characteristics remains a stumbling block to predictive power. Nevertheless, in many forests, enough experience and scientific knowledge do exist to carry out *relatively* nondestructive sustainable harvests in which much of the diversity is preserved (Putz et al. 2000, 2001; Grieser Johns 1997; Dickinson et al. 1996).

At the same time, there is basic agreement that *some* biodiversity will inevitably be lost, at least at the local level, through almost any process of timber harvesting (Szaro et al. 1999). There is disagreement as to how *much* will be lost and about which components—which species or species guilds—are likely to be lost. Crucially, there is disagreement about what the overall effect of this biodiversity loss is likely to be on the stability and functioning of the forest ecosystem as a whole. The argument for increasing research into harvesting methods that threaten a relatively small number of species only makes sense if the system as a whole can continue to function adequately in the local or regional absence of those species. Therefore, this argument depends on the organization and natural regulatory mechanisms of forest ecosystems. Unfortunately, these mechanisms are not well understood.

Forest Ecosystems: Regulation and Stability

Some have argued that the regulatory mechanisms of natural ecosystems, including tropical forest systems, are primarily top-down. According to this model, diversity is maintained and competitive effects are dampened by a relatively few species of top predators, which check the growth of herbivore populations and thus entrain pressure on plant populations (e.g., Terborgh 2001). If this is true, then it seems likely that virtually any tropical forest logging may put the functioning of the system at risk, not because the logging itself is necessarily so destructive, but because logging in tropical forests has so often been associated with patterns of road building, human intrusions and hunting of game animals, increased rates of invasion by exotic species, increased rates of destructive wildfire, and structural changes to the forest. Such cumulative effects are rarely considered in analyses because they are expressed over long time periods. In addition, the elimination of functionally important regulatory "keystone" species is expected to trigger a cascade of effects through trophic levels, which could produce extreme population swings and the extinction of some species at the local or regional level.

On the other hand, if regulatory mechanisms function preponderantly bottom-up, such that plant chemical and mechanical defenses principally limit the growth of herbivore populations and hence their predators, structural changes to the forest may not have as strong an effect on the system as a whole. Every natural system is without doubt an intricate, mutually modified, and continuously changing blend of bottom-up and top-down control mechanisms. Hence, the true dynamic in any tropical forest system is certainly far more complex and variable than either of these two relatively simple models indicates. The relevant point here is that since this question cannot be resolved quickly, management decisions must be made (as, of course, they are being made) in the presence of uncertainty.

The mechanisms by which forest ecosystems resist change in response to disturbance, or recover from disturbance when it has happened, are likewise uncertain. Maintenance of ecosystem stability is one of the "ecosystem services" hypothesized to rely at least partly on diversity (Tilman 1997; Kinzig et al. 2002). It is thought that diversity may be important for the maintenance of stability and productivity under changing conditions, providing an "insurance" against environmental fluctuations. Diverse ecosystems seem to be composed of many relatively weak ecological interactions within and between trophic levels, and fewer strong ones. ("Weak" interactions include such common interactions as predator-prey relationships in which the prey species is not the only component of the predator diet, and competition for resources among a group of species occupying "neighboring" but not identical ecological niches. "Strong" interactions include the relationship of a highly specialized predator to the single prey species that makes up its diet, and strongly coevolved pairs of species wholly dependent on each other for pollination or other reproductive services.) Weak interactions are thought to mask and neutralize the few strong interactions that may be present, stabilizing the system's functioning over time. According to this model, the loss of a large number of species would be likely to destabilize the system by weakening the overall diluting effect. Likewise, novel invasive species exerting strong trophic pressures might overwhelm the weak interactions and thus escape their modifying influence (McCann 2000). These ideas have been extremely difficult to verify experimentally, because of (1) the slowness with which environmental disturbances express themselves through population dynamics in natural systems (scale of years or decades), (2) the difficulty in separating out confounding effects of other variables, and (3) the uncertainty in applying results from one ecosystem or scale to others.

Some of the best-replicated and most detailed work on ecosystem productivity and stability has been done over a period of more than a decade on Minnesota grassland plots (Tilman et al. 2001; Kinzig et al. 2002). In some respects, this large project exemplifies the dilemmas faced by ecologists doing large-scale research with important conservation implications. The Minnesota grassland project has been one of the best-designed, consistently funded, and carefully executed long-term projects in the history of experimental ecology. The project has received extraordinary investments of time and energy from a large number of talented and dedicated scientists, and it has produced clear and unambiguous results in a number of areas. However, the grassland plots are necessarily small (scale of meters), and important questions have been asked about the applicability of results to systems with more complex structures, like forests (McCann 2000).

To cite a single example: the soil processes carried on by microbes, fungi, and invertebrates probably respond more to the chemical characteristics of individual plant species than to plant diversity per se. Yet the roles of the mutualistic interactions linking what goes on below and above ground are poorly understood (Loreau et al. 2001), especially in forests, where these interactions may be particularly important and variable.

In sum, community ecologists have traditionally treated species diversity as the result of the interactions of abiotic conditions (especially climate and soil type) with various ecosystem-level constraints, including competition for resources, productivity, and niche availability. More recently, the idea is being explored that species diversity may itself play an active role as a "potential modulator of processes" (Loreau et al. 2001: 808). Landscape-scale research such as that of Lawton et al. (2001) is producing evidence that the simplification of landscapes can have strong effects on abiotic processes regionally. These in turn can feed back upon regional biodiversity, showing that biotic and abiotic conditions are mutually influenced.

What Are the Practical Implications?

Do such questions matter to a forest agency faced with trying to decide whether to grant a concession to a logging company for a pristine area of tropical forest? Economic and political issues will undoubtedly loom large in the foreground of the decision process. However, if the agency has been charged with conservation of the forest patrimony, foresters would like to be able to predict, with some degree of confidence, what the medium- and long-term effects of logging will be on the forest system. They will find little guidance in the literature.

These kinds of unanswered questions are to a large extent contained within the set of scientific problems or imprecisions that we assume could be answered for local conditions, given a large enough research effort. Following Wynne (1992), we might call them "first-order risks" or uncertainties. Perhaps a bigger stumbling block to effective and secure forest management lies in predicting human behavior. In those places where management guidelines have been put in place, adherence to them has often been weak. Most critics of the sustainable forestry model warn that the real question is not *can* sustainable forestry be done but *will* it be done? In Wynne's terminology, these "second-order risks" or "indeterminacies" may be far less amenable to the conventional (reductionist) scientific approach.

Financial returns to SFM still appear to be, in most cases, inferior to those from unsustainable "timber mining" (Rice 2001; Howard and Valerio 1996; Kishor and Constantino 1994), although this partly depends on the scale of the analysis (Kremen et al. 2000). Financial returns to logging must in any case be seen in a broad political-economic context. Who receives credit from financial institutions and who does not, who controls the distribution and the security of timber concessions, how the regional and national timber market is structured, what degree of vertical integration the timber industry has achieved: these are examples of ways in which the political and economic environment will directly affect financial returns and, hence, the attractiveness of logging. Attempts to alter incentive structures at the level of the national or international timber trade have rarely been effective in promoting sustainable harvest and have often had the paradoxical effect of increasing the rate of unsustainable harvest, even of deforestation (Rice 2001; Brown et al. 2001). This is primarily because most high-level changes in forest policy are intended to increase the profitability of logging *in general,* rather than of sustainable logging *in particular.* "Unless loggers are forced to adopt SFM through strong government control, increasing the profitability of logging alone may in fact increase the pace and scale of conventional 'unsustainable' logging" (Rice 2001: 17). To this extent, it is the policy environment that determines the success or failure of attempts to promote sustainable development, rather than the quantity or quality of scientific and silvicultural know-how. "The most important factor affecting long-term trends will be the efficiency with which forestry rules are observed, and the ability of foresters to protect the integrity of the forest estate" (Grieser Johns 1997: 185).

Toward an Integrated Conservation Science: Integrating Uncertainty into Management Decisions

Uncertainty and indeterminacy are ineluctable components of ecological knowledge. However, this is anything but a recipe for inaction. To return for a moment to David Ehrenfeld's criticism, scientists interested in conservation must adapt their methodologies to the needs of the moment. Actions must be taken and policies defined that seek to protect the earth's biodiversity without excluding or underplaying uncertainty. On the contrary, policy prescription must explicitly integrate uncertainty. If we pursue Ehrenfeld's analogy of conservation biology to medical science, we see a parallel need for preventive strategies in both disciplines.

Uncertainties in system functioning and in chains of cause and effect, unpredictability of synergistic interactions within systems, call for strategies that *prevent* breakdowns rather than concentrating on trying to fix them once they have occurred. A preventive strategy "reaches out into that external world of forests and economies and climates, treats it as part of an enlarged world system of human activity, and tries to influence events before they have an impact on us. Therefore, we move from models of response and management to models of positive design" (Levins 1995: 55). If societies are to move toward some kind of positive design in their development strategies, they will have to avoid exogenizing uncertainty.

Conservation-oriented research programs, too, must escape the pattern of exogenizing those elements of uncertainty endemic to the battery of ecological research tools. We suggest that more research be directed specifically at the linkages and feedbacks between the economic and ecological systems. The mechanisms by which the human economy exerts effects on ecological systems are still little understood in detail. This is because such mechanisms must be studied within specific local-scale contexts. The effects on biodiversity con-servation of large trends such as economic globalization, for instance, are extraordinarily variable at the local level. Effects of trade liberalization policies or increased foreign investment are filtered through a complex network of social, legal, and historical relationships between local institutions and between institutions and the local biota.

As regards tropical forest management, there is little research on tropical forest resource use at the household, firm, and community levels. Decisions affecting many aspects of resource use are taken at those levels, while impor-tant consequences often sum to express themselves strongly at the regional level. Our understanding of decision making at the local level is inadequate (Kaimowitz and Angelsen 1998). For instance, the conventional economic understanding of environmental degradation is that it is a kind of *disinvestment* in the future, which must be countered by education and reincentivation. It has been clear for some time that this picture is flawed (see, e.g., Honadle 1993). On the contrary, there are many circumstances in which people seem to deforest, overhunt, or overharvest as part of a strategy to *invest* in the future (Wunder 2001). These circumstances may include opening access to markets for certain kinds of goods, land occupancy races, changes in tenure rights pat-terns, changes in labor markets, and anticipated regulatory changes. How regional or national regulatory environments affect such decisions is unclear in most cases. There are feedbacks between local, regional, and international poli-cies on forest management, each of which interacts as well with natural and human-induced ecological changes, social custom, and local history. Research that begins to pick apart the strands of these interactions would be useful. Comparative research programs using comparable methodologies in different places can do much to distinguish purely local phenomena from larger trends.

Likewise, it would be useful to understand more about the economic decision making of logging firms, if models could be constructed on the basis of enough data. In general, institutions are absent as parameters from most attempts to model forest resource use, although they are indispensable participants in every policy change (Kaimowitz and Angelsen 1998). This kind of detailed research, however, is slow and expensive, and although it can push back the boundaries of uncertainty about processes in a particular geo-graphical area, it is easy to criticize its conclusions as being inapplicable to other places and times. It cannot, therefore, precede or substitute for policy changes or management initiatives, but must complement them and even be a part of them.

Ideally, perhaps, research and natural resource management would no longer be seen as separate functions. "Ultimately . . . all management is exper-imental and all research involves managers; there is little distinction between management and research" (Sayer and Campbell 2001:32).

Below we sketch several aspects of the changes we see as necessary to the better functioning of conservation initiatives. These deal with the integration of the disciplines of ecology with economics, institution building, international funding systems for research and conservation, and education.

Integrating Ecology with Economics and Vice Versa

Conservation-interested scientists must push for the greater intellectual integration of ecology with economics. Far too many resource management decisions must now be made with economic caveats in mind; and far too many economic models, including influential ones, are constructed as though the human species had somehow escaped from its ecological dimension, as though economic growth could somehow be achieved without the consumption of real resources and the discharge of real wastes. The public debate about "the economy" is still carried on without mention of even the most basic biophysical concepts, including carrying capacity or resource and waste flows. Many of the fundamental principles of economics were articulated by the classical economists of the eighteenth and nineteenth centuries, who were primarily agricultural economists. They were explicitly concerned with the uses and properties of the land and its resources. As the world's leading economies have industrialized, the discipline of economics has distanced itself from its agricultural roots. The analytical procedures of twentieth-century "neo-classical" economics have largely avoided explicit reference to the constraints of the natural world, since these complicate many aspects of analysis. Nevertheless, the human economy is affecting natural systems on every scale. Scientists interested in the protection of biodiversity must debate the propositions of an establishment economics that so often ignores the wider implications of its own analyses.

One way in which this integration can be achieved is by explicitly treating the ecological and economic systems as a unit. Conceptual integration of complex systems can probably only be achieved by a "paradigm shift" away from attempts to direct systems toward a predetermined state and toward a focus on building resilience within systems. Groups such as the Resilience Alliance (http://www.resalliance.org/) and the Beijer International Institute for Ecological Economics in Sweden (http://www.beijer.kva.se/) have been attempting to outline this approach under the rubric of "ecosystem resilience." Resilience is here defined as the capacity of a system to experience disturbance without moving into a "different state and different set of controls" (Walker 2000: 5). Such efforts are still in the preliminary stages of theoretical working-out, but in combination with local-scale research and management, they hold considerable hope for more effective biodiversity conservation. Calls for the development of a "sustainability science" to focus on the interactions between nature and society (e.g., Kates et al. 2001) are setting broad agendas within which new approaches to conservation should flourish.

Institution Building

Institutions, both formal and informal, are required for nurturing and promoting local solutions. All over the world people have evolved strategies that are locally adapted to cope with uncertainty, but many traditional institutions for resource management have been considerably weakened due to commercialization and centralization. Thus, local institutions need to be strengthened or, in some cases, resurrected. Since the local institutions can influence conservation and management of resources at a relatively small spatial scale, strong and functional institutions are required at other levels as well. The inability of global and national environmental organizations to bring about institutional reforms, or to create and strengthen existing institutions, has been a major hurdle in the implementation of programs designed to reduce uncertainty in the conservation of biological diversity.

Incentives for Conservation

Conservation does not, and will not, pay for itself within the market economy. Most biodiversity-rich countries are economically poor. Many suggestions for the transfer of conservation funds from the developed to the developing world have been made; few have been successfully implemented, perhaps partly because they tend to be complex and difficult to monitor. Conservation initiatives linked to development projects (ICDPs) "fund" conservation in an indirect way, by attempting to add to or highlight the economic value of ecosystems, or to raise local incomes to the point at which people are no longer dependent on resource extraction. Economic incentives for such projects do exist, but benefits from projects accrue at the local level and (by some measurements) at the global level, rather than at the national level, where many of the decisions about larger-scale projects are made (Kremen et al. 2000). Another problem that must be confronted is the need for coordination between local-scale conservation programs and international or global-scale funding systems. Acknowledging these realities has led to proposals for direct payment schemes, which, if they can be supported politically, offer the hope of relatively streamlined transfer of funding for cooperatively administered conservation programs (Ferraro 2001; Ferraro and Simpson 2001). Direct payment contracts may help reduce some of the uncertainties associated with social behavior by simplifying the pathway between funding and conservation performance. They may clarify the relationship between donor expenditures and conservation objectives, which are often obscured in the more complicated conservation-and-development model. There are substantial difficulties with this kind of idea, not the least of which are the substantial institutional requirements for providing a legal and regulatory framework. Nevertheless, the institutional requirements are at least as significant in most of the indirect projects. Direct payment schemes may offer an efficient way of addressing head-on the global need for conservation initiatives that work without negatively affecting the livelihoods of local people.

International Funding for Research

Biodiversity-rich countries generally lack financial and human resources for adequate investments in research on issues related to conservation and natural resource management. Although a number of international initiatives have resulted in modest investments, much of the funding is highly concentrated in organizations based in the West, with meager flow of funds to appropriate institutions in the developing world. Equally important, the research paradigms developed by international organizations, often in isolation from realities on the ground, have failed to address critical issues. Efforts are now under way to develop fundamentally new paradigms that incorporate interdisciplinarity and integrated approaches to conservation of natural resources and at the same time address institutional issues (Sayer and Campbell 2001). Such efforts will yield results only if funding is substantially increased and directed to appropriate institutions.

Education

The institution of the university has at various times both nourished and obstructed dialogue between the disciplines. Many universities are presently struggling to realize ideals of interdisciplinarity. Conservation-oriented departments of biology and ecology must integrate the study of the human economy into their programs, so that graduate students do not leave without a strong appreciation of the inextricability of these issues. Despite nods in the direction of interdisciplinarity, most graduate programs in ecology and organismal biology still reward their students' ability to carry out programs of independent original research, rather than their ability to cooperate with researchers from other disciplines on interdisciplinary projects of the type that are needed to address current problems. Historically, the most effective conservation programs have been spearheaded by charismatic specialists with fortuitous talents for political and strategic thinking, communication, and action. We cannot afford to depend exclusively on the emergence of such individuals, but should be preparing a generation of scientists able to communicate effectively with specialists in other fields, with policy makers, and with nonscientists (see also comments by Jacobson and McDuff 1998; Cannon et al. 1996; Orr 1991; Noss 1997).

Climate Change

Climate change remains a variable with sweeping implications for the biodiversity question. Because of predicted increases in global surface temperatures, we can expect major changes in species distributions and in phenology (and hence in species interactions); some changes in physiology (particularly in plants and phytoplankton populations); and rapid evolutionary changes to species with short generation times and large populations, including parasites, bacteria and pathogens, and invertebrates. Many such changes are already being documented, although most data series are still too short to produce unequivocal statistical results (see Hughes 2000; McCarty 2001; Willott and Thomas 2001).

For forest management and biodiversity conservation, climate change presents a true wild card. If the distribution ranges of plant and animal species begin to shift across a landscape already fragmented by human activities, it is virtually impossible to say what the specific results will be. Under the conditions engendered by a 1 to 3.5°C temperature rise, many species will be forced into new ecological relationships with one another. Some will move geographically into new areas and interact within new community settings. Species interactions will loosen or break off because of changes in phenology and timing. It can be assumed that some marginal populations will be particularly stressed, whereas others will be favored. Parasites and infections will respond in idiosyncratic ways. The potential for longer-lived species to adapt to new and changing conditions is itself highly constrained by the ongoing fragmentation, degradation, and transformation of many habitats.

These facts throw even the most well-supported assumptions and hopes about biodiversity conservation into doubt. Even if we choose for a moment to ignore all the other sources and loci of uncertainty in biodiversity management, dealing with this one source alone will absolutely require the explicit integration of uncertainty into management policy. Climate change will accelerate and complicate ongoing human–induced changes in biological systems. "Conventional" ecological research, seeking to quantify and analyze these changes piecemeal, will not be able to keep up with this rate of change over the next century or more.

Concluding Remarks: Can Science and Policy Afford to Await the Reductionist Project?

One may agree with E. O. Wilson that reductionist methods can show us the way to reasonable and sustainable living or with Wendell Berry that such methods miss the very core of the problem, which is human nature and is irreducible (Berry 2001). Either way, given the uncertain margin of time and geography still separating us from the potential effects of irreparable ecological harm, one must question whether we are permitted the luxury of waiting. According to the best estimates of the overwhelming majority of people who are working on biodiversity issues, species are presently being lost to extinction at a rate unprecedented in the history of our planet. There are many different ways of measuring biological diversity, but none of them shows diversity as anything other than severely stressed worldwide. Management changes are being made in many places and contexts, but Ehrenfeld's "huge forces" continue to sweep biodiversity away in some places and to challenge it severely in many others. A large proportion of the earth's terrestrial and aquatic biodiversity inhabits tropical, populous, economically impoverished, and industrializing nations, in which policy changes are in some ways the hardest to make. Protecting the diversity of life will entail a concerted effort to implement integrated programs that depart from traditional approaches to reducing uncertainty. And unfortunately, these programs will cost money.

References Cited

Bawa, K. S., and R. Seidler. 1998. "Natural Forest Management and the Conservation of Biological Diversity in Tropical Forests." *Conservation Biology* 12: 46.

Berry, W. 2001. *Life Is a Miracle: An Essay Against Modern Superstition.* Washington, D.C.: Counterpoint.

Brandon, K., K. Redford, and S. Sanderson. 1998. *Parks in Peril: People, Politics, and Protected Areas.* Washington, D.C.: Island Press.

Brown, C., P. B. Durst, and T. Enters. 2001. *Forests Out of Bounds: Impacts and Effectiveness of Logging Bans in Natural Forests in Asia-Pacific (Executive Summary).* Bangkok: Asia-Pacific Forestry Commission, FAO-UN.

Cannon, J. R., J. M. Dietz, and L. A. Dietz. 1996. "Training Conservation Biologists in Human Interaction Skills." *Conservation Biology* 10: 1277–82.

Daily, G. C. 1997. *Nature's Services: Societal Dependence on Natural Ecosystems.* Washington, D.C.: Island Press.

Dickinson, M. B., J. C. Dickinson, and F. E. Putz. 1996. "Natural Forest Management as a Conservation Tool in the Tropics: Divergent Views on the Possibilities and Alternatives." *Commonwealth Forestry Review* 75, no. 4: 309.

Ehrenfeld, D. 2000. "War and Peace and Conservation Biology." *Conservation Biology* 14, no. 1: 105.

Ferraro, P. J. 2001. "Global Habitat Protection: Limitations of Development Interventions and a Role for Conservation Payments." *Conservation Biology* 15, no. 4: 990–1000.

Ferraro, P. J., and D. Simpson. 2001. "The Cost-Effectiveness of Conservation Payments. Resources for the Future. Discussion paper no. 00–31. Resources for the Future." Washington, D.C.: Resources for the Future.

Frumhoff, P. C. 1995. "Conserving Wildlife in Tropical Forests Managed for Timber." *BioScience* 45: 456.

Grieser Johns, A. 1997. *Timber Conservation and Biodiversity in Tropical Rainforests.* Cambridge: Cambridge University Press.

Halpin, P. N. 1997. "Global Climate Change and Natural-Area Protection: Management Responses and Research Directions." *Ecological Applications* 7, no. 3: 828.

Honadle, G. 1993. "Institutional Constraints on Sustainable Resource Use: Lessons from the Tropics Showing that Resource Overexploitation Is Not Just an Attitude Problem and Conservation Education Is Not Enough." In *Defining Sustainable Forestry,* edited by G. Aplet, N. Johnson, J. T. Johnson, J. T. Olson, and V. A. Sample, 90–119. Washington, D.C.: Island Press.

Howard, A. F., and J. Valerio. 1996. "Financial Returns from Sustainable Forest Management and Selected Agricultural Land-Use Options in Costa Rica." *Forest Ecology and Management* 81: 35.

Hughes, L. 2000. "Biological Consequences of Global Warming: Is the Signal Already Apparent?" *Trends in Ecology and Evolution* 15, no. 2: 56.

Jacobson, S. K., and M. D. McDuff. 1998. "Training Idiot Savants: The Lack of Human Dimensions in Conservation Biology." *Conservation Biology* 12, no. 2: 263–67.

Kaimowitz, D., and A. Angelsen. 1998. *Economic Models of Tropical Deforestation: A Review.* Bogor, Indonesia: CIFOR.

Kates, R., W. C. Clark, R. Corell, J. M. Hall, C. C. Jaeger, I. Lowe, J. J. McCarthy, H. J. Schellnhuber, B. Bolin, N. M. Dickson, S. Faucheux, G. C. Gallopin, A. Gruebler, B. Huntley, J. Jager, N. S. Jodha, R. E. Kasperson, A. Mabogunje, P. Matson, H. Mooney, B. Moore III, T. O'Riordan, and U. Svevin. 2001. "Sustainability Science." *Science* 292: 641.

Kinzig, A. P., S. W. Pacala, and D. Tilman. 2002. *The Functional Consequences of Biodiversity: Empirical Progress and Theoretical Extensions.* Princeton: Princeton University Press.

Kishor, N. M., and L. F. Constantino. 1994. "Sustainable Forestry: Can It Compete?" *Finance and Development* 31: 36.

Kremen, C. J. O. Niles, M. G. Dalton, G. C. Daily, P. R. Ehrlich, J. P. Fay, D. Grewal, and R. P. Guillery. 2000. "Economic Incentives for Rainforest Conservation Across Scales." *Science* 288: 1828.

Lawton, R. O., U. S. Nair, R. A. Pielke Sr., and R. M. Welch. 2001. "Climatic Impact of Lowland Deforestation on Nearby Montane Cloud Forest." *Science* 294: 584.

Levins, R. 1995. "Preparing for Uncertainty." *Ecosystem Health* 1, no. 1: 47.

Loreau, M., S. Naeem, P. Inchausti, J. Bengtsson, J. P. Grime, A. Hector, D. U. Hooper, M. A. Huston, D. Raffaelli, B. Schmid, D. Tilman, and D. A. Wardle.

2001. "Biodiversity and Ecosystem Functioning: Current Knowledge and Future Challenges." *Science* 294: 804.

Ludwig, D., R. Hilborn, and C. Walters. 1993. "Uncertainty, Resource Exploitation, and Conservation: Lessons from History." *Science* 260: 17, 36.

McCann, K. S. 2000. "The Diversity–Stability Debate." *Nature* 405: 228–33.

McCarty, J. P. 2001. "Ecological Consequences of Recent Climate Change." *Conservation Biology* 15, no. 2: 320–31.

Noss, R. 1997. "The Failure of Universities to Produce Conservation Biologists." *Conservation Biology* 11, no. 6: 1267–69.

Orr, D. 1991. "Politics, Conservation, and Public Education." *Conservation Biology* 5: 10–12.

Putz, F. E., G. M. Blate, K. H. Redford, R. Fimbel, and J. Robinson. 2001. "Tropical Forest Management and Conservation of Biodiversity: An Overview." *Conservation Biology* 15, no. 1: 7.

Putz, F. E., D. P. Dykstra, and R. Heinrich. 2000. "Why Poor Logging Practices Persist in the Tropics." *Conservation Biology* 14, no. 4: 951.

Rice, R. E., C. A. Sugal, S. M. Ratay, and G. A. Fonseca. 2001. "Sustainable Forest Management: A Review of Conventional Wisdom." *Advances in Applied Biodiversity Science.* Series no. 3. Washington, D.C.: Center for Applied Biodiversity Science, Conservation International.

Sayer, J. A., and B. Campbell. 2001. "Research to Integrate Productivity Enhancement, Environmental Protection, and Human Development." *Conservation Ecology* 5, no. 2: 32. http://139.142.203.66/pub/www/Journal/vol5/iss2/art32/index.html.

Szaro, R. C., J. A. Sayer, D. Sheil, L. Snook, A. Gillison, G. Applegate, J. Poulsen, and R. Nasi. 1999. *Biodiversity Conservation in Managed Forests.* Austria: CIFOR, Indonesia and International Union for Forestry Research Organizations.

Terborgh, J., L. Lopez, P. Nunez, M. Rao, G. Shahabuddin, G. Orihuela, M. Riveros, R. Ascanio, G. H. Adler, T. D. Lambert, and L. Balbas. 2001. "Ecological Meltdown in Predator-Free Forest Fragments." *Science* 294, no. 5548: 1923.

Tilman, D. 1997. "Biodiversity and Ecosystem Functioning." In *Nature's Services: Societal Dependence on Natural Ecosystems,* edited by G. C. Daily. Washington, D.C.: Island Press.

Tilman, D., P. B. Reich, J. Knops, D. Wedin, T. Mielke, and C. Lehman. 2001. "Diversity and Productivity in a Long-Term Grassland Experiment." *Science* 294: 843–45.

Walker, B. 2000. "Analysing Integrated Social-Economic Systems." Report on a workshop, 12–14 September. Stockholm: Royal Swedish Academy of Sciences.

Willott, J., and C. Thomas. 2001. "Implications of Climate Change for Species Conservation." Briefing paper, IUCN workshop, 19–21 February. Gland, Switzerland.

Wilson, E. O. 1998. *Consilience.* New York: A. Knopf.

Wunder, S. 2001. "Poverty Alleviation and Tropical Forests: What Scope for Synergies?" *World Development* 29, no. 11: 1817–33.

Wynne, B. 1992. "Uncertainty and Environmental Learning: Reconceiving Science and Policy in the Preventive Paradigm." *Global Environmental Change*, June, 111.

PART IV

Science in Governance
and Governance of Science

Environmental laws and policies and the agencies that administer them generally determine how environmental risks are studied, the boundaries of analysis,
the threshold for action (including the strength of scientific evidence needed),
and the range of plausible solutions to those risks. Environmental science is an
applied science, ultimately integrated into a political process. The precautionary
principle demands a more dynamic and transparent interaction between science and policy, accepting the uncertainty involved in assessing complex systems. It requires an acknowledgment that decision making under uncertainty is
often "messy"—that precise, point estimates may not be available—and that
values and differing worldviews do infiltrate the scientific process.

Yet, in many political systems, environmental science is considered to be
rational, value-neutral, and objective, cleanly separated from the political process. How is this so, given the range of uncertainties and ignorance in our
understanding of complex risks?

Decision making regarding hazards in the face of uncertainty is ultimately
complex, value laden, and contentious. Scientific information and procedures
play important roles in the process but cannot fully resolve difficulties and
uncertainties. As a result, we need to think about decision making like solving
a puzzle: gathering bits of incomplete information from many sources, looking for patterns, and making intelligent guesses to arrive at solutions.

Many of the current debates about the precautionary principle revolve
around what constitutes enough (and relevant) scientific information for decision makers to act. Debates between European countries (which espouse the
precautionary principle) and others on issues such as genetically modified
organisms and beef hormones demonstrate the divide between nations in how
science should be incorporated in policy. They reflect different understandings
of the ability of science to quantify and solve complex environmental prob-

lems; different beliefs on the roles of science, values, and government policy in environmental decision making; and very different regulatory and political systems.

The chapters in this part examine the science-policy nexus in detail. They provide perspectives on the application of science in precautionary policies from three different continents. They examine questions such as how does the relationship between science and precaution differ between countries and cultures; how can we use science more effectively in preventive policy; and what changes to law and policy are needed to support a more "precautionary" approach to science.

In chapter 13, David Gee and Andrew Stirling analyze the lessons learned from not taking precaution based on early warnings with regards to a variety of ecosystem and health risks. Their chapter is based on a landmark historical case-study analysis published by the European Environment Agency. They outline twelve lessons for improving policy making and the use of science under uncertainty.

Through an examination of the implementation of biodiversity conservation plans in developing nations, Reginald Victor analyzes the types of uncertainties involved in translating conservation goals into concrete actions. He argues in chapter 14 that greater examination of uncertainties in these plans is necessary, particularly those related to social, political, psychological, and other human factors.

Sheila Jasanoff examines the unique aspects of the U.S. regulatory system that lead to controversy over the precautionary principle and argues that the presumed dichotomy between the supposedly nonprecautionary U.S. approach and the precaution-based European approach is mistaken or greatly overstated. In chapter 15, she notes that in the United States precaution must be articulated within the context of homegrown American scientific, legal, and policy practices.

In chapter 16, Theofanis Christoforou presents a European perspective on the precautionary principle and science. He argues that science is critical to implementing precaution, but decisions must also reflect the public's willingness to accept risk (society's chosen level of protection) and the nature of uncertainties involved.

CHAPTER 13

Late Lessons from Early Warnings: Improving Science and Governance Under Uncertainty and Ignorance

David Gee and Andrew Stirling

The European Environment Agency (EEA) is an independent Agency of the European Community, which was founded in 1993. Its main task is to provide information to the policy-making bodies of the European Union (EU) and its member states that can be of direct use for improving decision making and public participation in the fields of environment and sustainable development. The main background against which these activities are set is the growing scale and pace of scientific and technological innovation, with the nature of the consequences increasingly outstripping any social capacity for prediction (WBGU 2000). Decision makers therefore often need information in situations of scientific uncertainty and ignorance. Herein lies the increasing relevance of the precautionary principle, enshrined, along with the principles of prevention and the polluter pays, in the Maastricht Treaty on European Union (European Commission 2000).

The precautionary principle is not just an issue for the EU: its potential impact on trade means that its application can have global repercussions. There are currently disputes both between and within the EU and the United States on the use and application of precaution to hormones in beef, genetically modified organisms (GMOs), and climate change. In 2002 the EEA published *Late Lessons from Early Warnings: The Precautionary Principle 1896–2000* to examine the use, or neglect, of information and precaution in protecting both

Table 13.1. Uncertainty and Precaution: Toward a Clarification of Terms

Situation	State and Dates of Knowledge	Examples of Action
Risk	"Known" impacts; "known" probabilities, e.g., asbestos causing respiratory disease, lung and mesothelioma cancer, 1965 to present	Prevention: action taken to reduce known hazards, e.g., eliminate exposure to asbestos dust
Uncertainty	"Known" impacts; "unknown" probabilities, e.g., antibiotics in animal feed and associated human resistance to those antibiotics, 1969 to present	Precautionary prevention: action taken to reduce potential risks, e.g., reduce/eliminate human exposure to antibiotics in animal feed
Ignorance	"Unknown" impacts and therefore "unknown" probabilities, e.g., the "surprises" of chlorofluorocarbons and ozone layer damage prior to 1974; asbestos mesothelioma cancer prior to 1959	Precaution: action taken to anticipate, identify, and reduce the impact of "surprises," e.g., use of properties of chemicals such as persistence or bioaccumulation as "predictors" of potential harm; use of the broadest possible sources of information and long-term monitoring

Source: EEA.

human and ecosystem health. The report was based on a historical approach to scientific uncertainty (Gee 1997). Since discussions on precaution are sometimes hampered by confusion about the meaning of terms used in the debate, we thought it was important to analyze and learn from past experiences in the use and nonuse of precaution in controlling hazardous technologies and to do so with a common approach and terminology. Analysis of these histories generated useful definitions of some of the key terms used in debates on precaution and prevention (table 13.1).

Late Lessons from Early Warnings: An Approach to Learning from History

In trying to understand the causes and impacts of not acting on early evidence, examinations of historical cases and their lessons such as that of asbestos (see box 13.1) have rarely been conducted. The *Late Lessons* report attempts to help fill that gap.

Fourteen case studies were chosen from a range of hazards to workers, the public, and the environment, where enough was known of their impacts for

Box 13.1. A Lesson from History

In 1898, U.K. factory inspector Lucy Deane observed, "The evil effects of asbestos dust have also instigated a microscopic examination of the mineral dust by HM Medical Inspector. Clearly revealed was the sharp glass-like jagged nature of the particles, and where they are allowed to rise and to remain suspended in the air of the room in any quantity, the effects have been found to be injurious as might have been expected" (Deane 1898:171–172).

One hundred years later, in 1998, the U.K. government decided to ban "white" asbestos, a decision that was echoed by the EU the following year. The current asbestos-induced death rate in the United Kingdom is about 3,000 deaths per year, and some 250,000 to 400,000 asbestos cancers are expected in Western Europe over the next thirty-five years, due to past exposures (Peto 1999).

---

conclusions to be drawn about how well they were dealt with by governments and civil society (see box 13.2). Obviously, there are other public health impacts and environmental disasters that were not looked at, such as thalidomide (James 1965), lead (Millstone 1997), and the Aral Sea (Small et al. 2001). These provide further lessons about unintended consequences and about the conflict between short- and long-term interests. The authors of the case studies were asked to structure their chapters around four key questions:

• When was the first credible scientific "early warning" of potential harm?
• When and what were the main actions or inactions on risk reduction taken by regulatory authorities and others?
• What were the resulting costs and benefits of the actions or inactions, including their distribution between groups and across time?
• What lessons can be drawn that may help future decision making?

Authors were also asked to base their conclusions on the information available at the time and not on the luxury of hindsight. The objective was to see what could be learned from the histories that could help prevent, or at least minimize, the impacts of current and future economic activities that may turn out to be harmful, and to do so without stifling innovation or compromising science.

The case studies and authors were chosen with a transatlantic audience in mind. Three chapters focus either on a North American issue (pollution of the Great Lakes) or primarily on the North American handling of issues that are

---

Box 13.2. Late Lessons from Early Warnings: The Case Studies
  Fisheries
  Radiation
  Benzene
  Asbestos
  Polychlorinated biphenyls
  Halocarbons and the ozone layer
  Diethylstilbestrol
  Antimicrobials as growth promoters
  Sulfur dioxide
  Methyl tert-butyl ether in petrol
  Chemical contamination of the Great Lakes
  Tributyltin antifoulants
  Hormones as growth promoters
  Mad cow disease (bovine spongiform encephalopathy)

---

also directly relevant to Europe (benzene and diesthylstilbestrol [DES] administered in pregnancy) and authored by scientists from North America. Three chapters cover issues of some conflict between North America and Europe (hormones as growth promoters, asbestos, and methyl tert-butyl ether in petrol), and all other chapters are as relevant to the environments and public health of North Americans as to Europeans. The case study authors are not without their own views, being for the most part active participants in the process of making the histories summarized in each chapter. However, as respected scientists in their fields, the authors attempted to adopt an objective position in answering the four questions put to them.

The case studies are all about false negatives in the sense that they are agents or activities that were regarded at one time as harmless by governments and others, at prevailing levels of exposure and "control," until evidence about their harmful effects emerged. While the authors wanted to include some examples of false positives, where unnecessary action was taken on the basis of a precautionary approach, such examples were difficult to find. Our attention was drawn to a U.S. publication, *Facts versus Fears* (Lieberman and Kwon 1998), which attempted to provide some twenty-five examples of false positives but on closer examination these turned out not to be robust enough for those who recommended them to accept the invitation to write case studies for the report. The challenge of demonstrating false positives remains: possible candidates that have been mentioned include the ban on dumping sewage sludge in the North Sea and the Y2K millennium bug.

An Early Use of the Precautionary Principle: London, 1854

The introduction to the Late Lessons report illustrates an early use of the precautionary principle in Europe: When in the midst of a cholera outbreak in 1854 London, the physician John Snow recommended removing the handle of the Broad Street Pump, based on observation of patterns of the disease. The removal of the pump, despite proof or a full understanding of the etiology of the disease, helped control the outbreak.

The story of John Snow and cholera contained several of the key elements of precautionary policy making:

• The long time lag between "knowing" about a hazard and its likely causes and "understanding" the chemical and other processes underlying the causal links
• A focus on the potential costs of being wrong
• The use of minority scientific opinions in policy making

There are many differences among cholera, asbestos (which came into use at about the time of Snow's action), and the other harmful agents in the case studies, not least being the time lag between exposure to the harmful agent and the health damage: hours in the case of cholera but decades in the case of asbestos and most of the other agents studied. Had governments adopted an approach to precautionary prevention similar to that of Dr. Snow, once the early warnings on asbestos had been published, much of the tragedy and the huge costs of asbestos exposure could have been averted.

Politicians today are working in similar conditions of scientific uncertainty and stress, but now made more difficult by the greater risks and uncertainties (economic, health, and ecological) of larger-scale activities (Beck 1992) and by greater pressure from the mass media (Smith 2000). They also work with more democratic institutions and are accountable to a better-educated and involved citizenry that has greater access to information. Globalization and free trade issues add further complications, as does the emerging science of complexity, which adds to the need for greater humility and less hubris in science. It is in these circumstances of trying to prevent potentially serious and irreversible effects, without disproportionate costs, that the precautionary principle can be useful.

Learning from History?

The European Scientific Technology Observatory (ESTO) project on technological risk and the management of uncertainty (Stirling 1999) provided an initial framing for this analysis. The ESTO project sets out a comprehensive structure for the consideration of issues relating to precaution. The most difficult question for the EEA case study authors to answer was, "What were the resulting costs and benefits of the actions or inactions?" This difficulty arose partly from the authors' lack of an economic background and partly because the incommensurable nature of different effects and their diverse social distributions militate against any definitive assessment of the general pros and cons of alternative courses of action.

In many of the EEA case studies, adequate information about potential hazards was available well before decisive regulatory advice was taken. However, the information was often either not brought to the attention of the appropriate decision makers early enough, or it was discounted by them and other stakeholders for one reason or another. It is also true that in some of the case studies, early warnings—and even loud and late warnings—were effectively ignored by decision makers because of short-term economic and political interactions (for example, asbestos, polychlorinated biphenyls [PCBs], the Great Lakes, and sulfur dioxide and acidification). Many of the late lessons therefore relate to the type, quality, processing, and utilization of information set within the context of a more participative and democratic process. The scale and scope of such an integrated and comprehensive process of hazard and options appraisal need to be related to the likely scale of the potential consequences (environmental, health, social, and economic) of the activity in question. For example, the global and generational consequences of GMOs would merit a more comprehensive options appraisal than the local consequences of a road bypass.

Twelve Late Lessons From Early Warnings

The report authors generated twelve particular lessons for improving science and governance about uncertain, complex risks.

1. Respond to Ignorance as Well as Uncertainty

A central lesson of the EEA report concerns the importance of recognizing and fully understanding the nature and limitations of our knowledge. What is often referred to as "uncertainty" actually hides important distinctions. All of the activities in the case studies were subjected to some form of (formal or informal) assessment of risk. What remained neglected, however, was the virtual certainty that there would be factors that remained outside the scope of the risk assessment. This is the domain of ignorance—the source of inevitable surprises, or unpredicted effects (see table 13.1).

Just as one basis for scientific research is the anticipation of positive surprises—"discoveries"—so it can yield the corresponding prospect of negative surprises. By their nature, complex, cumulative, synergistic, or indirect effects in particular have traditionally been inadequately addressed in regulatory appraisal. At first sight, responding to ignorance may seem impossible. How can strategies be devised to prevent outcomes that, by definition, are not known? Analysis of the case studies suggest that it is possible to do so.

- Use knowledge of the intrinsic properties of a substance or activity when assessing possible impacts; for example, a chemical substance that is persistent and bioaccumulates carries a significant risk of a long-term hazardous impact.
- Use a diversity of robust and adaptable technological options to meet needs. This helps to limit technological "monopolies" such as that of asbestos, chlorofluorocarbons [CFCs], PCBs, and thereby the scale of any "surprises."
- Use a variety of scientific disciplines as well as "lay" and "local" knowledge in risk assessments in order to more effectively gather whatever information is available in society.
- Reduce specific exposures to potentially harmful agents on the basis of credible early warnings of initial harmful impacts, thus limiting the size of any other surprise impacts from the same agent.
- Reduce the general use of energy and materials via greater ecoefficiencies, so as to reduce overall environmental burdens, thereby limiting the scale of future surprises.
- Use liability measures (e.g., legal duties and insurance bonds) to compensate for potentially harmful impacts and to provide an investment fund if no "surprise" occurs.
- Use prospective analyses and scenarios to help foresee unintended consequences.
- Use more long-term environmental and health monitoring to help detect "earlier warnings" of surprises.
- Use more and better research and more effective dissemination of "early warnings" to encourage earlier action to reduce risks.

2. Research and Monitor for "Early Warnings"

Well-planned research and long-term monitoring are essential to the systematic identification of areas of uncertainty and to increase the prospect of timely alerts to problems arising out of ignorance. Awareness of uncertainty and ignorance helps in posing appropriate research questions. The case studies illustrate that even "critical path" issues, identified at an early stage, were not necessarily followed up in a timely or effective fashion. For example, bovine spongiform encephalopathy (BSE) was first identified as a new disease in cattle in

1986, but research to verify its supposed absence of maternal transmission—important to the early position of the U.K. Ministry of Agriculture, Fisheries, and Food (MAFF)—was not initiated until 1989. Maternal transmission did occur. Experiments concerning the transmissibility of sheep scrapie to cattle (a favored hypothesis of the source of the disease) were not begun until 1996. No surveys of the number of infectious, but asymptomatic, cattle entering the food chain were conducted. The U.K. government continued prominently to cite the absence of evidence of harm, when very little evidence supporting this position was available or was even being sought.

As human economic activities become geographically more widespread and sometimes less reversible, the use of the "world as a laboratory" (the only one we have) requires more intelligently targeted, ecological, and biological surveillance. Research may reduce uncertainties and ignorance, but that will not necessarily be the case. There are examples in which research can compound uncertainty and reveal new sources of ignorance. For example, a Canadian mathematical model of the interactions between various fish species suggested that they became more unpredictable as progressively more biological data were incorporated into the model. Calls for more research should be as specific as possible about the scientific questions that need to be addressed, the time such research may take, and the independence of the relevant organization carrying it out.

3. Search Out and Address "Blind Spots" and Gaps in Scientific Knowledge

The confirmation of an Antarctic ozone hole in 1985 was a by-product of an experiment conducted for other purposes. A dedicated satellite observation program to monitor stratospheric ozone had earlier detected major depletion, but the results were considered suspect and set aside. Another blind spot can occur where technological change is purported to solve historic problems, even when the hazard is long term. For example, successive claims were made that earlier health impacts from asbestos were the result of working conditions that had been solved. Yet, even with improvements in working conditions, disease was still identified, and it took decades to identify the risks at each successively lower level of exposure. A more precautionary approach means systematically searching out such blind spots using multiple disciplines and other sources of knowledge that can help stimulate the interactions between disciplines that are more likely to expose uncertain assumptions and blind spots.

4. Identify and Reduce Interdisciplinary Obstacles to Learning

Hazardous impacts in a particular area (such as veterinary health) can lead to the regulatory appraisal becoming unduly dominated by a particular discipline,

such as medicine in the asbestos case and veterinary science in the antimicrobials case. This implicitly created a form of institutional ignorance. The
risk appraisal for methyl tert-butyl ether (MTBE) was based mainly on
knowledge concerning engines, combustion, and air pollution. Water pollution aspects associated with persistence and significant taste and odor problems were essentially disregarded, though the information was available. With
regard to mad cow disease, U.K. veterinary officials considered the possibility of BSE transmissibility to humans as acceptably slight. This contrasts with
the attitude in the United States, where the possible link between sheep
scrapie and human Creutzfeldt-Jakob disease (CJD) had been regarded as a
possibility since the 1970s, when the entry of infected animals into the food
chain was banned.

5. Ensure That Real-World Conditions Are Fully Accounted For

Real-world conditions can be very different from theoretical assumptions,
with serious consequences. In the real world, bioaccumulated metabolites of
PCBs were found to be more toxic than indicated by experiments using original commercial PCB formulations. It was also assumed that PCBs could be
contained within "closed" operating systems. This proved impossible, resulting
in releases into the food chain from accidents and poorly maintained equipment and illegal disposal. Optimistic assumptions about the performance of
halocarbon containment equipment and the efficiency of decommissioning
played a role in reducing the effectiveness of control measures. Scientific advisory committees on growth promoters only considered circumstances relating
to authorized use and assessments of individual growth promoters, rather than
in combination. There was a long delay before authorities acknowledged that
asbestos users (or even local residents around a factory), as well as manufacturing workers, could be at risk from exposure.

6. Systematically Scrutinize and Justify the Claimed Pros and Cons

The regulatory appraisal process generally does not systematically examine the
claims made about the benefits of a technology or product, including an identification and assessment of the real-world conditions under which the
claimed benefits could arise.

In the diethylstilbestrol (DES) case study, the data from 1953 trials
showed that DES was ineffective as a means of reducing risks of spontaneous
abortion in certain groups of mothers. It was not for another twenty to
thirty years that use of this drug was actually banned in some countries, as a
result of the discovery of an increase in a rare cancer of the vagina in daugh-

ters of treated women. Had greater critical attention been paid at the outset to the claims of efficacy, some of these second-generation cancers might have been avoided.

The need to formally justify the claimed benefits of a technology is rare. The ionizing radiation case study is a rare example of such a "justification principle," developed by the International Committee on Radiological Protection in the 1950s. This was in response to the growth of a variety of dubious or ineffective uses of radioactive materials, such as for fitting shoes for children, the cosmetic removal of hair, and the treatment of mental disorders. Yet, surveys of radiography practices over the past decade or so conclude that while doses have reduced considerably, a large proportion of medical X-rays are still of dubious clinical benefit.

Considerations of the full environmental and health costs and benefits of various options is also important. The failure, in cases like asbestos, halocarbons, and PCBs, to reflect full environmental and health costs in market prices gave these products an unjustifiable advantage in the marketplace. This in turn helped to keep technically superior substitutes off the market for longer than necessary. Although the mechanisms for the internalization of external environmental costs and the practical implementation of liability regimes are controversial, such measures are essential if both efficiency and equity objectives are to be addressed effectively.

7. Evaluate Alternatives and Promote Robust, Diverse, and Adaptable Solutions

Even when the pros are scrutinized alongside the cons, if attention is restricted simply to isolated technologies or products, then important practical insights may be missed. One concern is that once a technological commitment is made, a host of institutional and market processes act to reinforce its position, even if it is markedly inferior to potential alternatives.

For example, while in principle the function of MTBE might be substituted by alternative oxygenates such as bioethanol, by improved engine technology, or by an increase in the octane rating of the fuels themselves, little formal scrutiny of these alternatives appeared to have been undertaken at the time of the promotion of MTBE. The ozone-depleting properties of second-generation CFC substitutes were perhaps also unduly tolerated because of their relatively low ozone impacts when compared with the original substances. The existence of more benign substitutes, with less global warming potential, was not properly examined. Broader consideration of problems may give rise to more beneficial solutions than simple "chemical for chemical" substitution.

The promotion and production of alternatives need to take place within a culture of ecoefficiency, clean production, and closed-loop material flows so as to minimize the size of any future surprises in the use and impact of all technologies, including those considered to be safer those they are replacing.

8. Use Lay and Local Knowledge as Well as All Relevant Specialist Expertise

Such lay people may include industry workers, users of the technology, and people who live in the locality or those, who because of their lifestyle or consumption habits, are most strongly affected. The value of involving lay people lies not in assumptions that they are more knowledgeable or environmentally committed, but in the complementary value of perspectives that are broader in scope, more firmly grounded in real-world conditions, or more independent from the narrow professional perspectives that can accompany specialist expertise. The histories of asbestos and PCBs provide examples where workers were aware of what regulators subsequently recognized to be a serious problem.

Another form of lay knowledge concerns remedial measures. For example, although fishers can be less precautionary about stock depletion than others, there are many examples where fishers wish to act in a precautionary manner but are prevented from doing so because of a systems failure. There is an increasing emphasis in Canada and elsewhere on the need to involve fishers in management and to take full account of their knowledge and perspectives. Swedish farmers' knowledge of alternative animal husbandry techniques allowed them to promote animal health and growth without the large-scale use of antimicrobials. Not only did they bring valuable insights to the regulatory debate, but they were able to undertake voluntary controls in advance of regulatory requirements.

Lay knowledge should also receive critical scrutiny, because lay citizens are not immune to the pitfalls and difficulties noted in these conclusions about specialist expertise. One example is the "pensioners' party fallacy" among asbestos workers who pointed to the presence of healthy pensioners at the firm's Christmas party as evidence of the apparent harmlessness of asbestos, overlooking that those who had been harmed would, of course, not be able to attend a party.

9. Take Account of Wider Social Interests and Values

Social and political conflicts can be aggravated by a regulatory preoccupation with expert judgments and a lack of attention to public perspectives and values. This is critical to the wider assessment of the pros and cons. The implicit assumptions and values of specialists and interest groups need to be aired and shared. The Swedish farmers in the antimicrobials case study show how lay views can help ensure that the regulatory process reflects enlightened public and consumer ethical values.

While expert institutions tend to focus on scientific analysis, a public aversion to situations outside the bounds of normal experience ("common sense"), or at least a desire to proceed with caution, can be defended as a rational response to scientific uncertainty. A key feature of the public reaction to the emerging evidence of BSE was the surprised revulsion that ruminants were being fed on offal and bodily wastes. It seems likely that avoiding offal in ruminant feed would have at least significantly limited the scale of the subsequent BSE and CJD problems.

10. Maintain Regulatory Independence from Economic and Political Special Interests

There is evidence in the case studies that interested parties are often able to unduly influence regulators. As a result, decisions that might reasonably have been made on the basis of the available evidence were not taken. Benzene was demonstrated to be a powerful bone marrow poison in 1897; the potential for acute respiratory effects of asbestos was first identified in 1898; and the first cases of PCB-induced chloracne were documented in 1899, with effects on workers known by the late 1930s. Yet it was not until the 1960s and 1970s that significant progress began to be made in restricting the damage caused by these agents. Similarly, the temporary lifting of the ban on DES as a growth promoter in the United States in 1974 followed strong pressure from the farming lobby and occurred despite the availability of alternatives.

Independent information institutions are thus a key element of authentic regulatory independence and robust governance and appraisal. This is increasingly being recognized, for example, by the shifting of advisory committees from producer directorates in the European Commission (for example, agriculture) to the Health and Consumer Directorate. The setting up of independent food agencies in some member states and at the EU level also reflects this concern for more independent hazard appraisal institutions.

11. Identify and Reduce Institutional Obstacles to Learning and Action

The asbestos, benzene, and PCB case studies provide examples of how the short-term horizons of government and businesses can militate against social welfare in the medium and long term. However, institutional obstacles against timely protection of health and the environment can take other forms as well. The case studies illustrate three other areas: those resulting from periods of transition (for example, between succeeding elected administrations), those from tensions between different departments or levels of government and "their" agencies, or those from differing national approaches.

An official U.K. commission in 1979 recommended the setting of minimum processing standards in the rendering industries. A new administration later that year decided to withdraw the resulting proposed regulations, finding them to be an unnecessary burden on industry. It is not clear to what extent such tighter standards might actually have inhibited the later BSE outbreak, but it is notable that the implementation of standards of this sort featured prominently among that same government's later responses to the BSE crisis in 1996.

12. Avoid Paralysis by Analysis and Apply Precautionary Measures When There Are Reasonable Grounds for Concern

The general tenor of the lessons so far is to "know more." But how much information about potential hazards is deemed enough to trigger risk reduction measures? There is a danger of paralysis by analysis when either information overload or lack of political will lead to a failure of timely hazard reduction measures. One example is the antiprecautionary straightjacket imposed on U.S. benzene regulation by a Supreme Court decision, which required layer upon layer of additional information before regulatory action to reduce risks was possible, resulting in a ten-year delay in the occupational benzene standard.

Experts have often argued at an early stage that we know enough to take protective action. For antimicrobials, the U.K. Swann Committee in 1969 concluded: "Despite the gaps in our knowledge . . . we believe . . . on the basis of evidence presented to us, that this assessment is a sufficiently sound basis for action . . . the cry for more research should not be allowed to hold up our recommendations." Other case studies, such as asbestos and BSE, suggest that more, or better-targeted, research at an earlier stage would have helped minimize future costs. Similarly, for fisheries, the Ecosystems Principles Advisory Panel to the U.S. Congress concluded: "There will always be unmeasured entities, random effects, and substantial uncertainties, but these are not acceptable excuses to delay implementing an ecosystem-based management strategy."

The level of proof (or strength of evidence) needed to justify hazard reduction measures will vary with the size and nature of the claimed benefits of the economic activity, its likely costs, the significance of the uncertainties and types of ignorance involved, and the availability of alternatives. The choice of which level of proof to use is a political choice based on values and ethics as well as science.

The usual assumption in the case studies was that an activity was harmless until proven harmful by the public authorities. However, when activities are considered to be intrinsically harmful, as with pesticides and pharmaceuticals, the burden of demonstrating at least some evidence of harmlessness is placed on the proponents of the activity. The Swedish Chemicals Act of 1975 provides a clear illustration of both different levels of proof and different locations of the burden of proof in the same legislation. It requires the Public Authority to take precautionary action on a chemical substance based on a "scientific suspicion of risk" but then the burden of proof passes to the producer of the substance, who has to show that it is harmless "beyond all reasonable doubt." This example illustrates that a high level of proof is needed to show harmlessness when there is already evidence of potential hazard, whereas a lower level of proof is needed to demonstrate potential harm when harmlessness is assumed.

Burdens of demonstrating harmlessness can involve obligations to

- justify the technology in relation to the benefits claimed,
- show that alternative ways of meeting needs are likely to be more hazardous or disproportionately costly,
- monitor the impacts of the technology, and
- investigate "early warnings."

Conclusion

The EEA report provides a rich empirical history to underpin the twelve late lessons with which it concludes. Taken together, these lessons may help to minimize the future costs of being wrong about environmental and health risks. In the past, conventional scientific method has been biased toward avoiding the overstating of risks, sometimes at the expense of public safety and the environment. As Underwood (1999) concludes: "Typically there has been little concern about Type II (a 'false negative') error. The chances of erring in 'favor' of the environment (a Type I error) is deliberately kept small, whereas the chances of erring 'unfavorably' to environmental issues is not!" Participants at a recent European workshop of policy makers concluded, among other things, that "there should be a more efficient and ethically acceptable balance between the generation of 'false positives' and 'false negatives'" (Swedish Environment Ministry 2001). If implemented, the EEA's lessons should contribute both to fewer false negatives and to lower costs from any false positives that may occur.

A further key finding of the EEA report is that we need to be more precise and rigorous about what we mean by "uncertainty." Risk, uncertainty, and ignorance each warrant different treatment. More attention should be given to the handling of complexity, indeterminacy, ambiguity, and disagreement within or between technical disciplines (Stirling 1999, 2002; Wynne 2001). In short, risk assessment should become more reflective and humble. There is nothing scientific about the "pretence at knowledge" (von Hayek 1978). In particular, as the NRC (1996), other authorities in the United States (Presidential/Congressional Commission on Risk Assessment and Risk Management 1997), and the RCEP (1998) in the United Kingdom have concluded, the prior framing of risk science requires open public deliberation, addressing issues such as the questions to be addressed by science, the appropriate emphasis for different aspects of a given risk, and the balance to strike between comprehensiveness and specificity and the interpretation of uncertainty, ambiguity, and ignorance.

As long as these issues remain neglected, the "pretence at knowledge" in risk assessment has the effect of undermining the authority and credibility of the associated institutions, and even of science in general. As Stirling (1999, 2002) has pointed out, the precautionary principle has nothing to do with antiscience. Indeed, there are several respects in which it embodies a more rigorous and robust approach to the true nature of scientific uncertainty.

The tools for participatory approaches are in various stages of development, and the challenges are far from trivial. But this has to be set against traditional approaches, where the costs of failure can also be high for industry, as illustrated by the European public rejection of irradiated foods and the European response to many of the food applications of GMOs.

These considerations indicate the importance of acknowledging the interpenetration of facts and values. Popper pointed out long ago that is rationally impossible to derive a proposal for a policy from facts alone (Popper 1962). Policies that unduly emphasize the factual basis of decisions, without explicitly acknowledging and engaging with the value judgments that frame and constitute the relevant facts, are unlikely to lead to robust decisions or to achieve public acceptance (RMNO 2000).

Then there is the issue of precaution and innovation. The implementation of the EEA's twelve late lessons would stimulate application of a series of emerging tools for the fostering of more productive environmental innovations. By maximizing the breadth of available information and focusing on constructive solutions, approaches such as integrated environmental assessment (Dowlatabadi and Rotmans 2000; EFIEA 2000), multicriteria mapping (Stirling and Meyer 1999), constructive technology assessment (Rip et al. 1995), technology options analysis (Ashford 1981, 1994; Tickner 2000), alternatives analysis (O'Brien 2000), and "what-if" scenarios and participatory scenario development techniques can all assist in the management of ambiguities, uncertainties, and societal ignorance, while identifying a range of alternative courses of action.

Technological systems have a tendency to lock into particular configurations at a relatively early stage in their development, thus foreclosing other options and raising the costs of shifting to alternatives. The particular technologies that gain ascendancy in this way may do so for arbitrary reasons that may have little to do with intrinsic qualities and everything to do with chance and "first-leader" advantage. A technology can then virtually monopolize the meeting of some societal needs, as the histories of asbestos, CFCs, benzene, and PCBs illustrate. Application of these kinds of "precautionary" approaches to technology appraisal early in the innovation process offers a more effective way to promote environmentally beneficial innovation rather than relying on inefficient and disruptive social disputes at the conclusion of a closed risk process.

There is one final implication of the EEA's lessons for risk science and policy. Following episodes such as BSE in the United Kingdom, dioxins in Belgium, and HIV-contaminated blood in France, public trust in risk science is at a very low ebb in Europe. Governments are increasingly aware of this and are developing responses, such as the 2001 European Commission White Paper on European Governance. This includes recommendations for improving public participation in managing the interreactions among science, technologies, and society, issues that other public authorities outside Europe are also promoting (e.g., Ministry of the Environment, New Zealand 2001; National Oceans Office, Australia 2001) The need for broader engagement in risk decision making, highlighted in many of the EEA lessons, links directly with these wider imperatives for the democratizing of scientific expertise. The stakes are high, not just for public health and the environment, but for how we go about choosing our future technological pathways and for who has control.

References Cited

Ashford, N. 1984. "Alternatives to Cost-Benefit Analysis in Regulatory Decisions." *Annals of the New York Academy of Sciences* 363: 129–37.

———. 1991. "An Innovation-Based Strategy for the Environment and for the Workplace." In *Worst Things First: The Debate over Risk-Based National Environmental Priorities,* edited by A. Finkel and D. Golding. Washington, D.C.: Resources for the Future.

Beck, U. 1992. *Risk Society.* London: Sage.

Deane, L. 1898. "Report on the Health of Workers in Asbestos and Other Dusty Trades." In *HM Chief Inspector of Factories and Workshops, 1899, Annual Report for 1898,* 171–72. London: HMSO. See also the Annual Reports for 1899 and 1900, 502.

Dowlatabadi, H., and J. Rotmans, J. 2000. "Integrated Assessment." *Integrated Assessment* 1, no. 3. Special issue.

European Commission. 2000. "Communication from the Commission on the Precautionary Principle, COM 1." Brussels.

European Forum for Integrated Environmental Assessment (EFIEA). 2000. "Integrated Environmental Assessment in European Environment Agency Reporting." Report of the EEA/EFIEA special session, Copenhagen/Amsterdam.

Gee, D. 1997. "Approaches to Scientific Uncertainty." In *Health at the Crossroads: Transport Policy and Urban Health,* edited by T. Fletcher and A. J. M. Michael. London: Wiley.

Hanley, N., and C. Spash. 1993. *Cost Benefit Analysis and the Environment.* Cheltenham: Edward Elgar.

Harremoes, P., D. Gee, M. MacGarvin, A. Stirling, J. Keys, B. Wynne, and S. Guedes Vaz, eds. 2001. "Late Lessons from Early Warnings: The Precautionary Principle 1896–2000." Environmental issue report no. 22, Office for Official Publications of the European Communities (OPOCE), Copenhagen.

Hill, A. B. 1965. "The Environment and Disease: Association or Causation? *Proceedings of the Royal Society of Medicine* 58: 295–300.

James, W. H. 1965. "Teratogenetic Properties of Thalidomide." *British Medical Journal* 5469: 1064.

Kourilsky, P., and G. Viney. 1999. *Le Principe de Précaution.* Rapport au Premier Ministre, France.

Krohn, W., and J. Weyer, J. 1994. "Society as a Laboratory: The Social Risks of Experimental Research." *Science and Public Policy* 21: 173–83.

Levy, A., and B. Derby. 2000. "Report on Consumer Focus Groups on Biotechnology." U.S. Food and Drug Administration, Center for Food Safety and Applied Nutrition, Office of Scientific Analysis and Support, Division of Market Studies, Washington, D.C.

Lieberman, A. J., and S. C. Kwon. 1998. *Facts versus Fears: A Review of the Greatest Unfounded Health Scares of Recent Times,* 3rd ed., revised June 1998. New York: American Council on Science and Health. http://www.acsh.org.

Millstone, E. 1997. *Lead and Public Health.* London: Earthscan.

Ministry for the Environment, New Zealand. 2001. "Resource Management Act," text and background discussion. http://www.mfe.govt.nz/management/rma/rma11.htm.

National Oceans Office. 2001. Australia's Oceans Policy home page. http://www.oceans.gov.au/aop/main.htm.

O'Brien, M. 2000. *Making Better Environmental Decisions: An Alternative to Risk Assessment.* Cambridge, Mass.: MIT Press.

Peto, J. 1999. "The European Mesothelioma Epidemic." *British Journal of Cancer* 79: 666–72.

Popper, K. R. 1962. *The Open Society and Its Enemies,* vol. 1, 4th ed. London: Routledge and Kegan Paul.

Presidential/Congressional Commission on Risk Assessment and Risk Management. 1997. *Framework for Environmental Health Risk Management, Final Report.* Washington, D.C.: The White House.

Rip, A., T. Misa, and J. Schot. 1995. *Managing Technology in Society.* London: Pinter.
RMNO, Advisory Council for Research on Spatial Planning, Nature and the
Environment. 2000. *Willingly and Knowingly: The Roles of Knowledge about Nature
and the Environment in Policy Processes.* Utrecht: Lemma Publishers.

Small, I., J. van der Meer, and R. E. G. Upshur. 2001. "Acting on an Environmental Health Disaster: The Case of the Aral Sea." *Environmental Health Perspectives*
109, no. 6.

Smith, J. 2000. *The Daily Globe: Environmental Change, the Public and the Media.*
London: Earthscan.

Snow, J. 1849. *On the Mode of Communication of Cholera.* London.

Stirling, A. 1999. *On Science and Precaution in the Management of Technological Risk:
Volume I: A Synthesis Report of Case Studies.* European Commission Institute for
Prospective Technological Studies, Seville, EUR 19056 EN. ftp://ftp.jrc.es/
pub/EURdoc/eur19056en.pdf.

————. In press. "Risk, Uncertainty and Precaution: Some Instrumental Implications from the Social Sciences." In *Negotiating Change: Perspectives in Environmental Social Science,* edited by I. Scoones, M. Leach, and F. Berkhout. London:
Edward Elgar.

Stirling, A., and S. Mayer. 1999. "Rethinking Risk: A Pilot Multi-Criteria Mapping of a Genetically Modified Crop in Agricultural Systems in the United
Kingdom." Report to the UK Roundtable on Genetic Modification, Science
and Policy Research Unit, University of Sussex.

Swedish Environmental Ministry. 2001. *Bridging the Gap. Conclusions from the Conference on Sustainability Research and Sectoral Integration.* Stockholm: Swedish
Environmental Ministry.

Tickner, J. 2000. "Precaution in Practice: A Framework for Implementing the Precautionary Principle." Dissertation prepared for the Department of Work Environment, University of Massachusetts, Lowell.

Underwood, T. 1999. "Precautionary Principles Require Changes in Thinking
about and Planning Environmental Sampling." In *Perspectives on the Precautionary
Principle,* edited by R. Harding and E. Fisher, 254–66. Sydney: The Federation
Press.

U.S. National Research Council. 1996. "Understanding Risk." Report of an ad
hoc working party chaired by H. Feinberg. http://www.riskworld.com/Nre-
ports/1996/risk_rpt/html/nr6aa045.htm.

von Hayek, F. 1978. *New Studies in Philosophy, Politics, Economics, and the History of
Ideas.* Chicago: University of Chicago Press.

WBGU. 2000. *Welt im Wandel. Handlungsstrategien zur Bewältigung globaler Umwel-
trisiken* (World in transition: strategies for managing global environmental risks).
Jahresgutachten 1998 (Annual report 1998). Berlin: Springer.

Wynne, B. 1992. "Uncertainty and Environmental Learning: Reconceiving Science and Policy in the Preventive Paradigm." *Global Environmental Change* 6: 111–27.

————. 2001. "Managing and Communicating Scientific Uncertainty in Public Policy." Harvard University Conference on Biotechnology and Global Governance: Crisis and Opportunity. Boston: Kennedy School of Government.

Wynne, B., R. Grove-White, and P. Macnaghten 2001. *Wising Up.* Lancaster: Lancaster University.

Wynne, B., C. Marris, P. Simmons, B. De Marchi, L. Pellizoni, O. Renn, A. Klinke, L. Lemkow, R. Sentmarti, and J. Carceras. 2000. "Public Attitudes to Agricultural Biotechnologies in Europe." Final report of project PABE, 1997–2000. Brussels: DG Research, European Commission.

CHAPTER 14

Biodiversity Conservation in Developing Countries: Managing Uncertainty in Strategies and Action Plans to Support Precautionary Action

Reginald Victor

Biodiversity is defined as the variability among living organisms from all sources including, inter alia, terrestrial, marine and other aquatic ecosystems, and the ecological complexes of which they are part; this includes diversity within species, between species, and of ecosystems (Convention on Biological Diversity 1998). Biodiversity provides ecological, economic, recreational, cultural, and spiritual benefits to humanity, but it is adversely affected by human activities. The United Nations Convention on Biological Diversity, opened for signatures at the United Nations Conference on the Environment and Development (UNCED) in Rio de Janeiro, Brazil, in 1992 is a positive response to the threats on biodiversity. The main objectives of this convention are the conservation of biodiversity, the sustainable use of biological resources, and the fair and equitable sharing of benefits arising from the use of genetic resources.

The nations signatory to this convention are committed to producing strategies and actions plans and to implementing these to achieve the objectives of the convention. Strategies are thought-out plans and goals related to biodiversity, whereas action plans are optimal procedures to realize strategies. Both should strongly rely on precautionary science.

As a member of a national committee charged to produce the biodiversity strategy and action plan for the Sultanate of Oman, I had an opportunity to evaluate the main strategic goals and guiding principles of such an exercise as well as to review strategies and action plans submitted to the convention by some other developing countries. I found that, in general, these strategies and action plans followed the models of developed nations. Experts, who usually are short-term consultants with very little knowledge of local scenarios, supply the models. The precautionary elements used in any one model are not universal in application, and so these strategies and action plans are plagued with uncertainties.

None of the strategies and action plans I considered during this study dealt with uncertainties of any kind, whether they related to biodiversity conservation, sustainable development, or decision making. In this chapter, I analyze some of the lessons that can be learned from these planning exercises and argue for recognizing some specific types of uncertainty as inevitable components of strategies and action plans.

Discussions of scientific uncertainty in relation to the precautionary principle usually focus on data and the conclusions drawn from them. However, uncertainties related to social, political, psychological, and other human factors must be considered as well. If this kind of uncertainty is ignored, strategies and action plans become unrealistic "wish lists" that have little chance of being carried out. Effectively addressing these uncertainties, on the other hand, is likely to support precautionary decisions and actions.

Themes Used in Strategies and Action Plans

Biodiversity planning usually addresses the following themes:

• Conservation of natural resources (protected areas, endangered species, terrestrial and freshwater flora and fauna, marine life, and fisheries)
• Agriculture
• Energy, water, and mineral resources
• Industry and technology services, including biotechnology, biosafety, and tourism
• Urban environment
• Environmental emergencies
• Public, nongovernmental organization, and private sector participation
• Societal values, quality of life, and spiritual values

It is not imperative that all these themes be used in a single document. Each theme is addressed in four subcomponents: (1) key issues, (2) objectives, (3) options, and (4) priority actions. Key issues strongly influence the other subcomponents.

A Typical Example

This example is taken from one of the plans submitted by a developing country that is hailed for its thoroughness and vision. This document addressed protected areas—geographically defined areas designated or regulated and managed to achieve specific conservation objectives—as one of the subcomponents under the theme conservation of natural resources. Key issues identified around protected areas in this document included:

- Increasing pressure due to expansion in agriculture, fisheries, tourism, and urban development
- Incomplete implementation of management plans for existing protected areas
- Incomplete network of protected areas
- Poor public participation in the design and implementation of management plan
- Insufficient funding

In light of these realities, the strategy listed two objectives: (1) maintain and develop an integrated, representative, and sustainable network of protected areas that will ensure the protection of biodiversity and (2) conserve plant and animal diversity in existing and proposed areas.

The objectives are abbreviated and vague. The document addresses the issue of the network of protected areas, but the terms "integrated," "representative," and "sustainable" are buzzwords that are not explained anywhere in the document. The objective of conserving plant and animal diversity is redundant because achieving the first objective would do this. These vague objectives do not address key issues identified, such as the pressures imposed on protected areas caused by human activities, implementation of management plans, funding, and public participation. In other words, there were inadequate linkages between key issues and objectives, which in turn affect options and priority actions.

I believe that authors of these strategies and action plans are unable to make the connection between issues and objectives because of uncertainties relevant to environmental decision making (Shrader-Frechette 1995). However, that does not prevent them from suggesting a long list of options and priority actions, which, in my opinion, are mere wish lists fraught with more uncertainties.

Under the precautionary principle, uncertainties should not deter us from considering options and priority actions to achieve the objectives of the biodiversity convention. But wish lists are not precautionary if they have little or no chance of being implemented in a given time frame. Nevertheless, such wish lists are common in strategies and action plans produced by developing countries.

Accountability and Time Frame

In strategies and action plans presented by developing countries, accountability and time frame are flexible factors. The text of the convention advocates that nations take responsibility for preserving biodiversity, but the fine print says that it is not binding on nations to do so. In addition, the legalese of these documents provides loopholes permitting long delays in implementation. The failure to include accountability mechanisms and time frames accounts for much of the uncertainty and lack of realism apparent in these plans.

Strategies and action plans submitted by developing countries often generate priority matrices for issues specifying the nature of priority (low, high, etc.), agencies responsible (ministries, nongovernmental organizations, research institutes, etc.), funding (internal, international, etc.), and proposed performance indicators (assessment reports, database, red-lists of endangered species, etc.). However, none of the matrices analyzed during this study explicitly spelled out accountability criteria and time frame. High-priority issues often require the application of the precautionary principle, and delays are detrimental to the achievement of objectives. Specific accountability criteria and time frame required for their accomplishment must be indicated. Otherwise, the whole exercise is likely to be futile.

Consider the following real-life example. The issue concerns the assessment and recovery program of species at risk. The parties responsible are two ministries, two universities, and a private-sector organization. Performance indicators are specified: identification of species at risk and the number of recovery plans produced. But crucial questions are not addressed. For example, the document does not define accountability measures, time frame, or the specific duties of the five responsible parties.

A Study of Uncertainties and Time Frames

Figure 14.1 shows one of the results of a study conducted concerning four different issues chosen from a strategy and action plan submitted by a developing country. (The specific issues chosen are not revealed here to protect the identity of the country, and related but fictitious issues with broad similarities are used to explain the results.) Thirty environmental professionals attending an environmental workshop were randomly chosen and were presented with questions about four issues. They were asked to score the uncertainty of resolving each issue on a scale of 0 (no uncertainty) to 10 (maximum uncertainty) in a time frame ranging from five to twenty-five years. For convenience, all types of uncertainties (Lemons 1995; Shrader-Frechette 1995) were clumped into a single category. The five-year intervals reflect the five-year plans used by

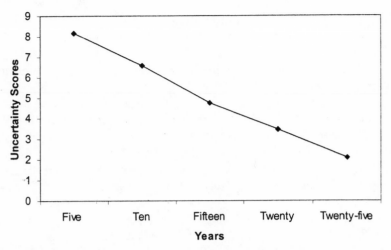

Figure 14.1. Relationship between time frame and mean uncertainty scores ($n = 30$) for issue 1 on legislation concerning protected areas.

the governments of several developing countries. The questionnaire offered complete anonymity.

Issue 1 (figure 14.1) was legislation concerning a protected area. The uncertainty of the legislation coming into effect decreased linearly with the increase in time. Issue 2 considered the possibility of public participation concerning the construction of a major tourist facility in a fragile ecosystem. The uncertainty scored was uniformly high without variability throughout the time frame of twenty-five years. Issue 3 concerned the reduction of fishing activities in a specific area. Here the high uncertainty decreased sharply in the first ten years and remained low for the next fifteen years. Issue 4 related to the realization of a very large dam project. The uncertainty in preventing its construction was very low in the first ten years, but rose sharply in the next fifteen years.

Role of Value Judgments

The major criticism of this exercise by three specialists in instructional development was that it was laden with biased, contextual, and methodological values. I completely agree but would argue that such value judgments are absolutely unavoidable and at times even needed for producing realistic and precautionary strategies and action plans. Biodiversity strategies and action plans are environmental policy documents full of uncertainties not only because they are based on underdetermined facts and data, but also because of value judgments planners must make. Value judgments are a critical component of a precautionary approach to environmental decision making.

Biased values occur in science due to intentional misinterpretation or omission of data (Longino 1990). In the production of biodiversity strategies and action plans, two kinds of bias may be introduced. The first is specialist bias, in which authors with expertise in a given area may give undue emphasis to one component of the conservation process. For example, biologists might give importance to the protection of specific plants and animals, while protection of habitat diversity receives less attention. The second is factual bias giving undue weight to known facts at the expense of unknown or uncertain entities. For example, flowering plants are better known than algae, so algae are completely ignored.

Of these, factual bias is more inimical to precaution, but specialist bias must also be considered. For example, if an herbivorous mammal is perceived as threatened, its protection receives serious attention, while other species sharing its habitat, including human beings, are ignored mainly because their interrelationship with the target species is poorly known.

Recognition of these biases is the first step to avoiding them. Assembling a diverse group of experts to develop strategies and action plans is essential. When obvious and important biodiversity components are targeted for conservation, all peripheral links and processes associated with them should be considered.

Many strategies and action plans suffer from what I term the "wrong template syndrome," in which an out-of-context country is used as a model. For example, one country prioritizes issues such as biotechnology and agrobiodiversity over conservation of water resources. However, in that country protection of water resources should have been a high priority because it is semiarid. Another country considers pollution from radioactive wastes at some length although it has no nuclear technology to speak of. If authors producing these strategies and action plans used sound contextual value judgments considering the natural, social, cultural, economic, and political environments of the countries concerned, these anomalies could have been avoided.

Recognizing Uncertainties

All strategies and action plans from developing countries I studied suffered from the perceived need for certainty. Authors of strategies and action plans find certainty appealing because they do not understand the probability of achieving the objectives in a given context. One result is that these documents are often "trimmed" and "cooked" (Jackson 1986; Shrader-Frechette and McCoy 1993). "Trimming" is a deliberate effort to make the documentation look accurate and precise while it is not, whereas "cooking" only retains data and information in favor of one action or interpretation and ignores others.

I believe that any attempt to evaluate uncertainties will be an underestimate, and any effort to account for them will be incomplete. Still, I would strongly argue for the transparent recognition of uncertainties in strategies and action plans. Acknowledging uncertainties may be construed as a weakness of these policy documents. However, in the case of developing countries, such admission of weaknesses will be helpful to justify precautionary environmental decisions. For example, the lack of expertise may be a reason for the uncertainty in the population estimate of an endangered species; this alleged weakness should justify the precautionary establishment of a large buffer zone to minimize human encroachment.

I propose the following procedures for evaluating and addressing uncertainties in biodiversity strategies and action plans.

• For each theme in the strategy, key issues and objectives should be linked. The "objectives" section should address all items listed in as key issues.
• Achievement of each objective should be linked to an estimated time frame. Considering the urgency of biodiversity protection and giving due allowance to the administrative processes, a maximum period of fifteen years seems reasonable. To put this in context, we should note that some developing countries have achieved grain, egg, and milk production targets in two to three five-year plans.
• Uncertainty assessment should be carried out for objectives of each theme. A broad range of environmental professionals, including scientists and decision makers familiar with the country, should be involved. Assessors should be asked to consider each objective carefully and score the uncertainty of achieving it in the next fifteen years on the scale of 0 (no uncertainty) to 10 (maximum uncertainty). They should then calculate the average score for each objective and the overall average for the whole theme. Following the assessment, the panel of assessors should be asked to discuss the results including the reasons for the uncertainty scores.
• After this assessment, options and priority actions can be formulated.

A real-life study conducted in a developing country is given below as an example:

THEME: Terrestrial Flora
OBJECTIVES: (i) Prevention of overharvesting of timber trees
(ii) Sustainable use of wild plants
(iii) Control of overgrazing
(iv) Control of invasive species
(v) Reforestation to reduce desertification
NUMBER OF ASSESSORS: 10

Table 14.1 Example of Uncertainty Assessment: Terrestrial Flora

Objective	Uncertainty scores for 15 years	Average
(i)	6; 8; 7; 6; 7; 8; 6; 6; 9; 7	7.0
(ii)	8; 9; 0; 8; 10; 10; 8; 7; 9; 8	7.7
(iii)	9; 10; 9; –; 10; 9; 9; 9; 9; 9	8.3
(iv)	5; 6; 6; 6; 6; 7; 5; 4; 6; 7	5.8
(v)	7; 7; 6; 6; 6; 7; 7; 8; 6; 7	6.7
OVERALL AVERAGE UNCERTAINTY SCORE:		7.3

Note: Objectives: (i) Prevention of overharvesting of timber trees; (ii) Sustainable use of wild plants; (iii) Control of overgrazing; (iv) Control of invasive species; (v) Reforestation to reduce desertification. Number of assessors: 10. The – sign means "no opinion."

In this example, the assessors considered average uncertainty scores for each objective first. The panel then examined the options available. The discussion was realistic, and priority actions suggested were thoroughly examined for feasibility. The overall average uncertainty score for the theme was considered as an indicator of difficulties associated with the issues. Reformulated options and priority actions were compared against the original options and priority actions recommended by the official strategy and action plan submitted by the country in question. The panel unanimously agreed that improvements were considerable. The new options and priority actions did not constitute a mere wish list, but appeared workable within constraints, in the next fifteen years.

To illustrate the differences between the above exercise and the official plan in existence, the objective of controlling overgrazing is given here. The official plan very briefly suggests culling of livestock and feral donkeys as a mitigation measure. In this exercise this objective had a high uncertainty score of 9.2 mainly because culling of livestock and feral donkeys is a complex social issue. The number of livestock is a symbol of wealth; meat and milk are secondary issues. Killing of feral animals that are not consumed or used in some way is not encouraged by the religion. The present exercise considered a seasonal restriction on imported meat and milk that would force the utilization of local livestock, especially with an offer of financial subsidies, privileged grazing rights for those with small herds, reestablishment of grazing-free traditional stone enclosures, supply of farm grown grass and alfalfa as feed alternatives, use of donkey meat as animal food and organic manure, and a selective male-sterilization program for donkeys in high impact areas. All these options when built into the priority matrix of the existing agricultural extension services appeared workable in the next fifteen years.

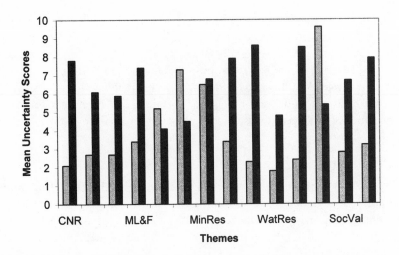

Figure 14.2. Bar graph comparing uncertainty profiles plotted as mean uncertainty scores (n = 45/country) for fourteen common themes used in the biodiversity strategies and action plans of countries A (light gray) and B (dark gray).

Uncertainty Profiles

Uncertainty profiles created by plotting average uncertainty scores for component themes in strategies and action plans may be useful for comparing strategies across countries and across areas of biodiversity conservation within a country. Themes with high uncertainty profiles should be scrutinized carefully to facilitate precautionary decision making. Comparisons of uncertainty profiles may reveal trends that can be used to group countries with common problems, thus making the precautionary approach and solutions common to countries with similar profiles. The information content in the profile is a qualitative indicator of the expected success of a strategy in a given time frame. These comparisons will also provide information on related success expectancy of strategies and action plans in different developing countries.

Figure 14.2 compares the uncertainty profiles for two developing countries sharing fourteen common themes in their strategies and action plans. Country A is under military rule, and Country B is a democracy. Average uncertainty scores plotted for themes were calculated from results obtained using teams of forty-five environmental professionals for each of the two countries. Each team was made up of twenty policy makers, twenty scientists, and five graduate students. All participants were well versed in these countries' environmental problems.

Discussion on causal reasons for differences between countries A and B is intentionally avoided. Some may link reasons of uncertainty to political philosophy, which is not a good predictor in matters of environmental protection (Arms 1994). In general, differences in uncertainty profiles could be attributed to several factors. Despite having common themes, the objectives and contexts may be different. Administrative efficiency, promulgation and implementation of regulations, scientific research base, education, and incentives for sound environmental practices are also factors affecting uncertainty profiles.

Conclusions

Strategies and action plans, even when produced with the best of intentions and not as a mere obligation to the international community, are inadequate to protect biodiversity loss. The rate at which these strategies and action plans are formulated and implemented is too slow to prevent damage to biodiversity. Strategies and action plans will not be of any precautionary value if they are not implemented in time. Uncertainty evaluations of these documents with stress on accountability and time frame should be used to influence precautionary policy decisions.

In developing countries, the political process and economic reality can work against precaution. Policy makers often question the scientific credibility of conservation that opposes development (Simberloff 1987), although conservation is "considered as the essential ingredient in development planning" (Oates 1999). Biodiversity strategies and action plans supposedly integrate conservation and development. If they are unrealistic and full of uncertainties, their implementation will be nearly impossible. Recognizing these uncertainties and dealing with them effectively will contribute to the success of these plans.

Note

Due to political sensitivities, the developing countries that participated in this study cannot be named.

References Cited

Arms, K. 1994. *Environmental Science,* 2nd ed. Philadelphia, Penn.: Saunders College Publishing.

Convention on Biological Diversity. 1998. Texts and Annexes. Secretariat of the Convention on Biological Diversity, Montreal, Quebec.

Jackson, C. I. 1986. *Honor in Science.* New Haven, Conn.: Sigma Xi.

Lemons, J. 1995. "The Conservation of Biodiversity: Scientific Uncertainty and the Burden of Proof." In *Scientific Uncertainty and Environmental Problem Solving,* edited by J. Lemons, 206–32. Cambridge, Mass.: Blackwell Science.

Longino, H. 1990. *Science as Social Knowledge.* Princeton: Princeton University Press.

Oates, J. F. 1999. *Myth and Reality in the Rain Forest: How Conservation Strategies Are Failing in West Africa.* Berkeley: University of California Press.

Shrader-Frechette, K. S. 1995. "Methodological Rules for Four Classes of Scientific Uncertainty." In *Scientific Uncertainty and Environmental Problem Solving,* edited by J. Lemons, 12–39. Cambridge, Mass.: Blackwell Science.

Shrader-Frechette, K. S., and E. D. McCoy. 1993. *Method in Ecology.* Cambridge: Cambridge University Press.

Simberloff, D. 1987. "Simplification, Danger, and Ethics in Conservation Biology." *Bulletin of the Ecological Society of America* 68: 156–57.

CHAPTER 15

A Living Legacy: The Precautionary Ideal in American Law

Sheila Jasanoff

Continental Drift

England and America, as George Bernard Shaw famously observed, "are two countries separated by the same language." Contemporary debates over environmental regulation in Europe and the United States call to mind this ironic remark. When it comes to managing today's complex problems of risk and sustainability, the countries of the industrialized world appear to be separated by a superabundance of common discourses. There are endless reasons why we should all see "our common future"[1] in the same light and make a common cause over how to secure it for generations still to come. Science, to which all industrial nations subscribe, offers universal benchmarks of knowledge; technology imposes common practical limits on what can be accomplished; economic realities promote a shared regard for innovation and progress; and commitments to fundamental human values, such as sustainability, underwrite legal mandates to prevent irreparable harm to the environment. Yet, listening to turn-of-the-century rhetorical exchanges on such issues as climate change, genetically modified crops and foods, and biodiversity, the naïve observer might well deduce that an unbridgeable philosophical divide has opened up somewhere along the Mid-Atlantic ridge, with destabilizing consequences for transnational legal and policy development. Much of the argumentative heat centers on *precaution,* a concept that is gaining ground in European law and jurisprudence while U.S. policy makers regard it with growing suspicion and ill-concealed impatience (see, e.g., Macilwain 2000).

To find a way out of this predicament, it is important to be clear about what is common to Western traditions of environmental regulation, as well as what is different about them. The commonalities lie in a set of shared presumptions about the obligation to prevent collective harms in rational, democratic societies, committed to principles of solidarity and to the acquisition and use of knowledge. The differences stem from the ways in which these basic expectations have been interpreted and operationalized within the constraints of divergent legal and political systems. In particular, the conflicts between the so-called risk-based approach and the so-called precautionary approach have arisen out of larger social and cultural contexts that include, as an important factor, the interaction of knowledge-making and regulatory practices.

Precaution, as this volume well illustrates, means different things to different actors involved in environmental regulation. In this chapter, I briefly reflect on the reasons for this ambiguity and then make three simple points about the theme of precaution in U.S. law. First, I show that the often-asserted dichotomy between the supposedly nonprecautionary U.S. approach and the precaution-based European approach is either mistaken or greatly overstated. Precautionary ideas are embedded in U.S. health, safety, and environmental law at many points. Second, although these ideas are present, they have been implemented within a distinctive legal and political framework whose biases are different from those of most European regulatory systems; precaution has therefore been differently understood in the U.S. and arguably kept from achieving some of the aims that U.S. environmentalists may once have hoped for. Third, and related, if one wants to reinvigorate the idea of precaution and give it new life in the context of U.S. law and policy making, then the way forward must come from inside America's own decision-making culture. Precaution cannot be seen as an exotic European bud to be grafted onto the solid rootstock of U.S. law; rather, the purposes of precaution have to be rearticulated and given renewed force within the context of homegrown U.S. scientific, legal, and policy practices.

The Meaning of Precaution

Much of the controversy that surrounds the precautionary principle (or, alternatively, the precautionary approach) has to do with the absence of a firm definition for the term, although it is mentioned or endorsed in a variety of international treaties and agreements, including the 1992 Rio Declaration and the Maastricht Treaty of the European Union.[2] A 1998 Wingspread conference in Racine, Wisconsin, produced one frequently cited statement of the principle: "When an activity raises threats of harm to human health or the environment, precautionary measures should be taken even if some cause and effect relationships are not fully established scientifically." Critics charge not only that this statement is too vague to be useful, but also that it rejects science and threatens innovation.

To make better sense of these criticisms, it helps to recognize that precaution collapses two kinds of normative obligations into a single word—the prudential and the moral. Prudential obligations are the kinds of *ought* statements that follow directly from what we know about the world; they tell us what are prudent ways to behave within the limits of available knowledge and experience. For example, it is prudent that one should carry an umbrella when it is raining or that one should not send a child to school when she has a severe cold. Moral obligations, by contrast, are the kinds of *ought* statements that tell us what are the correct ways to behave within a framework of highly valued social norms and conventions. The Hippocratic oath is a moral obligation that tells physicians to do no harm, regardless of what they may know or not know about a specific disease or patient. Other nearly universal moral obligations include the injunction not to hurt or to kill people and, increasingly, not to deprive them of fundamental human rights.

When talking about precaution, it is important to recall that no one seriously questions the moral obligation embedded in the concept. There is little disagreement today that we have a collective obligation as members of the human species to prevent harm to human health and the environment. Thus, even President George W. Bush, in his February 2002 speech outlining the unilateral U.S. approach to climate change, acknowledged that the U.S. has a commitment "to stabilize atmospheric greenhouse gas concentrations at a level that will prevent dangerous human interference with the climate" (*The Guardian* 14 February 2002). The conflicts over precaution concern the nature of the relationship between the norm's prudential and moral components. What responsibilities does the state have to gather knowledge about the environment, and are some methods more valuable than others? How much do we have to know before the in-principle obligation to do no harm is converted into an affirmative obligation? Should the capacity of current scientific knowledge define the baseline for moral action, so that we ought to act preventively only when we can scientifically foresee the harms that may ensue from inaction? And if there are differences on these questions, then who should decide and by what means?

All four questions are contested to varying degrees, but it is the third one that has drawn the most intense attention in transatlantic disputes over precaution. In a number of settings (e.g., negotiations over the biosafety protocol), the official U.S. position has tended to demand a close relationship, if not an actual identity of purpose, between the harms we can foresee using current analytic methods and the harms we should seek to avoid. Moreover, a particular analytic method has been advocated for trying to ascertain possible future harms, namely, science-based risk assessment. One consequence of this strategy has been to downplay hard-to-measure social and cultural impacts, and even long-term systemic effects, in favor of shorter-term environmental impacts that can be readily ascertained through the experimental sciences and formal modeling.

European policy on the whole has been reluctant to accept so strict an alignment between the prudential ought, particularly as defined by risk assessment, and the moral ought for purposes of environmental protection. On this reading of precaution, action may be required even when the harms sought to be prevented are not as yet precisely understood or subject to formal analysis. Regulation is expected to lead rather than follow analytic developments; it is "technology-forcing" in this sense. Equally, the views and preferences of lay citizens are considered highly relevant because expert knowledge is assumed to provide only a partial characterization of possible hazards. Not surprisingly, European policy makers have been more sympathetic to arguments that the precautionary approach should encompass risks to social as well as natural systems. All this is rejected by many U.S. commentators in politics and industry, but is their position consistent with the United State's own historical commitments in the field of environmental protection? And can U.S. regulatory policy find a positive role for precaution instead of simply dismissing the concept as unreasoning fear or covert protectionism?

A Precautionary Legal Tradition

Precaution, as I have already suggested, is not new to U.S. law. Indeed, its moral component lies at the heart of the ancient common law tradition—in the obligation of due care and the duty of private actors to behave as society has a right to ask of reasonable persons. Because of this preventive orientation, common law liabilities fall heavily on those who are careless or reckless of the safety of others. Long before there were environmental laws, the common law decreed that enterprises should not discharge pollutants into water so as to harm downstream users, blow smoke from a railroad so as to endanger the crops or cattle of neighboring farmers, or create public nuisances by raising livestock in densely settled residential areas. To be sure, the dangers in those days were easily apparent to the senses—sight, smell, and sound, for instance—but the limits of evidence back then should not distract us from the common law's basic prescription of foresight: to make actors think ahead and take responsibility for the likely consequences of their actions.

With modernization and the explosive industrial development of the twentieth century, the state became an increasingly important player in environmental affairs, sometimes directly, as in the large construction projects of the U.S. Army Corps of Engineers, and more often indirectly, by approving, authorizing, and regulating activities with significant implications for health, safety, and the environment. The nature of evidence and proof also changed during this time, with both common law and regulatory decisions increasingly recognizing that it was not enough to guard against imminent harms; longer-range problems, often amenable to technical analysis and control, also had to be avoided (Jasanoff 1995). As the law evolved to accommodate the changing roles of science, technology, and the state, precautionary concepts that had long guided behavior between private parties were gradually incorporated into the government's legal obligations. Although this process was unsystematic and not attributable to a single guiding precept such as the "precautionary principle," the result was a compelling requirement of foresight in the behavior of public agencies charged with regulatory responsibilities, particularly in the field of environmental protection.

Several important provisions were enacted into law by the mid-1970s. The most important for a time was the 1958 Delaney Clause (Section 409) of the Federal Food, Drug and Cosmetic Act, which forbade the addition to food of substances found to cause cancer in humans or in appropriate animal tests. Taking the then-available science of toxicology as a benchmark, the law stipulated that a positive cancer test should be treated as sufficient evidence of a risk to humans and that this risk should be absolutely avoided. This was in essence a precautionary statute, but not as many have argued simply because it prescribed a "zero risk" standard for carcinogenic food additives. Rather, the precautionary element of the Delaney provision lay in the fact that gaps in knowledge about human responses to carcinogens were not taken as a cause for inaction. The moral obligation not to place people at risk was triggered in this instance by surrogate markers, such as positive animal tests, not by actual evidence of harm to human beings.

We should remember as well that there was a defensible scientific argument for choosing the zero risk standard for suspected carcinogens in the late 1950s. Carcinogens, it was widely believed, were nonthreshold substances, capable of inducing cancer by causing cell mutations even at the most minimal levels of exposure. As a result, the standard toxicological practice of determining a "no observable exposure level" (NOEL) could not, in theory, work for carcinogens. Similarly, the regulatory practice of setting exposure limits at some fixed precautionary level below the NOEL—for example, at one-hundredth or one-thousandth of that amount—was also out of the question. The ban on carcinogenic food additives thus reflected more an extension of an existing precautionary norm than a wholly unprecedented approach to regulation.

Over the next forty years or so, the Delaney Clause was at the center of many scientific and policy debates, including a particularly heated controversy over the wisdom of applying the provision's zero risk standard to pesticide residues in processed foods (Brickman et al. 1985). Eventually, the 1996 Food Quality Protection Act replaced the Delaney Clause with a standard of "reasonable certainty of no harm" from exposure to pesticide residues. What changed in this period was the nature of the evidence required to trigger the uncontested moral obligation not to endanger public health. The 1996 revision acknowledged that more was known about the nature of environmental carcinogens in the 1990s than in the 1950s, including facts about naturally occurring carcinogens, mechanisms of cancer causation, and processes of DNA repair. More sophisticated techniques were therefore available to make better prudential judgments about the threshold of harm from residual exposure to carcinogens, and the revised law opened the door to the use of these new methods. Questions remain about whether the new law unacceptably diluted the precautionary mandate of the Delaney Clause. For our purposes, the more significant point to take from this history is that a regulatory system founded on precautionary principles *can* evolve and take new knowledge on board, although it may take decades to move from one scientific and political equilibrium to another.

Similarly, the National Environmental Policy Act of 1969 (NEPA), the first major piece of legislation in the modern environmental history of the United States, was very much a precautionary statute. Interestingly, this law (unlike the Delaney Clause) had no operational methods specified within its text, neither technical analytic methods nor even provisions for judicial review. All these had to be read into the law as it was gradually implemented. NEPA superbly articulated the precautionary obligation of the state; before federal agencies undertook actions that could have a significant impact on the environment, they were required to think hard about what those impacts might be. Such reflection was mandated even though no one knew at the time just how to carry out the required environmental impact assessment; once again the moral obligation ran ahead of the strictly prudential. Moreover, agencies were instructed not to conduct their analyses away from public view but to draw citizens into a combined analytic and deliberative process (Stern and Fineberg 1996).

NEPA also fostered a standard of judicial review that was known as the "hard look" doctrine. We can see this in retrospect as a kind of precautionary echo in the courts, reflecting the law's primary mandate to the executive agencies. In essence, reviewing judges said that their job was to make sure the decision-making agency had complied with NEPA by taking a hard look at the data before it. In this way, the act initiated a kind of trickle down effect of precaution throughout the U.S. decision-making system. Courts, of course, are essential to the implementation of U.S. public policy, so it was an important "downstream effect" of NEPA, if we may call it that, to make courts the partners of the regulatory agencies in a precautionary unfolding of the law. We now live in different ideological, as well as scientific and legal, times. Many things have changed, and NEPA is no longer the primary instrument of environmental protection that it once was. As regulatory agencies established their substantive and procedural good faith, the "hard look" doctrine, too, was abandoned in favor of greater judicial deference.[3] But for a formative period, federal courts adopted a rhetoric and style of review that were nontrivial in their impact on governmental practice, creating a high water mark for precaution in U.S. law.

A third example of precaution is the provision for setting primary ambient air quality standards under the 1970 Clean Air Act. Section 109 of that law requires the Environmental Protection Agency (EPA) to set standards protective of public health, allowing an "adequate margin of safety," for so-called criteria pollutants. This again was a precautionary provision, both because it did not, unlike many U.S. environmental laws, ask the EPA to balance the costs of regulation against the expected benefits and because it asked for a margin of safety in standard setting, presumably to be determined at the agency's discretion. Once more, Congress had instructed a powerful regulatory agency to move beyond the limits of existing knowledge in deciding how much protection to afford to public health. "Adequate margin of safety" was, after all, an indefinite concept; there were in 1970 no accepted statistical tools or analytic methods to give the term a precise quantitative meaning. The EPA's mandate was simply to approach the task of public health protection extra cautiously, taking account of known factors, but also leaving room for uncertainty, variability, and error. It was, more than anything, an injunction to give greater weight to the ethic of health protection than to scientific and economic calculations.

Risk and Property: Two Challenges to Precaution

Ideas, like precaution, however ingenious or benign, remain mere ideas unless they are inserted into social practice. Even the most carefully drafted legal principles are not self-executing but have to be unpacked, unfolded, and implemented in decision-making contexts that have torques and biases built into them. The United State's regulatory culture is no exception. It has enormous strengths, but sometimes those very strengths can operate as barriers against learning and ethical progress. In the generation since the enactment of NEPA, both the prudential and the moral component of the precautionary ideal have been substantially watered down in the United States, the former through bureaucratic responses to regulatory pressures and the latter through the very judicial system that had once been a staunch ally of environmentalism. On each front, we see that particular strengths of U.S. decision making—transparency and judicial responsiveness—have assisted in the decline of the precautionary ideal.

Transparency is one of the most important norms in U.S. public affairs, enshrined in countless laws and administrative practices, including the Freedom of Information Act, the Federal Advisory Committee Act, and most recently the 1998 "Shelby amendment," which required the public disclosure of all scientific data generated with federal financial support.[4] This proliferation of disclosure requirements is not in any sense accidental. Closed preserves of expert decision making, common to many European regulatory systems, would be inconsistent with the central tendencies of American democracy. Transparency is an essential value in a democratic, egalitarian society that is committed to leveling the playing field on which citizens interact with decision-making elites.

Yet even a norm that reflects such a high seriousness of purpose can prove, at its limits, suffocating. In the U.S. policy system, regulators are perpetually accountable to interested members of the public and hence vulnerable to questions about the basis for their policies. Quantification, with its powerful appearance of neutrality, provides a welcome refuge for beleaguered agency officials. By the same token, any decision-making strategy that cannot represent itself as objective, through quantification or other forms of analysis, risks becoming politically unsustainable (Brickman et al. 1985; Jasanoff 1986, 1992; Porter 1995). Yet, to the extent that precaution is a philosophical orientation that takes account of the limits of knowledge and analysis and, hence, marries prudential with ethical norms, the command to quantify it produces a kind of displacement. It requires a shift of gears into a register of managerial decision making that overemphasizes current awareness and understanding and, thus, imperils the core meaning of precaution.

By 1980, the precautionary aims of U.S. environmental policy were already giving ground to the preference for objectivity (or at least the appearance of it) and the sheer regulatory overload produced by the torrential legislation of the preceding decade. The U.S. Supreme Court in that year issued a very important ruling reviewing the standard for benzene exposure set by the Occupational Safety and Health Administration (OSHA). The "benzene decision" held that for OSHA to demonstrate that the standard was "reasonably necessary," as required by law, it would have to show that there was a "significant risk" in some quantitative sense, though not with "anything approaching scientific certainty" (*Industrial Union Department* 1980). The decision brought about an almost immediate transformation in U.S. risk analytic practices. Before 1980, most standard-setting agencies regarded risk assessment as a highly judgmental and largely qualitative exercise. Although pressure was mounting for techniques to speed an often unwieldy regulatory process, agency experts still resisted quantification and formal risk assessment as the appropriate management tools. The benzene decision, together with many recently enacted laws that required the government to compare the risks and benefits of regulation, provided a huge impetus for quantitative risk assessment. The only language in which agencies could credibly balance regulatory costs against the health of workers and citizens or the value of ecosystems was the language of numbers. Where the law led, universities and consultancies soon followed, creating within less than a generation a new professional field of risk analysis. In time, many came to regard this as the "science" that had to be invoked in addressing the prudential dimension of precaution.

A further retreat from precaution came with the resurgence of private property rights as a basis for challenging environmental policy. A 1994 Supreme Court decision, *Dolan v. City of Tigard,* challenged both the prudential and the moral aspects of environmental decision making by requiring a city planning commission to balance environmental gains against losses to private property. Unhappy with a requirement to set land aside for a greenway and a bicycle path, Florence Dolan, a would-be developer in Oregon, took her case all the way to the U.S. Supreme Court. In a 5–4 split decision, the Court held that the city had failed to show the necessary "reasonable relationship" between the requested set-asides and the proposed building. As in the benzene decision many years before, the justices refrained from demanding a precise mathematical calculation. But the majority made it clear that some form of explicit analysis would be needed to establish the constitutionally required "rough proportionality" between the impact of the proposed development and the cost of the environmental improvements requested by the planning commission.

A 2001 decision handed another partial victory to property rights claimants. In a case originating in Rhode Island, the Supreme Court held that the landowner's claim for uncompensated taking of property was not barred by the fact that the state had adopted the restrictive land-use regulations *before* the property was acquired (*Palazzolo v. Rhode Island* 2001). Future generations, the Court explained, with an unconsciously ironic twist on the language of sustainability, must have a right to challenge unreasonable limitations on the use and value of land. According to this line of reasoning, current precautionary judgments might have to be defended and modified in the future not only in the light of changing scientific knowledge but also in response to rising property values.

Even a conservative Supreme Court, however, has not been wholly oblivious to the normative commitments underlying environmental protection. Two recent decisions can be seen as placing limits on the developments described above. I have suggested above that the Supreme Court's decisions not to meddle in agency decision making in the 1980s put an end to the trickle-down effect of precaution throughout the decision-making process. More specifically, as we have seen, the "hard look" doctrine yielded to a judicial stance that gives agencies greater freedom to choose how prudential they want to be in their technical assessments, within relatively loose constraints on their discretion. But that laissez-faire judicial stance need not invariably lead to less precaution.

In *Whitman v. American Trucking Associations, Inc.*, capping what was arguably the most significant environmental litigation of the 1990s, the Court's deference to the EPA's technical judgment amounted in effect to upholding the Clean Air Act's precautionary approach. At issue in that case was the EPA's proposed new standard for fine particulate matter, which the agency claimed, on the basis of highly contested epidemiological research,[5] played a major role in causing respiratory illnesses such as asthma. The Court of Appeals for the District of Columbia Circuit invalidated the particulate standard partly on constitutional grounds, holding that Section 109 had delegated de facto legislative power to the EPA and that the agency had failed to cure this unbounded grant of authority by providing any "intelligible principles" to guide its decisions. The Supreme Court, however, backed away from this position. Statutes, the Court announced with uncharacteristic unanimity, do not have to provide determinate criteria for agencies to decide how much harm is too much. Rejecting industry's arguments to the contrary, the Court also reaffirmed that Section 109 does not permit the EPA to consider the costs of implementation in setting primary air quality standards.

The expansion of property rights also received a check from the high court in a decision of April 2002 (*Tahoe Sierra Preservation Council, Inc.* 2002). Once again landowners had contested a planning decision on the ground that it constituted an uncompensated taking of their property. The focus of controversy was a thirty-two-month moratorium on development announced by the Tahoe Regional Planning Authority while it formulated a comprehensive land-use plan for the celebratedly beautiful Lake Tahoe basin. Owners protested that this delay deprived them of all viable economic uses of their land. In a 6–3 opinion, the Supreme Court turned back this challenge, stressing that temporary restrictions on land use should not be conflated with permanent takings and that a moratorium was an appropriate instrument to guide informed and balanced planning decisions.

Resurrecting the Precautionary Ideal

If we believe that precaution has lost ground in U.S. environmental policy, and are convinced moreover that the debate over precaution has been couched in overly simplistic, binary terms—such as good science versus protectionism or analytic rigor versus political motivation—what antidotes can we offer? I have suggested that proposals for resurrecting the precautionary ideal will have to address both the prudential and the moral commitments that this term incorporates. As a practical matter, these proposals will also have to respect the traditions and possibilities of U.S. legal and political culture.

On the side of prudential obligations, precaution in the American context should not be viewed simply as a substitute for risk assessment. So long as precaution is forced into this conceptual straitjacket, it risks being treated as just another way to answer the age-old regulatory question, "How safe is safe enough?" If we look at precaution as an end-of-pipe principle, merely replacing risk-benefit balancing with a different set of technical methods, then it will fail in its objectives again, for all the reasons outlined above. The pressures to discipline precaution, to transform it into easily performable bureaucratic routines that can survive judicial review and meet citizens' demands for objectivity, will all prove irresistible. This instrumental version of precaution will quickly lose the active engagement with the uncertain and the unfathomable, and the commitment to collective moral reflection, that gives the term its political edge. Precaution has to be seen instead as a habit of thought that applies up and down through a production system, not just as a technique for controlling known negative externalities at the system's end. Only then will we be able to develop innovative ideas, such as upstream risk assessment,[6] for how to operationalize the concept within the constraints of a living, but changing, regulatory process.

On the moral front, the discourse around precaution has to be reoriented so that it is not identified, immediately and automatically, with a Luddite rejection of technology. The most frequent criticism of precaution is that it is nothing more than a blocking strategy, designed either to gain competitive space for one's own enterprises or to keep innovation at bay. To liberate precaution as a productive ideal in policy making, it is essential to reject the easy association of this fundamentally care-taking concept with fear, selfishness, greed, or political opportunism. Precaution has to be translated into a principle of production rather than of regulation, and the best place to begin is *within* the processes of innovation. Producers have to be challenged with implementing the idea of precaution so as to tap into their unquestioned reserves of knowledge and ingenuity.

This will not be a matter of rhetoric alone but of creating new institutional structures that promote precaution. Key to such a development is the recognition that, for both democratic and epistemological reasons, closed decision-making processes are likely to be more blinkered and less precautionary than open ones that make room for multiple critical viewpoints. Yet even within America's admirably transparent and accessible political system, there remain large pockets of decision making that are insulated from effective participation and review, whether because citizens lack the knowledge and resources to intervene or because the legal and corporate culture of production excludes the very possibility of public intervention.

Some years back, for example, Monsanto Company devised a controversial plan to invest in the "Terminator technology," a process for developing genetically modified plants with sterile seeds. The company did not consider it prudent, let alone ethically necessary, to discuss its plan with farmers, consumers, and social activists who might eventually have had a stake in the consequences of this commercial strategy. Monsanto (unlike many politicians and even regulatory agencies) consulted no focus groups, carried out no surveys, and held no town meetings to see how people might feel about the economic, social, or environmental risks and benefits of this new technology. As a result, the company embarked without any form of structured, deliberative process upon a course of innovation that might have radically altered the ways in which billions of people around the world meet their most basic needs. When, in response to intense public pressure, Monsanto announced that it would not pursue this particular technological option, that decision too lacked the kind of openness and analytic rigor that should have guided a decision with so many unknown human and environmental consequences.

Last, but not any the less important, let us return to the role of the law in crafting new precautionary practices. The project of reviving precaution in the U.S. will very likely require the recalibration of the law as an instrument of progressive action. There are precedents for this in the history of U.S. environmentalism. In the 1960s, when much of the present architecture of environmental protection was first conceived, the law was widely seen, and used, as an instrument of progressive social change. Now in its third century, a legal tradition that blends common law concepts of care with constitutional guarantees of fundamental rights still offers a wealth of terms for rearticulating the discourse of precaution in a genuinely American idiom. Here are a few of the most compelling:

- The *duty of care,* on which negligence law is built.
- The constraint of *rationality,* which guards against arbitrary and unreasoned decision making.
- The idea of *accountability,* which underlies all governmental regulation.
- The commitment to *participation,* which is central to the functioning of democracy.
- The concept of *environmental equity,* which corrects for past injustice.
- The instrument of *compensation,* with which bad decisions can be remedied and deterred.

Resonant terms such as these provide doctrinal entry points through which the law can participate in developing new societal practices of precaution. Only this time the beneficiaries of legal creativity should not be U.S. citizens alone, tenaciously protecting the safety of their own backyards against encroachment by dirty, unsafe industries. This time U.S. legal ingenuity should be put in service of a global ethic of precaution, corresponding to the global spread of environmental hazards and corporate capital. How to take up this task is a challenge for progressive legal thinkers at this formative moment in human history.

Endnotes

1. This was the title of the agenda-setting report of the World Commission on Environment and Development in preparation for the Earth Summit held in Rio de Janeiro in 1992 (WCED, *Our Common Future* [Oxford: Oxford University Press, 1987]). A puzzle for analysts is why some concepts embraced by that report (such as "sustainability") took hold in international policy discourse, whereas others (such as "precaution") remained contested.

2. Some of the confusion follows from the absence of the word "precaution" in significant environmental and legal texts. Another difficulty is that varied terms have been used even in documents that explicitly speak of precaution—thus, either the "precautionary principle" or a "precautionary approach." In this paper, I do not seek to resolve these debates over the correct lexical choice but attempt rather to clarify how precaution has worked as a concept and a norm in U.S. law.

3. See in particular *Baltimore Gas and Electric Company v. Natural Resources Defense Council*, 462 U.S. 87 (1983); *Chevron USA, Inc. v. Natural Resources Defense Council*, 467 U.S. 837 (1984).

4. Pub. L 105-277 (1998). The Office of Management and Budget responded to scientists' concerns about the cost and sweep of the amendment by restricting it to studies used in policy making.

5. Much of this research was carried out at the Harvard School of Public Health in the so-called Six Cities Study. Attempts by opponents of the standard to obtain the raw data generated by these studies led to the Shelby amendment discussed at note 4.

6. Although it remains as yet only a phrase on the drawing boards of policy, this concept refers to the fact that current risk assessment practices address problems only very far along in the process of industrial production, at a time when many consequential choices have already been made and cannot be reviewed or revisited. Upstream risk assessment would insert precautionary principles closer to the point of innovation, a point that is currently hedged around with legal protections, such as intellectual property rights.

References Cited

Brickman, R., S. Jasanoff, and T. Ilgen. 1985. *Controlling Chemicals: The Politics of Regulation in Europe and the United States.* Ithaca, N.Y.: Cornell University Press.

Guardian, The. 2002. "George Bush's Global Warming Speech." *The Guardian*, 14

February 2002. http://www.guardian.co.uk/globalwarming/story/0,7369, 650821,00.html.

Jasanoff, S. 1986. *Risk Management and Political Culture.* New York: Russell Sage Foundation.

———. 1992. "Science, Politics, and the Renegotiation of Expertise at EPA." *Osiris* 7: 195–217.

———. 1995. *Science at the Bar: Law, Science, and Technology in America,* 36–39. Cambridge: Harvard University Press.

Macilwain, C. 2000. "Experts Question Precautionary Approach." *Nature* 407: 551.

Porter, T. 1995. *Trust in Numbers: The Pursuit of Objectivity in Science and Public Life.* Princeton: Princeton University Press.

Stern, P. C., and H.V. Fineberg, eds. 1996. *Understanding Risk: Informing Decisions in a Democratic Society.* Washington, D.C.: National Academy Press.

Cases Cited

Industrial Union Department, AFL-CIO v. American Petroleum Institute, 448 U.S. 607 (1980).

Dolan v. City of Tigard, 512 U.S. 374 (1994).

Palazzolo v. Rhode Island, 533 U.S. 606 (2001).

Whitman v. American Trucking Associations, Inc., 531 U.S. 457 (2001).

Tahoe Sierra Preservation Council, Inc. v. Tahoe Regional Planning Agency, U.S.—(2002).

CHAPTER 16

The Precautionary Principle in European Community Law and Science

*Theofanis Christoforou**

The precautionary principle is about scientific uncertainty. It permits and in some cases requires regulatory authorities to take action or adopt measures in order to avoid or reduce risk to health, safety, or the environment, if necessary by erring on the side of safety. In the European Community (EC; also referred to as the European Union or EU), considerations of health and environmental protection are paramount, and the precautionary principle is a legal norm that can be deployed to ensure that the societal values and policy choices on health and environmental protection are fulfilled. It serves three main functions in EC law: (a) to allow or compel precautionary action by the regulatory authorities when necessary to achieve the chosen level of protection; (b) to allow *ex ante* and *ex post* administrative and judicial control of regulatory discretion, that is, before regulatory action is taken or after it has been applied; and (c) to place the burden on the manufacturer or other interested party to demonstrate safety.

This chapter provides a brief account of the history, content, and role of the precautionary principle in EC law, with particular reference to how the European Commission views the relationship of science and policy in applying precaution in risk regulation.

---

*All views presented are strictly those of the author.

241

The 2000 European Commission Communication

The European Commission, after much study and discussion, issued on 2 February 2000 the *Communication on the Precautionary Principle* (European Commission 2000). The *Communication* had four aims: to outline the commission's approach to using the precautionary principle; to establish commission guidelines for applying it; to build a common understanding of how to assess, appraise, manage, and communicate risks that science is not yet able to evaluate fully; and to avoid unwarranted recourse to the precautionary principle as a disguised form of trade protectionism. The *Communication* also sought to contribute to the ongoing debate on this issue, both within the EC and internationally, since relevant international law and approved agreements and conventions are part of EC law and, thus, influence the development and application of the precautionary principle in the EC and its member states. The *Communication* has indeed had considerable impact on public debate and academic research in the broad areas of science, regulation, and precaution.

The commission's *Communication* was endorsed by the December 2000 Nice European Council *Resolution,* which called on the commission to systematically apply its guidelines, making allowance for the specific features of the various areas in which they may be implemented, and to incorporate the precautionary principle, wherever necessary, in drawing up its legislative proposals and in all its actions. The European Council also called on the member states and the commission to develop scientific expertise and the necessary institutional coordination; ensure that the precautionary principle is fully recognized in the relevant international health, environment, and trade fora, particularly at the World Trade Organization; ensure that the public and the various parties involved are informed as fully as possible about the state of scientific knowledge, the issues at stake, and the risks to which they and their environment are exposed; and, finally, to work actively for international partners' commitment to reaching an understanding on the application of the principle.

The work of the EC institutions on the precautionary principle, including the numerous resolutions of the European Parliament and the active participation of nongovernmental organizations, intensified an already lively public debate that was influenced by recent food and health crises and the ongoing debate on biotechnology and genetically modified products at the national, community, and international levels.

Definition and Status

The principle that responsible governments should act on the basis of precaution to protect health and ecosystems despite scientific uncertainty is so widely accepted that it is now in the process of becoming, or has already crystallized as, a rule of customary international law in the areas of health and environmental protection. Consequently, some divergence in the terminology used ("principle" or "approach" or "measure") in the various international conventions and agreements is of no legal significance.

Stripped of all its peripheral and functional elements, the precautionary principle deals with the question of what appropriate actions should be taken to protect health and the environment in the face of scientific uncertainty. In accordance with generally accepted theory of international law, it can therefore be argued that the precautionary principle, understood in that sense, has already become a principle or customary rule of international law, because all the requisite elements of general state use practice and scholarly legal opinion exist and have been met with quite strong, consistent, and widespread acceptance. Any differences in the formulation of the principle in the available definitions relate rather to its application, such as the nature or extent of risk identified or the need to conduct a risk assessment or a cost-benefit analysis before taking the measure, which do not as such affect the core and basic rationale of the principle.

In any case, in EC law the precautionary principle has the status of a mandatory treaty principle. The EC regulatory authorities (and those of the member states acting in the area of EC law) are *obliged* to consider the application of the precautionary principle *when this is necessary to achieve the chosen level of health or environmental protection.* Article 174(2) of the EC Treaty, as modified by the Maastricht Treaty in 1992, provides that the "Community policy on the environment shall aim at a high level of protection. . . . It shall be based on the precautionary principle. . . ." The EC Treaty did not provide a definition of the precautionary principle. However, the following passage from the European Court of Justice judgment in a recent case on bovine spongiform encephalopathy (BSE, or mad cow disease) contains all the necessary elements of a general definition of the precautionary principle that can be applied in all areas of EC law:

> Where there is uncertainty as to the existence or extent of risks to human health, the institutions may take protective measures without having to wait until the reality and seriousness of those risks become fully apparent. (*BSE* [1998], at paragraph 63, repeated in subsequent judgments)

This passage lays down three basic conditions that may trigger application of the precautionary principle in EC law: uncertainty, risk, and lack of proof of direct causal link. These are explained in more detail below.

Chosen Level of Protection

The difficulty of providing a generally applicable and universally acceptable definition of the precautionary principle stems not from any uncertain or imprecise nature of its basic rationale but, rather, from the fact that its application is context and case specific, depending on the level of risk a society considers acceptable for a specific substance or activity at a given moment in time. This, in turn, is a function of many considerations and factors, such as the understanding experts, regulators, and lay people have of science and its role as a tool to identify, analyze and predict risk, the nature and extent of the risk (irreversibility, magnitude, preventability, necessity, etc.), and the confidence of the public in regulation. It should be noted that although the acceptable level of risk can be defined both in qualitative and quantitative terms, in the EU it is practically never expressed in a precise quantitative manner, such as a one-in-one-million risk of death. It is interesting to note that European law speaks in terms of *high level* of health or environmental protection to be achieved, not in terms of *significant risk* to be avoided, which is essentially the U.S. practice. However, there is no doubt that even a qualitative expression (e.g., significant or serious risk) of the acceptable level of risk includes or implies a chosen level of health or environmental protection. The precautionary principle is not automatically applied whenever there is uncertainty, as the potential harm may be considered to be acceptable to the regulatory authorities and the public in a specific case or compatible with a previously chosen level of health or environmental protection established by statute.

The level of protection does not always have to be chosen in advance or in the abstract. It may be decided on a case-by-case basis at the time of taking a specific regulatory measure. In this latter situation, a requirement to apply *consistency* in the choice of the level of protection could provide some means of controlling the discretion that EU regulatory authorities normally enjoy in the design and application of precautionary measures. Nonetheless, no legal system can ensure regulatory consistency in a strict sense across the board and over time for any kind of risk.

Science has always been the basis of risk regulation in the EC. This was necessary to establish the internal market and to resist national protectionism. When the aim to achieve a "high level" of health and environmental protection was given the status of a general objective in the EC Treaty, Article 100a of the EC Treaty was amended in 1992 to dispel any doubt that this should be based "on scientific facts" (Article 100a[3]). Consequently, the member states of the EC were allowed to adopt stricter standards in health and environmental protection (Article 130t) only on condition that the stricter national measures will be "based on new scientific evidence relating to the protection of the environment or the working environment" (Article 100a[5]). The EC Treaty provisions also prohibit explicitly the adoption of national stricter measures that are "a means of arbitrary discrimination or a disguised restriction on trade between member States and whether or not they shall constitute an obstacle to the functioning of the internal market" (Article 100a[6]). Moreover, in a series of seminal judgments, the European Court of Justice had clearly examined the scientific basis of many national or EC measures purporting to regulate risk to human health or the environment. The Angelopharm case (C-212/91, [1994] ECR I-171) made consultation of the relevant scientific committees mandatory. The Cassis de Dijon case (120/78, [1979] ECR 649), the German Beer case (178/84, [1987] ECR 1227), and the Danish Bottles case (302/86, [1988] ECR 4607) and the Fedesa case (C-331/88, [1990] ECR I-4023) have all made clear that any measure purporting to regulate risk should be based on scientific evidence and respect the principle of proportionality. It should also be pointed out that, contrary to conventional wisdom (Breyer and Heyvaert 2000; Vogel 1995), the stringency of the control by judges in the European Union as well as in the United States is not that much different in substance, especially since the creation of the Court of First Instance in 1992. In both systems the courts, in solving a specific legal dispute, are only required to ensure that the authorities have not used their regulatory discretion in an arbitrary and unjustifiable manner (see the Angelopharm case, supra, and the Pentachlorophenol [PCP] case, C-41/93, [1994] ECR I-1829). But the courts are not required nor are they capable epistemically of resolving the underlying basis of scientific uncertainty (Christoforou 2000).

If this is so, how then can one explain the constant claim by the United States that many European regulatory measures lack scientific basis and constitute disguised protectionism (e.g., meat hormones, recombinant bovine somatotropin [rBST], genetically modified organisms [GMOs], etc.)? There are many reasons and factors that explain the current divergence in the regulatory approach of the two systems. They range from social, economic, legal, scientific, cultural, ethical, tradition, political, and regulatory policy choices (Echols 2001; Pollack and Shaffer 2001). They all interact and play an important role, although the relevance of one or the other of these factors may be different depending on the circumstances of each case. Two factors, however, appear to play a predominant role: the Europeans' desire to achieve and maintain a high level of health and environmental protection, on the one hand, and the Americans' greater reliance on economic cost-benefit and market-oriented values, on the other (Christoforou 2001; Vogel 2002). The application of precaution has also played a role in addressing the Europeans' risk aversity. Discussing the precautionary principle, therefore, helps underscore the fundamentally divergent understandings in the two systems of what science is and its role in risk assessment and risk regulation.

Uncertainty and the Assessment of Risk

Although there are many discussions of different forms of uncertainty, in this chapter I will adhere to several definitions that have influenced interpretations of the precautionary principle in the EU. I will distinguish scientific uncertainty from risk and situations of ignorance (Stirling 1999; EEA 2001). These contribute to a lack of proof of direct causal links between an activity and a particular outcome—the reason for the precautionary principle. These distinctions are important in the context of risk assessments.

In the EU, nearly any substance or technological activity that may potentially have an adverse effect or impact on health, the environment, or safety in the workplace is subject to a risk assessment requirement. Detailed provisions exist on how to conduct such an assessment, especially in the areas of dangerous substances, medicinal products, food additives, contaminants, pesticides, GMOs, product standard-setting procedures, and environmental impact assessment.[1]

Risk is a function of at least two variables: the likelihood (or probability) of an adverse effect and its severity or magnitude (Codex Alimentarius Commission 2000). A formal definition of risk, therefore, is a condition under which it is possible to describe the probabilities of occurrence of most possible outcomes and their magnitude.

Scientific uncertainty exists when there is no adequate theoretical or empirical basis for assigning possibilities to a possible set of outcomes. This may be due to the novelty of the substance or activity concerned or because of complexity or variability in its context (Stirling et al. 1999; Stirling 2000). From a systematic point of view, this type of scientific uncertainty usually results from five sources of error in the scientific methods used to describe information and data: the variables chosen, the measurements made, the samples drawn, the models used, and the causal relationships employed (Walker 1991).

Uncertainty should also be distinguished from ignorance, where some of the possible outcomes, at the time of assessing the activity or substance, are completely unknown or unknowable and, thus, fail entirely to be assessed (EEA 2001). As it has rightly been argued, studying uncertainty and causality to death often can result in the death of those the regulatory authorities are supposed to protect (Infante 1987, 2001).

However, allowing fears from ignorance and indeterminacy to guide any risk regulation is likely to halt technological progress (Dupuy 2002). On the other hand, the mistake in the past has been (probably too often) to require scientific certainty before deciding to take restrictive or protective action (EEA 2001; Raffensperger and Tickner 1999). Two factors have led to such mistakes. The first is a positivist view that considers science to be a powerful and neutral tool capable of predicting risk and causality. This has been demonstrated to be wrong in several cases, because the experts' judgments appear to be prone to many of the same mistakes and biases as those of the general public, particularly when experts are forced to go beyond the limits of available information and data and rely on assumptions and intuition (Fischhoff et al.

1981; Slovic 1987, 1996; Jasanoff 1998). Second, existing risk-assessment methodologies are inherently biased in favor of avoiding too stringent regulatory measures (i.e., the inclination is to avoid false positives), for fear of imposing undue costs on technological progress and on society (Breyer 1993; Cranor 1993; Graham 1996, 2001; EEA 2001).

In a risk assessment exercise, instead of adopting a positivist approach to science, it is probably more accurate to view science in general and scientific research in particular as the products of a technical and sociological process that involves both objective and subjective observation, interpretation, verification, and discourse. The EC seems to be increasingly guided by this type of consideration in its risk assessment and risk management processes.

Most legal systems and courts tend to acknowledge scientific uncertainty (and the potential for error) in the presence of incomplete, inconclusive, or contradictory evidence coming from credible and reliable sources (even if they are minority scientific views). Courts cannot and should not judge science, but employ legal principles (e.g., the standard of proof) and the classic tool of dialectic analysis to resolve the legal dispute only, not to judge the scientific accuracy of the theory on which the measure regulating risk to health or the environment is claimed to be based (Brewer 1998; Christoforou 2000). As the EC's *Communication* has pointed out, varying degrees of scientific uncertainty may result from information that is insufficient, inconclusive, or contradictory—from gaps in knowledge or a state of controversy on existing data that renders problematic an estimation of the adverse effect (European Commission 2000). However, when uncertainty is properly analyzed and explained throughout the scientific assessment process, the regulatory authorities should be provided with sufficient scientific basis for a decision. That is the basis for the valid claim that the precautionary principle is firmly based on science.

Factors in Precautionary Risk Management

The decision to take action on risks cannot depend on scientific certainty. Moreover, regulatory bodies must make these decisions, rather than courts, because uncertainty and lack of clear evidence of causality normally undercut the ability to prove negligence in courts (Wiener 2001). What EU regulatory bodies strive for is to understand scientific uncertainty and take precautionary risk management action, which takes into account the chosen level of protection and other factors, including costs, to a greater or lesser degree.

Determining Acceptable Risk

Research has demonstrated that risk means more to people than the expected number of fatalities based on probabilistic quantitative assessments (Fischhoff et al. 1981; Slovic 1987). The perception people have of risk is wider than that of experts and reflects a number of legitimate concerns (e.g., familiarity with risk, catastrophic potential, irreversibility of harm, threat to future generations, risk control possibilities, whether exposure is voluntary), which are frequently omitted from an expert risk assessment (Fischhoff et al. 1981, 1978; Slovic et al. 1985; Slovic 1987). This has been elegantly described by the World Trade Organization (WTO) Appellate Body in the *Hormone Beef* case as follows:

> It is essential to bear in mind that the risk that is to be evaluated in a risk assessment under Article 5.1 is not only risk ascertainable in a science laboratory operating under strictly controlled conditions, but also risk in human societies as they actually exist, in other words, the actual potential for adverse effects on human health in the real world where people live and work and die. (at paragraphs 187 and 194)

Risk managers, instead of trying to patronize consumers with positivist views on science, should take into account their legitimate concerns and perceptions. This is indeed the risk management philosophy in the European Union.[2] Detailed studies indicate that people tend to view current levels of risk as unacceptably high for most activities and substances. Studies have also shown that the gap between perceived and desired risk levels suggests that people are not satisfied with the ways in which the market and regulatory authorities have balanced risks and benefits (Slovic et al. 1980; Slovic 1987). Therefore, being able to define appropriately the acceptable level of risk (or chosen level of health or environmental protection) is fundamental in risk management and the application of the precautionary principle. In simple terms, therefore, the objective is to discover how safe is safe enough for the people. This should be decided on the basis of transparency and public dialogue with all interested parties.[3]

It is generally agreed that defining the level of acceptable risk is a normative decision that belongs to the democratically elected and accountable institutions of a state.[4] Regulation of risk entails making important decisions about how much health and safety people want and can afford. As this touches upon the basic functions and mission of a democratic system of government, which is to protect the life and health of its people and the environment, decisions about the level of acceptable risk cannot be made by scientists or other experts who are unaccountable to the public. This is why in the EC legal system, as in many other systems, the opinions of technical and scientific committees are of advisory nature only, which means that their opinion is a necessary but not sufficient condition for risk regulation.

As a general rule, people and regulatory authorities normally pursue policies that seek to avoid risk to health or the environment, unless this becomes clearly too high a burden on them or the society to bear (Slovic 1987). Pursuing zero risk policies, that is, the goal of eliminating particular risks altogether, is common in many legal systems, and that right has been upheld explicitly both by national and international courts and tribunals.[5]

Weighing Costs and Benefits

Unlike U.S. law,[6] EC law has no general guideline that obliges regulatory authorities to analyze systematically the economic impact or cost of risk management measures. However, nothing prevents regulatory authorities from, whenever feasible, measuring and reporting on the economic impact of their decisions, so as to inform themselves and the public. Indeed, in its December 2000 Nice *Resolution* on the precautionary principle, the European Council stressed that

> the measures adopted presuppose examination of the benefits and costs of action and inaction. This examination must take account of social and environmental costs and of the public acceptability of the different options possible, and include, where feasible, an economic analysis, it being understood that requirements linked to the protection of public health, including the effects of the environment on public health, must be given priority. (at paragraph 20)

Regulatory authorities in the EC sometimes make, consciously or unconsciously, gross estimates of first-level, direct costs and benefits of their decisions. However, considerations of the level of economic impact or cost from adopting a future precautionary action do not play a decisive role in determining *whether* to adopt a measure, but only in the actual *choice* or *design* of the measure to be taken. Moreover, the European Court of Justice has held several times that, in a risk management and balancing exercise, considerations of health should take precedence over economic or commercial considerations. (See, for example, Order in *UK v. Commission* [1996] at paragraph 93; Order in *Farmers' Union* [1996] at paragraph 105; *Affish* [1997] at paragraph 43.)

Some have argued in favor of adopting a detailed cost-benefit analysis in nearly all risk management decisions in the EC, based on the multirisk nature of our world and on reasons of efficient allocation of resources (Wiener 2001; Majone 2001). These arguments are not only misconceived and flawed but also potentially dangerous. Cost-benefit analysis and other influences can lead to undue delays in precautionary action and further losses (see chapter 13 for examples).

The fact that in our technologically complex society there are multiple sources of risk, including risks to which people voluntarily expose themselves, does not cancel out the legitimate objective to aim, whenever possible, to eliminate risks to health or the environment (Geistfeld 2001). Moreover, voluntary exposure to risk by *some* must not be balanced against unintended, involuntary exposure to the same or other types of risks by *others*. The right to life and health is the most fundamental of all human rights, and no restriction should be placed on it without proper consideration. Indeed, as a matter of principle, reasons of justice, fairness, and morality militate against a balancing exercise based on broad considerations of efficient allocation of resources (Sen 1986; Dworkin 1987; Rawls 1999; Nussbaum 2000).

In the EC legal order the principle of proportionality is used to check the balance between the health or environmental objective pursued and the restrictive effects of the precautionary measure. The ECJ has defined the principle of proportionality in EC law as follows:

> It must be recalled that the principle of proportionality, which is one of the general principles of Community law, requires that measures adopted by Community institutions do not exceed the limits of what is appropriate and necessary in order to attain the objectives legitimately pursued by the legislation in question; when there is a choice between several appropriate measures recourse must be had to the least onerous, and the disadvantages caused must not be disproportionate to the aims pursued. (BSE [1998] at paragraph 60)

Science, Precaution, and the Courts

The problem of understanding and defining uncertainty in the context of a risk assessment can be large, complex, and nearly intractable, unless the analysis is structured into small and simpler concepts for each stage and component of the risk analysis. It is of paramount importance for risk assessors to explain in detail any kind of scientific uncertainty they encounter in every step of their analysis and the techniques, assumptions, and values they employ to eliminate or reduce it. These uncertainties should then be communicated in clear and understandable language to both risk managers and the people. Residual uncertainties, however, will often remain.

Precaution should be applied both by the scientists completing the risk assessment, on the basis of science policy guidelines, and by the regulatory authorities themselves who have to draw the necessary implications. Both risk assessors and risk managers attribute at any given time different subjective values to available scientific data, the risks, and the nature of possible adverse effects. Precaution applied by scientists in a risk assessment does not, therefore, eliminate the need for risk managers to apply precaution to the same agent, activity, or process when taking regulatory action. Risk assessors' *technical* precaution (when developing hypotheses, modeling, and interpreting evidence and data) is, therefore, distinguishable from the risk managers' *regulatory* precaution (when taking normative regulatory action). This proposition is forcefully denied by the United States internationally, basically for reasons of economic competition, trade policy considerations, and general litigation and negotiation tactics.[7]

Because of the EC's objective to achieve a "high level" of health and environmental protection, the precautionary principle provides the means for affected parties to control, if necessary by means of action before the courts, the way risk management institutions make decisions under uncertainty. The precautionary principle, in situations of scientific uncertainty, should not only allow but in certain cases may *oblige* the regulatory authorities to err on the side of caution when this is necessary in order to achieve the chosen level of protection (necessity test). Thus, one of the functions performed by the precautionary principle is to put constraints on how regulators act under uncertainty. This entails both *ex ante* and *ex post* control of measures taken to regulate risk.

Allocating the Burden of Proof

The EC, like other countries, applies a prior approval procedure for many industrial and technological products, substances, or processes (additives, contaminants, medicinal products, veterinary drugs and growth promoters, GMOs, etc.). It also places the burden of proving safety or lack of harm on the applicant manufacturer or operator (burden of proof = burden of producing evidence + burden of persuasion). The candidate products, substances, or processes are generally deemed to be dangerous unless and until the interested manufacturer carries out the necessary scientific work and demonstrates to the satisfaction of the authorities the *safety* or *lack of harm,* compared to the chosen level of protection, which may be decided case by case. The amount of evidence required may vary. As the commission's *Communication* has suggested, measures based on the precautionary principle may be adopted when there are "reasonable grounds for concern" or when there are "valid reasons to consider" that there may be a risk.

The standard of proof applied by courts when reviewing cases may also vary. Some may call for a preponderance of evidence; others, for proof beyond reasonable doubt or clear and convincing evidence. In EC law, in the context of products or substances requiring approval before being marketed, the burden of proof is clearly on the manufacturer, who must usually demonstrate safety "adequately or sufficiently," which is comparable to the "proof beyond reasonable doubt" standard applied in common law jurisdictions.

There is also an implied initial burden on regulatory authorities to demonstrate the existence of a risk. However, this carries a relatively low threshold of producing evidence and burden of persuasion, especially in case of emergency or safeguard measures. Otherwise, the very purpose of applying the precautionary principle risks being defeated.

In cases where no prior authorization procedure is applied, for example, in existing chemicals, the burden normally rests on the public and their representatives to demonstrate existence of a risk from a product, process, activity, or project. However, in cases of scientific uncertainty, precaution can compel authorities to take specific regulatory measures to reverse this burden and place it upon the producer or importer when this is necessary to meet the chosen level of protection.

The Interface of National Laws with International Rules on Precaution

A basic traditional assumption is that countries accede to international agreements and treaties only when they think they are going to be to their interest. Recently, however, countries tend to accede to various international agreements and treaties, especially to environmental protection agreements, for a broader range of reasons. It has been argued that one of those reasons is that these agreements lack efficient implementation and enforcement mechanisms (Brown Weiss and Jacobson 1998). Conversely, agreements with binding and legally enforceable dispute settlement mechanisms, if properly applied and implemented, should normally discourage free riders. The practice of the EU and of the United States in the areas of international agreements and conventions in the areas of environment, health, and trade has been accurately described and analyzed by scholars elsewhere (see Vogel and Kessler 1998).

The 1947 General Agreement on Tariffs and Trade and its successor the World Trade Organization play an important role in the application of precaution at the international level. No country can afford today to stay outside the WTO Agreement because of the breadth of its coverage, embracing practically all aspects of international trade in goods, services, and intellectual property.

The meat hormones dispute under GATT (1989) and ten years later under the WTO Agreement on Sanitary and Phytosanitary Measures (1998) between the EC and the United States typifies the approaches these trading partners adopt to regulation under uncertainty. In the hormones case, described as "the mother of all food safety trade disputes" (Josling 1998: 3), a country was forced to pay compensation for its regulatory choice aiming to achieve a very high level of health protection in its territory (Vogel 2002). It typifies the trans-Atlantic policies and differences in the role of science in risk regulation.

The introduction in 1989 of a total ban on the use of hormones for animal growth promotion (Council Directive 88/146/EEC) clearly marked a conscious departure by the European Union from the United States' standard of protection despite the visible trade tensions this policy was expected to raise. A first measure, which restricted only domestically the use of hormones for growth promotion, was adopted in 1981 (Directive 81/602/EEC) and had, in addition to scientific uncertainty, a strong ethical and moral background because it sprang from scandals (use of illegal substances and high concentrations of residues in baby food) in Germany and Italy (where very young children exhibited serious symptoms of early puberty) (Fara et al. 1979; Chiumello et al. 2001; Perez Comas 1982; Saenz et al. 1982). However, the subsequent total ban on the use of hormones for growth promotion in 1998 was applied on a national treatment basis without discrimination to animals and meat treated with hormones in the EC and from third countries.

The WTO Appellate Body in the Hormones case reversed an earlier WTO panel finding and accepted that the EC legislation was not a disguised restriction on international trade because the record showed that it was motivated by the

> depth and extent of the anxieties experienced within the European Communities concerning the results of the general scientific studies showing the carcinogenicity of hormones, the dangers of abuse (highlighted by scandals relating to black-marketing and smuggling of prohibited veterinary drugs in the European Communities) of hormones and other substances used for growth promotion and the intense concern of consumers. (WT/DS26/AB/R and WT/DS48/AB/R, at paragraph 245)

The resistance of European consumers and farmers to meat hormones and other modified food products (e.g., rBST, GMOs, etc.) reflects a certain unease with artificial growth of agricultural production by new technologies, the harmless nature and long-term effects of which have not been clearly established scientifically (demonstrating a general risk aversity). After the findings of the WTO Appellate Body in 1998 (Walker 1999), the relevant scientific committees of the EC found, in three subsequent risk assessments carried out in 1999, 2000, and 2002, that meat hormones pose a number of potential adverse effects, in particular cancer, and that they appear to pose greater risks to children (Anderson 2001; Zhu and Conney 1998).

Much of this evidence was not available previously, but when the United States authorities evaluated these hormones in the 1960s and 1970s, they did not find that these hormones are risk free but rather that they do not pose a "significant" risk (Hertz 1977; Epstein 1998: 585; Schell 1985). On the basis of those risk assessments, the United States allowed the use of six meat hormones and continues to do so today. Conversely, on the basis of its latest risk assessments, the European Commission decided again to propose the maintenance of the total ban on the use of these meat hormones both within the EC and on imports from third countries (EC OJ No C 337, 28.11. 2000, p. 163, and EC OJ No C 180, 26.6.2001, p. 190).

The same regulatory attitude has been observed in other cases, for example, the hormone rBST to enhance milk production. In 1990, the EC introduced a moratorium on the use of rBST, for the same reasons as in the meat hormones case, that is, uncertainty, lack of knowledge on long-term effects, animal welfare, and artificial growth of agricultural production. In 1999, the EC introduced a permanent ban on the use of rBST within its territory (but not on the very tiny quantities of imports from third countries) when clear evidence showing detrimental effects on animal health and welfare became available (case C-248/99P, *France v. Monsanto* (BST), judgment of 8 January 2002). In September 2000, the European Council decided also not to fix a maximum residue limit for rBST under another EC regulation (Council Regulation 2377/90/EEC) on the grounds of scientific uncertainty about possible risks to human health and the precautionary principle (EC Council doc. 11307/00 of 21 September 2000). It should be pointed out that Canada, on the basis of a similar risk assessment, decided also in 1998 to withdraw the authorization of rBST on grounds of animal health and welfare. The United States is one of the very few countries that continues to allow the use of rBST.

There is no doubt that despite the costs of the regulatory choices made by the EU (e.g., hormones in meat, rBST, etc.) on its industry, farmers, and consumers, these measures were imposed in the first place and are now maintained in order to achieve a high level of health or environmental protection, not for protectionist reasons. It should also be pointed out that the United States does not deny that precautionary measures or a precautionary approach may be adopted to regulate risk (Shaw and Schwartz 2002; Boisson de Chazournes 2002). What it has been contesting since the early 1990s in nearly all international fora (e.g., WTO, Codex Alimentarius Commission, Cartagena Protocol on Biosafety, Organization for Economic Cooperation and Development) is the existence or emergence of a precautionary principle that can trump provisions in existing agreements (WTO Appellate Body report in Hormones, at paragraph 122). It is the status of precaution under international law as such that explains the United States' resistance, as well as the wider discretion a general principle or rule of customary law will provide to cautious states to apply lawfully strict regulation even when there is no positive proof of harm.

A further related question is whether, given the normative status of the precautionary principle in national and international law, regulatory bodies are entitled to apply precaution even when there is no explicit or implicit legal reference made to the principle in the international agreement. I would argue that the normative value of the precautionary principle in national or international law is to compel *ex ante* and *ex post* consideration and application of precaution when this is necessary to achieve the chosen level of health, safety, or environmental protection (see Fisher and Harding 2001; Fisher 2001).

Conclusion

Every society is free to choose the level of acceptable risk to health or the environment. The precautionary principle provides a basis both to the regulatory authorities and the affected communities for implementing this choice. It is firmly based on science because its application is warranted only when uncertainty is scientifically established. It enables the regulatory authorities to take action when there is scientific uncertainty and risk but a direct causal link cannot be established. It can also oblige them to take such precautionary action when this is necessary to achieve the already chosen level of health or environmental protection. This is the most important normative function of the principle in EC law and, increasingly so, in international law and science on health and environmental protection.

Postscript

The Court of First Instance of the European Communities (CFI) handed down two seminal judgments on 11 September 2002, which upheld the European Community ban on the use of certain antibiotics in animal farming (i.e., virginiamycin, bacitracin zinc, spiramycin, and tylosin phosphate). The cases are important because for the first time the Court discusses in great detail the scope and conditions of application of the precautionary principle in community law.

In the decisions—which are closely in line with the analysis in this chapter and reaffirm the European Commission's Communication on the Precautionary Principle of February 2000—the Court holds that it is possible to take preventive measures (even on a class of similar hazards) without having to wait until the reality and seriousness of the risks perceived become fully apparent. In the Court's view, the concept of risk entails some probability that the negative effects will occur. In its judgments, the Court puts emphasis on the conditions with which the public authority must comply in its risk assessment. It places particular importance on the essential role of scientists in this context and concludes that the view of the competent scientific committees must be obtained, even if their opinion is only advisory or even if this is not specifically provided for by legislation, unless the public authority can ensure that it is acting on an equivalent scientific basis. However, the Court points out that the decision to ban a product is not a matter for the scientists to decide but rather one for the public authority to whom political responsibility has been entrusted and which can claim democratic—as opposed to scientific-legitimacy in risk regulation. The Court concludes that, despite uncertainty as to whether there is a link between the use of those antibiotics as additives and the development of resistance to them in humans, the ban on the products is not a disproportionate measure by comparison with the objective pursued, namely the protection of human health. An appeal against the judgments before the European Court of Justice, the highest Court of the European Communities, is still possible within a period of two months.

Endnotes

1. There is an extensive and still rapidly growing regulatory framework in the EC that lays down the general guidelines and detailed provisions about risk assessment methodologies and risk management decision-making. See E.Vos, *The Institutional Frameworks of Community Health and Safety Regulation: Committees, Agencies and Private Bodies,* 1999.

2. See, e.g., Article 3(12) of Regulation (EC) No 178/2002, of 28 January 2002, laying down the general principles and requirements of food law, establishing the European Food Safety Authority, and laying down procedures in matters of food safety, which defines risk management as follows: "Risk management means the process, distinct from risk assessment, of weighing policy alternatives in consultation with interested parties, considering risk assessment and other legitimate factors, and, if need be, selecting appropriate prevention and control options" (OJ No L 31, 01.02.2002, p.1).

3. For example, the United States has taken the position that the level of protection can be decided "by any means available to a Government under its law, including by a referendum." See US Statement of Administrative Action for WTO/SPS Agreements (1994): 103d Congress, 2d Session, H.D. 103-316, p. 745 (27.9.1994).

4. For instance, in the context of international trade it is accepted that defining the acceptable level of risk is the sovereign or autonomous right or prerogative of each state.

5. See, e.g., Council Directive 96/22/EC concerning the prohibition on the use in stockfarming of certain substances having a hormonal or thyrostatic action and of beta-agonists. For a similar example in the U.S. legal system, the Delaney Clauses (21 U.S.C.A. 2000) established a level of zero risk based on the assumption that if substances (food additives, color additives, and new animal drugs) induce cancer in animals, they pose some risk of cancer to humans.

6. See *Economic Analysis of Federal Regulations, Executive Order No. 12866,* 58 Fed. Reg. 51735 (30 September 1993).

7. See, e.g., U.S. Food and Drug Administration and Department of Agriculture,

United States Food Safety System: Precaution in U.S. Food Safety Decision Making: Annex II to the United States' National Food Safety System Paper (3 March 2000) found at http://www.foodsafety.gov/~fsg/fssyst4.html. For a critical review of the U.S. paper, see V. R. Walker, *Some Dangers of Taking Precautions without Adopting the Precautionary Principle: A Critique of Food Safety Regulation in the United States,* 31 Environmental Law Reporter 10040 (2001).

References Cited

Anderson, A-M., K. M. Grigor, E. Rajpert-de Meyts, H. Leffers, and N. E. Skakkebaek. 2001. *Hormone and Endocrine Disrupters in Food Water: Possible Impact on Human Health.* Copenhagen: Munksgaard.

Arrow, K. 1963. *Social Choice and Individual Values,* 2nd ed. New York: Wiley.

Boisson de Chazournes, L. 2002. "Le Principe de Précaution: Nature, Contenu et Limites," in C. Leben and J. Verhoeven, eds., *Le Principe de Precaution: Aspects de Droit International et Communautaire.* Paris: Patheon Assas, L. G. D. J. Diffuseur.

Brewer, S. 1998. "Scientific Expert Testimony and Intellectual Due Process." *Yale Law Journal* 107: 1535.

Breyer, S. 1993. *Breaking the Vicious Circle: Toward Effective Risk Regulation.* Cambridge, Mass.: Harvard University Press.

Breyer, S., and V. Heyvaert. 2000. "Institutions for Regulating Risk." In *Environmental Law, the Economy, and Sustainable Development,* edited by R. Revesz, P. Sands, and R. Stewart, Cambridge: Cambridge University Press.

Brown Weiss, E., and H. K. Jacobson. 1998. *Engaging Countries: Strengthening Compliance with International Environmental Accords.* Cambridge: MIT Press.

Chiumello, et al. 2001. "Accidental Gynecomastia in Children." In *Hormones and Endocrine Disrupters in Food Water: Possible Impact on Human Health,* edited by A.-M. Anderson, K. M. Grigor, E. Rajpert-de Meyts, H. Leffers, and N. E. Skakkebaek, 203. Copenhagen: Munksgaard.

Christoforou, T. 2000. "Settlement of Science-Based Trade Disputes in the WTO: A Critical Review of the Developing Case Law in the Face of Scientific Uncertainty." *N.Y.U. Environmental Law Journal* 8: 622.

———. 2001. *The Precautionary Principle, Risk Assessment, and the Comparative Role of Science in the European Community and the United States.* Unpublished paper.

Codex Alimentarius Commission. 2000. *Procedural Manual,* 11th ed., p. 48. Rome: FAO/WHO Standard Programme.

Cranor, C. 1993. *Regulating Toxic Substances: A Philosophy of Science and the Law.* New York: Oxford University Press.

de Sadeleer, N. 2001. "Le statut juridique du principe de précaution en droit communautaire: du slogan a la règle." CDE 91.

Dworkin, R. 1987. *Taking Rights Seriously.* Corrected 5th impression. London: Duckworth.

Dupuy, J.-P. 2002. *Pour Un Catastrophisme Eclaire: quand l'impossible est certain.* Paris: Seuil.

Echols, M. A. 2001. *Food Safety and the WTO: The Interplay of Culture, Science and Technology.* London: Kluwer.

Epstein, S. S. 1998. *The Politics of Cancer Revisited,* appendix 11. New York: East Ridge Press.

European Commission. 2000. *Communication on the Precautionary Principle.* Brussels.

European Council. 2000. *Resolution on the Precautionary Principle.* Brussels.

European Environment Agency (EEA). 2001. *Late Lessons from Early Warnings: The Precautionary Principle 1896–2000.* Environmental issue report no. 22, P. Harremoes, D. Gee, M. MacGarvin, A. Stirling, J. Keys, B. Wynne, and S. Guedes Vaz, eds. Copenhagen: Office for Official Publications of the European Communities (OPOCE).

Fara, G. M., G. Del Corvo, S. Bernuzzi, A. Bigatello, C. Di Pietro, S. Scaglioni, and G. Chiumello. 1979. "Epidemic of Breast Enlargement in an Italian School." *Lancet* 2: 295.

Fischhoff, B., S. Lichtenstein, P. Slovic, S. L. Derby, and R. L. Keeney. 1981. *Acceptable Risk.* New York: Cambridge University Press.

Fischhoff, B., P. Slovik, L. Lichtenstein, S. Read, and B. Combs. 1978. "How Safe Is Safe Enough? A Psychometric Study of Attitudes Towards Technological Risks and Benefits." *Policy Science* 9: 127–52.

Fisher, E. 2001. "Is the Precautionary Principle Justiciable?" *Journal of Environmental Law* 13: 315, 324.

Fisher, E., and R. Harding. 2001. "From Aspiration to Practice: The Precautionary Principle in Australia." In *Re-interpreting the Precautionary Principle,* 2nd ed., edited by T. O'Riordan, J. Cameron, and A. Jordan. London: Cameron and May.

Funtowicz, S., and J. Ravetz. 1992. "Three Types of Risk Assessment and the Emergence of Post-Normal Science." In *Social Theories of Risk,* edited by S. Krimsky and D. Golding, 251–73. Boulder: Westview Press.

Geistfeld, M. 2001. "Implementing the Precautionary Principle." *ELR News and Analysis* 31: 11326.

Graham, J. D. 1996. "Making Sense of Risk: An Agenda for Congress." In *Risks, Costs, and Lives Saved: Getting Better Results from Regulation,* edited by R. W. Hahn, 181–205. New York: Oxford University Press.

———. 2001. Opening Remarks, "U.S., Europe, Precaution and Risk Management: A Comparative Case Study Analysis of the Management of Risk in a Complex World." Conference held in Bruges, Belgium, 1 November. http://www.uspolicy.be/issues/Biotech/precprin.011502.htm.

Hertz, R. 1977. "The Estrogen-Cancer Hypothesis with Special Emphasis on DES." In *Origins of Human Cancer,* edited by H. H. Hiatt, J. D. Watson, and J. A. Winston, 1665-82, vol. 4 of Cold Spring Harbor Conference on Cell Proliferation, Cold Spring Harbor Laboratory.

Infante, P. F. 1987. "Benzene Toxicity: Studying a Subject to Death." *American Journal of Industrial Medicine* 11: 599–604.

———. 2001. "Benzene: An Historical Perspective on the American and European Occupational Setting." In *Late Lessons from Early Warnings: The Precautionary Principle 1896–2000,* 38–51. Copenhagen: European Environment Agency.

Jasanoff, S. 1998. "Contingent Knowledge: Implications for Implementation and Compliance." In *Engaging Countries—Strengthening Compliance with International Environmental Accords,* edited by E. Weiss Brown and H. K. Jacobson, Cambridge, Mass.: MIT Press.

Josling, T. 1998. "EU-US Trade Conflicts over Food Safety Legislation: An Economist's Viewpoint on Legal Stress Points That Will Concern the Industry." Paper presented at the Forum for U.S.-EU Legal-Economic Affairs, Helsinki, September 16–19. Sponsored by the Finnish House of Estates.

Luce, R., and H. Raiffa. 1957. *Games and Decisions.* New York: Wiley.

Majone, G. 2001. *The Precautionary Principle and Regulatory Impact Analysis.* Mimeo available from the author.

National Research Council. 1994. *Science and Judgment in Risk Assessment.* Washington, D.C.: National Academy Press.

Noiville, C., and N. de Sadeleer. 2001. "La gestion des risques écologiques et sanitaires a l'épreuve des chiffres—Le droit entre enjeux scientifiques et politiques." *R.D.U.E.* 389.

Nussbaum, M.C. 2000. "The Costs of Tragedy: Some Moral Limits of Cost-Benefit Analysis." *Journal of Legal Studies* 29: 1005.

OECD, Joint Working Party on Trade and Environment. 2000. *Uncertainty and Precaution: Implications for Trade and Environment.* COM/ENV/TD(2000)114, 20–21 November.

Otsuki, T., J. S. Wilson, and M. Sewadeh. 2001. *A Race to the Top? A Case Study of Food Safety Standards and African Exports.* Working paper of the World Bank-Country Economics Department, published under the Policy Research Working Paper Series (no. 2563). Washington, D.C.: World Bank.

Perez Comas, A. 1982. "Precocious Sexual Development: Clinical Study in the Western Region of Puerto Rico." *Boletin-Asociacion Medica de Puerto Rico* 74: 245–51.

Pollack, M. A., and G. C. Shaffer. 2001. "The Challenge of Reconciling Regulatory Differences: Food Safety and GMOs in the Transatlantic Relationship." In *Transatlantic Governance in the Global Economy,* edited by M. A. Pollack and G. C. Shaffer, 153–78. Lanham: Rowman and Littlefield.

Raffensperger, C., and J. Tickner, eds. 1999. *Protecting Public Health and the Environment: Implementing the Precautionary Principle.* Washington, D.C.: Island Press.

Rawls, J. 1999. *A Theory of Justice.* Oxford: Oxford University Press.

Saenz, C.A., M. Toro-Sola, L. Conde, and N. P. Bayonet Rivera. 1982. "Premature

Thelarche and Ovarian Cyst Probably Secondary to Estrogen Contamination." *Boletin-Asociacion Medica de Puerto Rico* 74: 16–19.

Schell, O. 1985. *Modern Meat.* New York: Vintage Books.

Sen, A. 1986. *On Ethics and Economics.* New York: Basil Blackwell.

Shaw, S., and R. Schwartz. 2002. "Trade and Environment in the WTO: State of Play." *Journal of World Trade* 36, no. 1: 129–54.

Slovic, P. 1987. "Perception of Risk." *Science* 236: 280–85.

———. 1996. "Risk Perception and Trust." In *Fundamentals of Risk Perception and Trust,* edited by V. Molak. New York: CRC/Lewis Publications.

Slovic, P., B. Fischoff, and S. Lichtenstein, 1985. "Characterizing Perceived Risk." In *Perilous Progress: Managing the Hazards of Technology,* edited by R. W. Kates, C. Hohenemser, and J. X. Kasperson, 91–125. Boulder: Westview Press.

Stirling, A. 2000. "Sciences et risques: aspects theoriques et practiques d'une approach de precaution." In *Le Principe de Precaution: Significations et Consequences,* edited by E. Zaccai and J. N. Missa, 73–103. Brussels: Editions de l'Universite de Bruxelles.

Stirling, A., O. Renn, A. Klinke, A. Rip, , and A. Salo. 1999. *On Science and Precaution in the Management of Technological Risk.* Sevilla: European Commission, Institute for Prospective Technology Studies. Report available at ftp://ftp.jrc.es/pub/EURdoc/eur1905en.pdf.

Sunstein, C. R. 2001. "Cost-Benefit Default Principles." *Michigan Law Review* 99: 1651.

U.S. Statement of Administrative Action for WTO/SPS Agreements. 1994. 103rd Congress, 2nd Session, H.D. 103–316. 27 September 1994.

Vogel, D. Forthcoming. "Risk Regulation in Europe and the United States." *Yearbook of European Environmental Law,* volume 3. Oxford: Oxford University Press.

Vogel, D., and T. Kessler. 1998. "How Compliance Happens and Doesn't Happen Domestically." In *Engaging Countries: Strengthening Compliance with International Environmental Accords,* edited by E. Brown Weiss and H. K. Jacobson, 19–37. Cambridge, Mass.: MIT Press.

Walker, V. R. 1991. "The Siren Songs of Science: Toward a Taxonomy of Scientific Uncertainty for Decision Makers." *Connecticut Law Review* 23: 567.

———. 2001a. "Consistent Levels of Protection in International Trade Disputes: Using Risk Perception Research to Justify Different Levels of Acceptable Risk." *Environmental Law Reporter* 31: 11317.

———. 2001b. "Some Dangers of Taking Precautions without Adopting the Precautionary Principle: A Critique of Food Safety Regulation in the United States." *Environmental Law Reporter* 31: 10040.

Wiener, J. B. 2001. *Precaution in Multi-Risk World.* Mimeo available from the author.

Zhu, D. T., and A. H. Conney. 1998. "Functional Role of Estrogen Metabolism in Target Cells: Review and Perspectives." *Carcinogenesis* 19: 1–27.

Cases Cited

120/78, *Cassis de Dijon* [1979] ECR 649
Case 174/82, *Sandoz BV* [1983] ECR 2445
Maine v. Taylor, 477 U.S. 131, 148 (1986)
178/84, *German Beer* [1987] ECR 1227
302/86, *Danish Bottles* [1988] ECR 4607
C-331/88, *Fedesa* [1990] ECR I-4023)
Pacific Northwest Venison Producers v. Smitch, 1994 U.S. App. LEXIS 6028 (20 F.3rd 1008)
C-41/93, *Pentachlorophenol (PCP)* [1994] ECR I-1829
C-212/91, *Angelopharm* [1994] ECR I-171
C-180/96R, *UK v. Commission* [1996] ECR I-3903
C-76/96R, *Farmers' Union* [1996] ECR I-3903
C-183/95, *Affish* [1997] ECR I-4315
Case C-157/96, *BSE* [1998] ECR I-2211
WTO Appellate Body Report in *EC Measures Concerning Meat and Meat Products (Hormones)*, WT/DS26/AB/R, WT/DS48/AB/R, adopted 13 February 1998
WTO Appellate Body report in *Australia—Measures Affecting Importation of Salmon ("Australia-Salmon")*, WT/DS18/AB/R, adopted 6 November 1998

Suggested Readings

Applegate, J. S. 2000. "The Precautionary Preference: An American Perspective on the Precautionary Principle." *Human and Ecological Risk Assessment* 6: 413.

Bourdieu, P. 2001. *Science de la Science et Reflexivity.* Paris: Raisons d'Agir Editions.

de Sadeleer, N. 1999. *Les Principes du Polluer-Payer, de Prévention et de Précaution.* Brussels: Bruylant.

Environmental Protection Agency. 1995. *Science Policy Council: Guidance for Risk Characterization.* Washington, D.C.: Environmental Protection Agency. Available at epa.gov/osp/spc/rcguide.htm (last visited site 8/12/02).

Funtowicz, S., and J. Ravetz. 1990. *Uncertainty and Quality in Science for Policy.* Dordrecht: Kluwer.

Godard, O., ed. 1997. *Le principe de précaution dans la conduite des affaires humaines.* Paris: Maisons des Sciences de l'Homme.

Hermite, M.-A., and V. David. 2000. "Evaluation des risques et principe de précaution." *Petites Affiches* 239: 13.

Molak, V. 1997. *Fundamentals of Risk Analysis and Risk Management.* New York: CRC/Lewis Publications.

Morgan, M., M. Henrion, and M. Small. 1990. *Uncertainty: A Guide to Dealing with Uncertainty in Quantitative Risk and Policy Analysis.* Cambridge: Cambridge University Press.

Noiville, C. 2000. "Principe de precaution et gestation des risques en droit de l'environnement en droit de la sante." *Petites Affiches* 239: 39.

Sunstein, C. R. 2002. "Beyond the Precautionary Principle," University of Chicago, John M. Olin Law & Economics Working Paper No. 149. Preliminary draft, April 2002.

Vogel, D. 1995. *Trading Up: Consumer and Environmental Regulation in a Global Economy.* Cambridge, Mass.: Harvard University Press.

Zaccai E., and J. N. Missa, eds. 2000. *Le Principe de Précaution: Significations et Conséquences.* Brussels: Editions de l'Universite de Bruxelles.

PART V

Science for Solutions: A New Paradigm

The role of science in decision making has traditionally been viewed as assessing, characterizing, and quantifying hazards. Under this view, science serves to provide warnings about and estimates of potential harms or demonstrating damage that has already taken place. Even in discussions about the precautionary principle, science is viewed as a means for characterizing uncertainties and developing a multidisciplinary picture showing when there is enough information to act. This forward-looking role for precaution in stimulating a search for alternatives and solutions to prevent harm—inherent in the original German formulation, *Vorsorgeprinzip*—has been lost in most enunciations of the principle.

The authors in this part argue for a new role for science: seeking solutions. This requires a more effective linkage between scientists studying risks and those analyzing prevention opportunities and restoration. They address such questions as, What are new directions for science that can support the development of solutions? How can a preventive approach to science be better integrated into decision-making structures? How do we promote these new directions in education, the research agenda, and funding?

Joel A. Tickner in chapter 17 presents the concept of precautionary assessment as a method for integrating a more holistic, interdisciplinary approach to science, with an examination of alternatives and participative structures. He argues that the acceptability of a risk must be a function of the ability to prevent harm in the first place.

Mary O'Brien argues in chapter 18 that we need to rethink the role of environmental science to conduct research not only in response to identified hazards (finding the level of harm that requires precautionary action) but also to achieve expressed values (e.g., respect for all beings) and ambitious goals.

Terry Collins, a leader in the green chemistry movement, argues in chapter 19 that we need to develop chemical technologies that are more compatible with natural cycles, to design problems out of technologies before they happen. He further notes that we need to train students in chemistry and engineering to understand the potential implications of the products they create.

CHAPTER 17

Precautionary Assessment: A Framework for Integrating Science, Uncertainty, and Preventive Public Policy

Joel A. Tickner

This chapter presents a framework and set of procedures—precautionary assessment—to implement the precautionary principle in environmental and health decision making in the United States and elsewhere. While definitions and implementation of the principle have been debated in Europe for two decades, discussions about the principle are at their early stages in the United States. Given substantial differences in the American environmental regulatory system, there is a need to translate precaution into a U.S. context. Under the European system, public and social values and politics are highly integrated into decision-making processes under uncertainty (see chapters 15 and 16).[1] In addition, political agencies are often separate from the technical agencies that conduct science for policy. Thus, European decision makers have no pressure to justify what are often political decisions in the artificially rational language of science or economics (Brickman et al. 1985).

Precautionary assessment can be a critical step toward making precaution an overarching guide to environmental and health decision making in the United States, overcoming some of the key barriers to its implementation in this country. The precautionary principle is conventionally understood to include two main components: action in the face of uncertainty and placing the burden of proof on proponents of potentially harmful activities. These elements lead to interpretation of the precautionary principle as reactive, based

265

on analyzing and responding to problems rather than proactively seeking solutions. Participants in the Wingspread Conference on the Precautionary Principle added two elements to restore the original spirit of the principle: assessment of alternatives and democratic decision-making structures (see Raffensperger and Tickner 1999). These elements refocus environmental policy on seeking prevention opportunities and increase the information base and legitimacy of decision-making processes. Precautionary assessment attempts to build all four components into administrative (and private) decision making.

The Fundamentals of Precautionary Assessment

Precautionary assessment integrates prevention and care in environmental health policy. The goal is not to replace existing decision-making structures but rather to reorient them to better support preventive, precautionary decisions in the face of uncertain, complex risks. Central to this framework are flexibility, continuous feedback and learning, and a diverse portfolio of information, constituencies, and scientific and policy tools used in the decision-making process. This "portfolio" or "heuristic" approach focuses attention on the bulk of accumulated experience and understanding (e.g., of similar activities), in addition to the details of particular hazards. This could be called a "qualitative Bayesian" approach to decision making in which knowledge is continuously updated (Malakoff 1999). It can facilitate preventive decision making on a chemical-by-chemical, activity-by-activity basis as well as by broad categories of hazards.

Precautionary assessment requires the following changes to current environmental health decision-making processes:

1. *Precautionary assessment redirects the questions asked in environmental decision making.* Instead of asking "How safe is safe" and "What level of impact can a human or ecosystem assimilate without showing any obvious adverse effects?" we must first ask such questions as "To what degree can impacts be reduced while maintaining societal values" and "What safer alternatives might achieve the desired goal?" In its simplest sense, precautionary assessment moves the initial focus of decision making from characterizing problems to characterizing solutions. This requires tools and expertise to comprehensively analyze not only risks but also the feasibility of alternative technologies and products.

2. *Precautionary assessment alters the basic assumptions of environmental and health decision making.* Rather than assume that specific substances or activities are safe until proven dangerous, precautionary assessment makes presumptions in favor of protecting the environment and public health under uncertainty. This places the responsibility for developing information, regular monitoring, demonstrating relative safety, analyzing alternatives, and pre-

venting harm on those undertaking potentially harmful activities. It also allows government to prevent harm and take action regarding potentially harmful activities. It lowers the amount and strength of evidence needed before preventive action can take place, in addition to allowing government agencies to institute "deterrent signals" to potentially harmful activities.

3. *Precautionary assessment modifies environmental decision making to permit a more careful consideration of technologies and activities.* For new activities, the framework establishes barriers, which may slow but do not stop the development process. For example, a tiered permitting process (where the activity is allowed to proceed slowly as different types of evidence are presented) might be instituted for a new activity with poorly understood impacts. For both new and existing activities, precautionary assessment involves more careful, ongoing consideration of all available evidence on impacts and detailed analysis of the least hazardous ways to achieve a specific purpose. Evidence of potential harm from various disciplines, magnitude of potential effects, uncertainty, and availability of alternatives and preventive opportunities are considered together to determine precautionary courses of action.

4. *Precautionary assessment expands the range of participants in risk decisions.* Environmental decisions tend to be primarily policy decisions, because of high scientific uncertainty. They are also public decisions, affecting human health or public resources. The framework more effectively incorporates those potentially affected by substances and activities in the decision-making process. This requires transparent decision-making processes and structures for increasing citizen control in all phases of science and technology decisions.

5. *The framework reconfigures the science used for public policy.* Precaution needs to be embedded in all phases of science, including the research agenda. Incorporating the precautionary principle in environmental science requires an a priori commitment to taking care and providing information to inform preventive policy. This leads to changes not only to the questions asked but also to the methods of science. These changes are discussed throughout this volume.

Precautionary assessment expands on the tools central to current environmental decision-making structures, such as risk assessment and cost-benefit analysis, but does not use them as the sole basis for decisions. Instead of using them to quantify "acceptable" risks, decision makers can use them to characterize hazards and risks, compare alternatives to an activity (or to establish priorities), and better understand trade-offs inherent in environmental decision making (see Tickner 2000).

Precautionary assessment provides a structural approach to agency priority setting by ranking hazards based on evidence of harm, accumulated experience and understanding, and opportunities to prevent harm. By focusing on alternatives, it reorients agency attention to what can be done, rather than what cannot be done due to limited resources. This can result in efforts to establish goals for prevention and "master plans" that array actions to be taken to achieve certain outcomes.

Applying Precautionary Assessment

Precautionary assessment represents a framework and set of procedural steps designed to embed precaution in both the science and policy of environmental decision making. It incorporates broad problem framing, thorough examination of alternatives, and an approach to science that expands the considerations, disciplines, and constituencies involved in the collection and weighing of scientific evidence and ultimate decision-making process.

Precautionary assessment incorporates a process flow that emphasizes flexibility. This is substantially different from the more rigid, formulaic four-step approach to risk-assessment and management set forward by the U.S. National Research Council (National Research Council, 1993), yet is consistent with approaches to sound decision making proposed in many business texts (Hammond et al. 1999). There are four reasons an iterative process flow is more useful than the prescriptive rules currently used in environmental decision making:

• Each decision is different—with different types of evidence, uncertainty, affected communities, and availability of alternatives.
• A more generalized approach permits a wider range of information to be used in the decision-making process and allows for more qualitative judgments in the face of uncertainty and complexity.
• A process flow does not oversimplify or narrow the decision-making process. Rather, it lays out a series of procedural steps that should be considered in all sound environmental and health decision-making processes.
• Since many environmental health decisions are made in the face of great uncertainty, yes/no or quantitative determinations are often not warranted by the available data and thus a broader range of options and considerations must be included.

A complete precautionary assessment would include all six of the procedural steps outlined in figure 17.1. It is not necessary for each component to be completed in order, and components may overlap or be repeated at several stages. However, it is clearly important to begin with a holistic definition of the problem as that will affect each of the following steps. Though it may appear

I. Problem Scoping
- Broadly frame and define problem - Outline the range and types of plausible impacts, including potential disproportionate impacts - Identify research and information needs about health impacts and alternatives

II. Participant Analysis
- Determine who should be involved, at what points during the decision-making process and what roles are. - Determine if participatory decision-making structures are triggered, the type/level of citizen participation to address the problem, and resources needed.

III. Burden/Responsibility Allocation Analysis
- Determine who has resources and access to information on the problem. - Consider burdens, duties, and responsibilities in existing laws and opportunities for discretion. - Determine appropriate burdens/duties/presumptions to apply to different actors given nature of the problem and available information.

IV. Environment and Health Impact Analysis
- Hazard analysis: weigh strength of evidence of plausible impacts. Broadly examine evidence of hazards from multiple sources and disciplines and prioritize concerns. Consider quality of studies. - Exposure analysis: Examine potential for exposure from various sources. Consider nature and intensity of exposure and who is exposed. Consider potential for cumulative and interactive exposures. - Magnitude analysis: Examine magnitude and severity of potential impacts including spatial and temporal scale, susceptible sub-populations; reversibility, connectivity - Uncertainty analysis: Examine amount and type of uncertainty as well as feasibility of reducing uncertainty and potential impacts on outcomes. - Consider weight of evidence on association, exposure, magnitude together to determine potential threat to health or environment. Develop narrative with rationale, limitations in studies, and research needs.

V. Alternatives Assessment
- Examine/understand impacts and purpose of activity. - Identify wide range of alternatives. - Conduct detailed comparative analysis of alternatives (pros/cons, economic, technical, h&s) - Select "best" alternative and institute implementation and follow-up plan.

VI. Precautionary Action Analysis
- Determine level of precaution needed based on level of threat of harm, uncertainty, and availability of alternatives - Determine appropriate actions based on level of precaution needed - Determine "precautionary feedback" regime to minimize unintended consequences and for continuous improvement

Figure 17.1. The steps of precautionary assessment.

cumbersome, the process should be thought of as a heuristic device and normative considerations to guide sound, preventive environmental decision making rather than an inflexible set of steps that must be completed in a particular way. In most cases, certain steps can be completed relatively quickly. Depending on the nature of the problem and evidence other steps can be completed in a relatively rapid fashion, without extensive quantitative analysis. Often, steps can be bypassed or the extent of analyses reduced, as when strong evidence of safer alternatives exists or there is an established presumption of potential harm. Three important steps of precautionary assessment are described below.

Environmental and Health Impact Analysis

A centerpiece of precautionary assessment is the Environment and Health Impact Analysis (EHIA), in which the science of hazards and exposures is weighed. In this step, evidence of risks and uncertainties are examined to determine the possibility (and plausibility) of a significant health threat and the need for precautionary action. Because many environmental risks are complex and highly uncertain, such an analysis must involve both the totality and individual pieces of the evidence for plausible indications of effects. The goal is to build a coherent picture of potential impacts—a "story." In precautionary assessment, this analysis is completed using a "research synthesis" (Stoto 2000) or weight of evidence approach.

Research synthesis is a formal method to summarize and integrate studies, drawing conclusions on the whole of the evidence but also uncovering variability, as well as consistencies between studies (Cooper and Hedges 1994). It can be considered a synthesis by careful thinking and analysis. While not well defined in the literature, research synthesis is commonly practiced by government agencies in developing regulatory policies. It is also commonly used by scientific bodies in examining a broad set of literature on a particular uncertain environmental hazard—such as the assessments of the Intergovernmental Panel on Climate Change (see chapter 10). Some attributes of a systematic research synthesis include:

- Development of an explicit protocol for study identification inclusion and exclusion
- Methods for both qualitative and quantitative review, examining individual study attributes such as bias, confounding, and study design
- A process for presenting and interpreting results and estimating an overall association or magnitude of effect (quantitatively or qualitatively)
- Methods for identifying research gaps and uncertainties requiring more primary research (Mosteller and Colditz 1996)

As the quality of data available in research syntheses varies widely, some scientists advocate taking the best available evidence, using professional judgment to support the evidence, and drawing the best possible conclusions (Mosteller and Colditz 1996).

In its review of the health and environmental impacts of persistent and bioaccumulative substances in the Great Lakes, the U.S.-Canada International Joint Commission (IJC) defined the weight-of-evidence approach as follows (IJC 1994):

> The approach takes into account the cumulative weight of the many studies that address the question of injury or the likelihood of injury to living organisms. If, taken together, the amount and consistency of evidence across a wide range of circumstances and/or toxic substances are judged sufficient to indicate the reality or a strong probability of a linkage between certain substances or class of substances and injury, a conclusion of a causal relationship can be made.

The IJC definition answers the question, "How and when do we know there is sufficient evidence or accumulated knowledge so that a reasonable person will conclude that policy makers should act?"

Precautionary assessment adjusts the notion of causality. It acknowledges that the determination of a causal relationship is a qualitative, case-by-case, and subjective judgment based on an analysis of the weight of all the available evidence taken as a whole. Such judgments can be guided by qualitative criteria for how the evidence should be weighed in evaluating the hazard.

In 1965 statistician Sir Bradford Hill proposed a set of considerations that scientists and policy makers could use in determining whether an observed association is likely to be causal in nature (Hill 1965). Now widely used (and frequently misinterpreted) by scientists and decision makers, these include strength of association, consistency across studies, specificity of effects, dose-response, temporality of effects, plausibility of effects, coherence with other knowledge, evidence from experiments, and analogy based on experience. Hill advised broad interpretation of the evidence with respect to these considerations to ensure that associations were not discounted simply because there was insufficient evidence or understanding about a hazard at a particular point in time. He concluded: "What I do not believe—and this has been suggested—is that we can usefully lay down some hard-and-fast rules of evidence that must be obeyed before we accept cause and effect" (Hill 1965: 299). According to Hill, causal judgments must not require perfect information and causality must be considered in the context of available knowledge and a responsibility to prevent impacts to health. In other words, the determination of whether the evidence is strong enough to act is not only a function of the strength of evidence of a causal relationship but also prior experience, uncertainty, and the need to prevent harm.

In arriving at a determination of the "significance of the threat" to health or the environment, Environment and Health Impact Analysis incorporates consideration of the strength of the evidence of a causal association (criteria such as biologic plausibility, analogy). These criteria can guide the collection and analysis of information, as well as the questions asked by scientists and decision makers. However, under EHIA, these criteria are defined broadly and cautiously. For example, recent evidence has demonstrated that for developmental toxicity, timing as well as dose are critical considerations (see Colborn et al. 1996). Biologic plausibility can be defined in terms of an ability to cause effects in one or more organ systems in humans (and not only a specific one or through a particular mechanism). In some cases, it is appropriate to make presumptions as to whether particular causal considerations have been satisfied.

The EHIA is a categorical approach.[2] Categorical presentations of evidence are supplemented by concise yet detailed narratives describing the nature of evidence upon which the categories were determined (e.g., limitations in studies). Experience on the U.S. Institute of Medicine's Committee on Agent Orange—where a research synthesis approach was used to determine whether an association existed between human exposure to herbicides used during the Vietnam War (including dioxin contamination) and adverse effects in humans—showed that when evidence is limited and uncertain, information is most appropriately presented in terms of categories of evidence rather than continuous, quantitative estimates of risk. Nonetheless, quantitative estimates may be integrated into the categories (Institute of Medicine 1993).

The categorical, or graded classification, approach has two important benefits over traditional "continuous" risk variable approaches for analyzing and presenting uncertain information. First, it provides greater accountability by providing clarity about the nature of the available evidence and choices in the analysis, instead of hiding behind a single number based on multiple assumptions that may be hidden. Numerical determinations often "crumple" information into a single value, losing track of nuances and qualitative details about that information. Further, by definition, the concept of "risk" requires that probabilities of occurrence are fairly well understood, whereas in most environmental health decision making available information and uncertainties do not allow for such precision.

Second, it opens up greater opportunities for prevention and intervention. Unpacking information on hazard, exposure, magnitude, and uncertainty provides greater flexibility, understanding of the nature of potential impacts, and opportunities for preventive interventions in decision making.

The EHIA should include consideration of the wide range of sources of information and plausible harms and impacts identified during problem scoping. Who should conduct this analysis will depend on the nature of the problem (e.g., small localized decisions versus decisions on whole classes of substances). Evidence of potential impacts and uncertainties should be gathered from as diverse an array of disciplines and constituencies as possible, including observational studies, worker case histories and case reports, toxicological studies, wildlife and domestic animal studies, cellular studies, ecological assessments, epidemiologic studies, community health studies, structure activity analyses, modeling, and monitoring. Impacts examined in the analysis should include human and ecosystem health impacts, acute and chronic effects, interactive and cumulative effects, direct and indirect impacts, and socioeconomic, historical, and aesthetic impacts. Since the list of plausible impacts might be very large, it is useful to prioritize by impacts of greatest concern from a scientific and political point of view.

While the studies and other information should be evaluated for their quality—strength of the methods, questions asked, source, bias, confounding, and peer review—anecdotal information, including single case reports or case series, and "lay" collected data should also be considered. If there are conflicts in the results and conclusions of individual studies, the strength of the studies becomes an important consideration in weighing evidence.

The four steps of EHIA include:

- *Hazard analysis:* The purpose of this step is to understand the strength and quality of the evidence that there is or could be a detrimental effect. Studies and potential impacts are examined individually and as a whole. When possible, meta-analyses can be performed to provide more detailed information. Inherent properties in the activity or substance that could lead to adverse impacts are considered.
- *Exposure analysis:* Evidence of actual or potential exposure is gathered from various sources. The nature (direct, dispersive, controlled, closed-system) and intensity of exposure are analyzed as well as when and to whom exposure occurs, including the potential for cumulative and interactive exposures.
- *Magnitude analysis:* Evidence on the seriousness of potential impacts is examined, including spatial and temporal scales of effects, potential catastrophic impacts, susceptible subpopulations, reversibility of adverse effects, and degree of connectivity of effects. When the potential magnitude of effects is large, weaker evidence provides a cause for concern.
- *Uncertainty analysis:* This step includes both a qualitative and quantitative assessment of gaps in knowledge. Uncertainty should be analyzed broadly in terms of type (parameter, model, systemic, ignorance), sensitivity to changing assumptions, and feasibility of reducing uncertainty.

The results of these subanalyses are combined into a final EHIA. Here, the weight of the evidence of potential or actual harm for a particular hazard or group of hazards is presented as one of five categories (based on analyses of hazard, exposure, and magnitude) as well as a concise, detailed narrative outlining the rationale for the categories, the evidence on which the determination was based, and other quantitative and qualitative considerations.

The narrative should be clear about what is known, not known, and can be known about the threat (and suspected); limitations of scientific studies to understand the threat; and gaps in information, including research needs. The narrative should also indicate the extent to which uncertainty, and particularly ignorance, can be reduced through additional research. Quantitative evidence such as uncertainty analyses and quantitative assessments of risk should be included in this narrative and final categorical determination. The plausibility and probability of various outcomes should also be considered (i.e., the sensitivity of the results). Finally, the strength of the evidence and categorical recommendation should be outlined. *The analysis provides a determination, based on the weight of evidence, as to whether an activity is associated with or may cause harm, and the potential severity of that harm.*

Alternatives Assessment

The other centerpiece of precautionary assessment is a thorough evaluation of alternatives to prevent or minimize harm. Alternatives assessment is the heart of the solutions-oriented approach of the precautionary principle and central to sound, forward-looking environmental decision making. This focuses decision-making attention on opportunities rather than simply the hazards associated with a narrow range of options (O'Brien 2000). Reasonableness of a risk must be a function not only of hazard and exposure but also of uncertainty, magnitude of potential impacts, and the availability of alternatives or preventive options. Availability of a safer alternative can obviate the need for a costly, contentious, and potentially misleading quantitative risk assessment.

The goal of alternatives assessment is to identify and examine opportunities to prevent environmental and health impacts from an activity. A secondary goal is to drive innovation toward more environmentally friendly and sustainable technologies, products, and practices. Thus, alternatives assessment should consider not only existing, easy, and feasible options, but also those that can be developed—that are "on the horizon." Critics of the precautionary principle have argued that it paralyzes innovation. However, the use of alternatives assessment in a precautionary context can embrace and encourage development of innovative, cleaner technologies—thus redirecting innovation. The most effective alternatives assessments start with a broad problem definition and address multiple risks at once (e.g., multiple chemicals, media, or facilities). Alternatives should be considered in terms of broadly defined substitution (not just chemical for chemical), modifications to an activity that would prevent impacts (prevention opportunities), as well as stopping an activity or preventing its initiation. Alternatives assessments often have the most impact when undertaken early in a decision-making process—in the development phase.

Nonetheless, alternatives assessment requires tools to comprehensively analyze not only risks but also feasibility of alternative technologies and products.

A variety of methodologies exist—including trade-offs analysis and health and environmental impact analysis—to evaluate technology and policy alternatives and to identify potential unintended consequences (see Tickner 2000). The steps of an alternatives assessment should include:

- *Examination and understanding of the impacts and purpose of the activity.* The purpose of this step is to better understand the "service" that the activity provides (and whether that service can be provided in a less damaging way), how hazardous materials are used (materials accounting), and potential impacts and benefits of the activity.
- *Identification of a wide range of options.* Here, a diverse group of stakeholders should brainstorm a wide range of options that could lead to multiple risk reduction opportunities.
- *Comparative analysis of alternatives.* The goal of comparative options analysis is to thoroughly examine and compare technical feasibility and economic, environmental, and health and safety impacts and benefits from the existing or proposed activity and identified alternatives.
- *Alternatives selection.* This step should include a narrative of the identified options, results of the analysis, and criteria on which the alternative was chosen. The alternatives plan should contain an analysis of the selected alternative, how it will be implemented (including how barriers will be addressed), and a plan for follow-up, continuous improvement, and monitoring for potential adverse impacts. It may be possible to institute interim alternatives while long-term alternatives with greater environmental and health benefits are being developed.

Precautionary Action Analysis

The last part of precautionary assessment is to determine the appropriate courses of action. This could be considered the "risk management" phase of the decision process, yet it is fully integrated into all of the previous steps. Precautionary action analysis involves weighing the information gathered earlier to determine how much and what type of precaution should be taken. Policy tools for implementing precautionary action and preventing harm, ranging from further study to banning the activity, are chosen on the basis of the entire evidentiary base. Finally, a feedback and monitoring scheme is developed to measure benefits and provide early warning of potential problems. The determination of actions is not based on a specific threshold for action but rather considers all of the available evidence in determining the most health-protective, yet reasonable, course of action. Precautionary assessment may also result in a decision that an activity is unlikely to cause harm or that its impacts would be minimal—in which case institution of a monitoring scheme may be the most appropriate action step.

Based on the results of previous analyses, the level of precautionary action required may range from weak to strict—a function of the serious of potential harm, the amount of uncertainty about impacts, and the availability of alternatives (see Tickner 2000). As uncertainty, seriousness of impacts, and availability of alternatives increase, so does the level of precaution that should be applied. For example, a determination of strict precaution would be made for an activity for which there is high uncertainty, significant threat of harm, and widely available alternatives. A weak precaution determination might be made in the case of an activity that had minimal evidence of potential harm yet low potential impacts, moderate uncertainty, and limited availability of alternatives. In the case of a threat with high uncertainty yet some evidence of potential significant impacts (irreversible or widespread harm or harm to sensitive members of a population), it would be prudent to take strong precautionary action before greater evidence accumulates, particularly if alternatives are available.

The selection of actions will depend on the level of precaution established in the previous step and the nature of the particular decision, and may depend heavily on the legislation or regulation under which a particular activity is being addressed. The most preventive and flexible decisions will apply an array of actions. For example, pollution prevention planning requirements could be combined with public right to know. Or, a permitting process could be stepwise, contingent on testing and providing evidence of relative harmlessness at each step. The final choice of actions should integrate all the considerations—technical, political, social, ethical, and economic—that form part of sound environmental decision-making practices. It should also consider the ability to monitor for early warnings, adaptability, and ease of compliance/enforcement.

Undertaking precautionary actions need not only include bans. Some possible actions to apply precaution in chemicals policy include requirements for pollution prevention and cleaner production; establishment of long-term goals and deterrent signals, including interim targets and objectives; phase-outs of particularly dangerous activities; health-based exposure limits; extended producer responsibility requirements; assurance bonding; premarket review and study requirements; environmental and health impact statement requirements; ecological taxes; right to know/disclosure requirements; monitoring and surveillance requirements; liability for damages (polluter pays principle); and adaptive management (see Tickner 2000).

Decisions made under a precautionary assessment should not be considered permanent, but part of a continuous process of increasing understanding and reducing overall impacts. Once precautionary actions have been chosen, follow-up and monitoring schemes for the activity should be developed. This type of feedback is critical to understanding the impacts of precautionary actions, as well as to provide early warnings of harm, thus helping to avoid unintended consequences. It also stimulates continuous improvement in environmental performance and technological innovation. Follow-up tools include periodic assessment, audit, or prevention planning requirements; regular reporting of environmental impacts (e.g., toxics use reporting); short- and long-term health and exposure monitoring; toxicological testing; and impact statements any time a major change is made to a product, process, or activity.

Conclusion

This chapter has outlined precautionary assessment, a framework for implementing precautionary principle in environmental health decision making. It is not meant to provide rigid rules for invoking the principle but rather a structure for stimulating consistent and thoughtful application of precautionary thinking. In its most basic sense, precautionary assessment establishes a new role for scientific research and public policy that expands the constituencies, disciplines, and considerations involved in decision making under conditions of uncertainty and complexity. Such a process—one that that examines the whole of evidence from various sources, examines a full range of alternatives, and injects sensible judgment and values—produces sound decisions.

Only to the degree that effective procedures are built and readily used by government agencies may "cultural precaution"—a value underlying all decision making affecting public health and environment—be achieved. Achieving a culture of precaution, however, will require more than procedures. It will require capable institutions that comprehensively, consistently, and fairly apply precautionary procedures. It will also require mandates, regulations, and guidelines so that the framework and procedures are broadly applied throughout environmental decision-making processes, addressing different media and different types of impacts. It will require modifications to the discretion courts provide in reviewing agency decisions. And, ultimately, it will require important changes in the conduct of environmental science and policy, many of which are detailed throughout this book.

Endnotes

1. These differences include the limited flexibility in the U.S. regulatory system to address broad or novel risks, with its prescriptive, media-specific rules, supplemented by numerous procedural requirements from statutes and executive branch orders; and the increasingly rigorous judicial review of agency decisions (including their scientific basis), which results in the need for agencies to defend decisions quantitatively using a "rational," "expert- and science-based" approach (see Tickner 2000).

2. See Tickner 2000 for further discussion. These categories are much like those
used by the International Agency for Research on Cancer. Each of the four
subanalyses of EHIA have a set of categorical representations. For example,
those used in hazard analysis include sufficient evidence of a hazard,
limited/suggestive evidence of a hazard, inadequate/insufficient evidence to
determine whether a hazard exists, and limited/suggestive evidence of no
hazard.

References Cited

Brickman, R. S. Jasanoff, and T. Ilgen. 1985. *Controlling Chemicals: The Politics of Regulation in Europe and the United States.* Ithaca: Cornell University Press.

Colborn, T., J. P. Myers, and D. Dumonoski. 1996. *Our Stolen Future.* New York: Dutton Books.

Cooper, H., and L. Hedges. 1994. "Research Synthesis as a Scientific Exercise." In *The Handbook of Research Synthesis,* edited by H. Cooper and L. Hedges, 3–14. New York: Russell Sage Foundation.

Hammond, J., R. Keeney, and H. Raiffa. 1999. *Smart Choices: A Practical Guide to Making Better Decisions.* Cambridge, Mass.: Harvard Business School Publishing.

Hill, Austin Bradford. 1965. "The Environment and Disease: Association or Causation." *Proceedings of the Royal Society of Medicine* 58: 295–300.

Institute of Medicine. 1993. *Veterans and Agent Orange: Health Effects of Herbicides Used in Vietnam.* Washington, D.C.: National Academy Press.

Malakoff, D. 1999. "Bayes Offers a New Way to Make Sense of Numbers." *Science* 286: 1460–64.

Mosteller, F., and G. Colditz. 1996. "Understanding Research Synthesis: Meta-Analysis." *Annual Review of Public Health* 17: 1–23.

National Research Council. 1983. *Risk Assessment in the Federal Government: Managing the Process.* Washington, D.C.: National Academy Press.

O'Brien, M. 2000. *Making Better Environmental Decisions: An Alternative to Risk Assessment.* Cambridge, Mass.: MIT Press.

Raffensperger, C., and J. Tickner. 1999. *Protecting Public Health and the Environment: Implementing the Precautionary Principle.* Washington, D.C.: Island Press.

Stoto, M. 2000. "Research Synthesis for Public Health Policy: Experience of the Institute of Medicine." In *Meta-Analysis in Medicine and Health Policy,* edited by D. Stangl and D. Berry, 321–57. New York: Marcel Dekker.

Tickner, J. 2000. "Precaution in Practice: A Framework for Implementing the Precautionary Principle." Doctoral dissertation, Lowell, Mass.: University of Massachusetts, Lowell.

U.S.-Canada International Joint Commission (IJC). 1994. *Applying the Weight of Evidence: Issues and Practice.* Windsor, Ont.: International Joint Commission.

CHAPTER 18

Science in the Service of Good: The Precautionary Principle and Positive Goals

Mary O'Brien

The precautionary principle is often discussed as a means to avoid harm. For example, the Second North Sea Declaration invokes the precautionary principle as a guide "to protect the North Sea from possibly damaging effects of the most dangerous substances" (Ministerial Declaration 1987). A climate change conference invokes it "to anticipate, prevent, or minimize the causes of climate change and mitigate its adverse effects" (Framework Convention 1992). In this use of the precautionary principle, the fundamental role of science is to provide "credible evidence" that particular proposed or ongoing human activities cross some *threshold* of harm, triggering precautionary action.

I suggest in this chapter that it can be more useful to treat the precautionary principle as *a means to achieve positive public and environmental health goals.* When the precautionary principle is used primarily to achieve positive goals, for example, "sustainable development" (Bergen Declaration on Sustainable Development 1990), then the role of science expands to provide evidence not only about the potential harms, but also the potential benefits of various activities that might be used to achieve those goals. Precaution in this context is not simply a process that is or is not triggered. Instead, it exists as an omnipresent *decision-making screen* through which goals, potential alternative courses of action, and options for midcourse corrections are examined.

The difference between two English translations of the German word *Vorsorgeprinzip*, "precautionary principle" and "forecaring principle" (Pollan

2001), points toward these two different ways of using the precautionary principle, the first as a triggered brake, the second as an ongoing screen.

Avoiding Harm versus Seeking Positive Goals: Asking Different Questions

Consider a fishery that has been depleted. We can ask a harm-driven question, such as, "Is this particular harvest rate or fishing technology likely to drive a particular fish species into unrecoverable populations?" To answer the question, scientists will be asked to investigate the effects of a particular harvest rate or technology. If the effects appear to be capable of threatening the existence of a particular fish species, precaution will be considered.

Alternatively, we can ask a goal-driven question, such as, "What alternatives do we have for insuring a robust, biologically diverse fishery from which humans can draw sustenance?" To answer this question, scientists may be asked to assemble information on such matters as the biological diversity that is possible, the results of various restoration steps that have been taken elsewhere, and/or the documented or likely recovery or maintenance of biological diversity using various harvesting methods and rates. Precaution will play a role in deciding which alternatives to promote, because attaining positive environmental and public health goals may be delayed or impossible if certain unnecessary harmful activities are undertaken.

The current, dominant paradigm for implementation of the precautionary principle is one of a harm-driven process (figure 18.1). However, a positive, goal-driven process for implementation of the precautionary principle (figure 18.2) is more likely to change fundamentally the terms of debate about harmful activities and to lead to consideration of ambitious, innovative, public interest-based beneficial activities. It is useful to examine these two processes separately, because goals, harm, science, precaution, alternatives, and public involvement play different roles in each.

Harm-Based Process

The *goal* of most harm-driven precautionary principle processes (figure 18.1) is to protect public health and/or the environment from significant or irreversible harm posed by some human activity or activities. This is similar to the ostensible goal of most risk assessments. Technically, a risk assessment is a process of estimating damages that may be occurring or that may occur if an activity is undertaken. A "risk assessment" can be formal or implicit; it can be quantitative and/or qualitative. But risk assessments use data, estimates, formulas, and/or models to provide an estimated answer to the question, "How

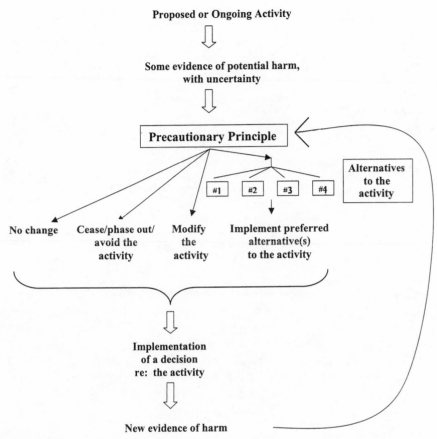

Figure 18.1. Harm-driven process.

much of this harmful activity or substance will not cause significant or unacceptable harm?" (O'Brien 2000).

Similarly, a harm-driven precautionary principle process begins with a proposed or ongoing activity that appears to pose harm to the environment or public health. The basic question is, "Is this activity harmful enough to be banned, altered, or replaced?"

This is thus fundamentally a risk assessment question of whether a particular threshold of harm has been or might be crossed by a particular harmful activity. This assessment of risk may not be quantified, it may not be formally called a risk assessment, but it has the essential features of risk assessment, namely, beginning with a hazardous or potentially hazardous activity, irrespective of its necessity, and proceeding to the narrow question of whether too

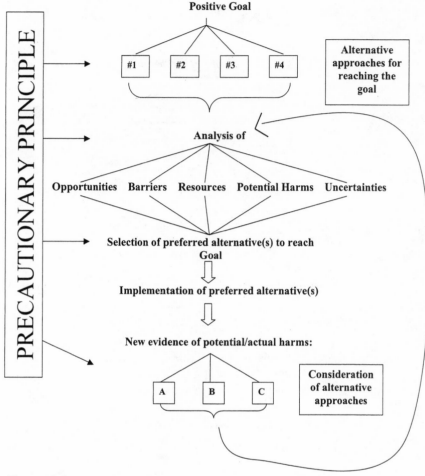

Figure 18.2. Goal-driven process.

much harm is sufficiently likely that some action should be taken, in this case invocation of the precautionary principle.

In Norway, for instance, an expert commission found that existing evidence on human health risks of genetically engineered food is not strong enough to trigger use of the precautionary principle, though a number of studies suggest that a new gene in food may be transferable to mammalian cells (see chapter 4).

Defenders of the potentially harmful activity almost inevitably claim that a threshold of significant harm or its likelihood has *not* been crossed, and precautionary action is thus not necessary. The proponents of the activity press for incorporating formal risk assessment into the process, that is, quantifying how much harm is how likely, how much harm is proven, how much harm is acceptable.

Role of Science

The science involved in harm-driven decision-making processes is primarily the science of biological or ecological harm: the potential of a given activity to cause death, extinction, developmental damage, or depletion of native species, for example. The science addresses a fairly narrow question, such as the effect of a particular amount of a particular industrial chemical on brain development of a human fetus. An ecological question may be equally narrow, for instance, the ability of a salmon species to survive passage through hydroelectric dam turbines or the contribution of a particular activity to global warming.

Scientific problems attendant to most risk assessments (e.g., lack of consideration of cumulative effects, cultural effects, or data gaps) (O'Brien 2001) are likewise frequently present in answering whether enough "credible evidence" of harm is present to trigger precaution with respect to a particular activity.

It is at this point, at the triggering threshold, that proponents of the activity in question try to wrench the harm-driven decision-making process back into conventional risk analysis and "sound science." This phrase, "sound science," ostensibly refers to objective, quantitative data. It is generally a euphemism for "enough smoking guns or dead bodies to provide proof that harm has happened and a call for some change in activities is unavoidable." In other words, the proponents of a harmful or potentially harmful activity try to reduce the precautionary principle to a mere nudge of risk analyses (which blur the distinction between quantification and reality), even though these same proponents characterize risk analyses as conservative and health protective, even prior to use of the precautionary principle (Graham 1999).

Invoking Precaution

In the harm-driven decision-making process, precaution may be invoked *after* scientific and other evidence has been gathered regarding an ongoing or proposed harmful activity. If scientific evidence suggestive of harm is not valued as credible, or if credible scientific information does not suggest harm of sufficient magnitude or probability to satisfy the decision makers, then precautionary action will be pronounced unnecessary.

Inevitably, a harm-driven approach, whether risk-based or precautionary principle–based, will sometimes trigger restrictive actions. The question arises whether the restrictive action is going forward in the absence of fully established cause-and-effect relationships (i.e., use of the precautionary principle), or is going forward only after harm and causation *have* been conclusively established, and the action is simply being taken in the *name* of the precautionary principle.

A recent "interim" report by a National Research Council committee concluded that the U.S. Fish and Wildlife Service did not have a sound scientific basis for recommending that more water should be allowed to remain in southern Oregon's Klamath Lake (rather than be taken for agricultural irrigation) than had been retained during 1990–2000, in order to provide for the continued existence of two endangered sucker species (Committee on Endangered and Threatened Fishes in the Klamath River Basin 2002). The committee based its finding primarily on the fact that three years of especially low water level during the past ten years did not result in fish kills from excessive algal growth and consequent poor water quality.

The Klamath Tribe, in objecting to this and other findings in the report, noted that the committee wrongfully expected a simple connection between low water level and fish kills when in fact climatic factors (specifically, unusually windy conditions, and a cooler spring one year) had intervened to partially mitigate for what could have been a catastrophic combination of low lake levels, lack of wind, and favorable climatic conditions for abundant algae (Kann and Dunsmoor 2002). In other words, in supporting the continuation of 1990–2000 irrigation drawdowns of Klamath Lake, the committee deferred to a lack of fish kills (i.e., lack of demonstrated, direct cause and effect) rather than scientific studies showing what could be expected to go wrong if low water in Klamath Lake coincided with climatic conditions that periodically occur and that are favorable to algal growth and still water.

Examining Alternatives

In the harm-driven process, alternatives are generally considered only after the precautionary principle has been invoked. Indeed, they may not be considered at all, if there is a decision to allow the activity to continue unchanged or with minor restrictions or modifications (as in most permitting processes for toxics).

When alternatives *are* considered, the tendency (often backed by powerful corporate or government pressure) is to consider slot-in alternatives: activities that approach as closely as possible the same specifications of the harmful activity that prompted the process. For instance, major reductions in commercial old-growth logging of Pacific Northwest public lands during the 1980s were forced by overwhelming public opinion and threats to rare and endangered animal species. Under the resulting 1994 Northwest Forest Plan, however, old-growth logging is allowed to continue in the spaces ("matrix") between old-growth "reserves," and the approximately 500 plant and animal species expected to suffer under this continued loss of old-growth were to be "surveyed and managed."

In a U.S. Forest Service contract paper on the then-upcoming Northwest Forest Plan, forest analysts noted the plan "does not escape the historic dependence on late-successional forest and old-growth as the source of harvest volume. How publicly acceptable this policy will be remains to be seen" (Johnson et al. 1993).

When the Forest Service failed to follow through on their "survey and manage" rules, a 1999 federal court decision required the Forest Service to survey their 1998 and 1999 old-growth sales (ONRC v. USFS). In January 2001, the Bush Administration dropped 72 species from the survey and management rules (USDA and USDI 2001); 90 more species will be dropped soon (Doug Heiken, ONRC, personal communication, 30 March 2002).

As in this example, the harm-driven process tends to ignore the question of whether the activity in question inherently runs counter to the natural limits of ecosystems or humans or whether far more important community and social goals might be accomplished by rethinking goals and alternatives.

A discussion of fundamental goals and social-change options is generally lacking in harm-driven processes. Further, the discussion of alternatives does not generally consider integrated systems alternatives or simultaneous or linked modifications of activities in multiple sectors of society.

Public Involvement

The initial challenge to a harmful activity generally arises from citizens or workers responding to such signals as a neighborhood cluster of children with brain cancer, mudslides on a clearcut forest slope, or a crop dying following applications of hazardous waste–laced fertilizer. However, once the issue has been formulated as the narrow question, "Is there enough credible evidence that a significant harm is significantly likely?" most of the public is excluded, with only those who are scientifically armed, unusually politically active, and organized able to weigh into the fray.

Later, the public has to weigh in again, first to transform even overwhelming evidence of significant harm into precautionary action and then influence what precautionary action(s) are taken. Generally, regulated industries apply intense pressure on public regulators to minimize the change that will qualify as precautionary, proposing mere tinkering that can be likened to "rearranging the deck chairs on the Titanic." State and federal agencies will propose better fish hatcheries to supplement dwindling native fish runs rather than arranging for restoration of degraded river habitat; industry and government will agree on better emissions controls on a polyvinyl chloride factory rather than facilitate the phase-out of industrial uses of chlorine; the U.S. Department of Agriculture will require that chicken egg-laying cages be slightly larger rather than require free-range feeding.

If alternatives involving systemic or significant change are considered at all in this process, they are usually proposed by the public. However, the burden then falls on the public and public interest scientists to prove that tinkering with the existing activity will continue to pose enough harm to warrant rejection or, in the case of revised technologies, will not adequately reduce harms. Making this case requires gathering "credible evidence," such as documentation of "dead body counts" that directly threaten a species, habitat, or population (see, e.g., Ludwig et al. 1993). Unfortunately, given wide ranges of natural variability in plant, human, and other animal populations and likely multiple causation, such direct, unequivocal evidence can be essentially impossible to produce.

Ultimately, then, it is upon the public that most key burdens of a harm-driven decision-making process fall: initially challenging and publicizing a harmful activity, subsequently arguing that the scientific evidence of harm is both credible and sufficient, and finally advocating for adequately precautionary changes. After precautionary measures have been adopted, the public will generally have to defend the measures against being weakened.

Goal-Based Process

A goal-driven precautionary principle process (figure 18.2) begins with a positive environmental health goal, such as net gain of topsoil, asthma-free children, or zero annual highway mortality. Basic questions addressed in this process include, " Which complex of activities is both feasible and likely to achieve our goal?" and "Are the activities we are undertaking moving us toward our goal?" These are largely benefits assessment questions, rather than risk assessment questions. They are questions about whether particular activities or complexes of activities are actually maintaining or enhancing environmental health.

It is interesting to consider, for instance, the Swedish Government's stated objective of "hand[ing] over to the next generation [i.e., in twenty-five years] a society in which the main environmental problems have been solved" (Sweden Ministry of Environment 1997/98). In its Environmental Policy for a Sustainable Sweden (Sweden Ministry of Environment 1997/98), fifteen environmental quality objectives have been established (including no eutrophication or unnatural acidification, a magnificent mountain landscape, and a protective ozone layer) to accomplish the overall environmental objective. The fifteen Swedish environmental goals were the outcome of extensive public discussion and debate (Per Rosander, Kemi and Miljo AB, Sweden, personal communication, 22 September 2001).

Each goal has multiple targets that, if met, will supposedly result in the larger goal being met. The "no eutrophication" goal, for instance, includes both qualitative and quantitative targets (Sweden Ministry of Environment 1997/98):

• Releases of nutrients to coastal waters, lakes, and watercourses and to groundwater should in the long term not exceed levels that cause an adverse impact on health, biological diversity, or the possibility of versatile use.
• Discharges of nitrogen from Sweden to the Baltic Sea south of the Sea of Åland must be reduced by 40 percent compared with the baseline year 1995.
• Protective areas should be established for water catchments and their most important areas of influence.

By stating positive, twenty-five-year goals for environmental quality (which is inherently linked with global environmental quality), Sweden has been almost inevitably led to discuss the global context of Sweden's use of resources:

> Calculations indicate that the *use of resources* in our part of the world needs to be reduced radically if the earth's ecosystems are to be able to support a growing world population while allowing developing countries to improve their living standards. A measure of the reductions that need to be made is *factor 10,* a concept that has been launched within the UN and signifies that resource utilization in the industrialized countries must be reduced within a generation by a factor of 10 (Sweden Ministry of Environment 1997/98, 2; emphases in original).

The U.S. Endangered Species Act (1982) includes positive goals: the restoration of endangered and threatened plant and animal species. Section 7(a)(1) mandates that public agencies use all possible means for recovering animal and plant species that have been federally listed as threatened or endangered, so that they no longer need to be listed. Section 7(a)(2) of the Endangered Species Act, on the other hand, allows agencies to issue permits for activities that will result in "takings," that is, the direct killing or indirect impairment of essential activities of individuals or populations of the endangered or threatened species. Such "takings" are to be allowed only if the agency estimates they will not jeopardize the survival of the species. It is of interest to note that court cases and regulations for implementing Section 7(a)(2), that is, continued diminution of the species, have been fully developed and that extinction has taken more species off the endangered or threatened list than recovery (Hitt 2002). On the other hand, the two agencies that administer the act, the U.S. Fish and Wildlife Service and the National Marine Fisheries Service, have never developed regulations for restoring species. When the agencies address restoration at all, it is generally through recommendations

(Hitt 2002). The development of implementing regulations for Section 7(a)(1) will most likely occur as a result of citizen-initiated, public interest litigation.

In the Klamath Basin case described earlier, the National Research Council committee focused only on whether low lake levels would jeopardize the existence of the sucker fish. The committee wrote, "The work of the NRC committee is tightly focused . . . by the inherent requirement of the Endangered Species Act, which prohibits federal actions that jeopardize continued existence of listed species through interference with their survival or recovery" (Committee on Endangered and Threatened Fishes in the Klamath River Basin 2002). The question of whether increased lake levels would positively restore these endangered fish was not on the table.

Examining Alternatives

A goal-driven process, then, is a debate about how to accomplish a worthy environmental health goal that has been adopted via some social process. In this debate, development of alternatives is the *first* step after a positive goal has been articulated. The most difficult, and yet most essential, part of this process, as with any public social process, is ensuring that all reasonable alternatives reach the table for consideration. This is resisted socially for some of the same reasons we resist it in our personal lives. Individually and socially, we are often entrenched in narrow patterns of behavior compared with the full range of behaviors we could undertake to achieve positive goals we have set for ourselves.

At the societal level, however, corporate power and market position are powerful added deterrents to consideration of a full range of reasonable alternatives. Those who have a stake in business-as-usual are likely to resist the surfacing (let alone selection) of alternatives that are innovative, decentralized, amenable to public influence or control, collaborative, or long-term as well as alternatives that diminish the power of or short-term financial returns to those benefiting from current practices.

The likelihood that a meaningful range of alternatives will be considered is increased dramatically if the public is encouraged to contribute to the range of options to be considered for achieving goals. While Sweden's fifteen environmental quality goals were set following extensive public debate, specific targets (i.e., means) for reaching the goals were set largely by government agency staff (Per Rosander, personal communication, 22 September 2001). Indeed, as the Swedish legislation itself directed, "It will be the task of the Government to fix targets that will facilitate achievement of the adopted goals" (Sweden Ministry of Environment 1997/98).

Major problems with this narrow-input approach include a reduced diversity of paths considered or employed for reaching the goal and reduced social involvement in or "ownership" of the selected paths. The means considered for reaching each goal are no broader or more innovative than the information, skills, and political courage possessed by the particular government staff assigned to the task of laying out a path achieve a particular goal.

The positive U.S. Clean Water Act national goals of eliminating the discharge of pollutants into "navigable waters" by 1985 and prohibiting discharge of "toxic pollutants in toxic amounts" (Federal Water Pollution Control Act 1972) have not been reached by 2002, because the basic paradigm of discharging toxic pollutants into water has continued. Amounts deemed not "toxic" are estimated through risk assessments. States and federal agencies work together to establish "standards" for continuing discharges of toxics into rivers, aquifers, lakes, and other waters.

However, the most basic commitment that is needed to restore clean water to any nation is quite different than establishment of "standards" for continued use of water bodies as open sewers and waste repositories. What is needed is to set a positive goal of zero discharge, with timelines, and then consider all technical, social, economic, cultural, and incentive options for reaching as close to that goal as possible. As law professor Oliver Houck (1991) has written:

> We will never, of course, by these means or any other, see zero discharge. The exceptions to even the most stringent prohibitions have ways of bending the rule, and nonpoint and other contributions of toxics to the nation's waters are, at present, less manageable. But in this life we manage what we can. We can abate point source toxic discharges with certainty, with fairness to industries wherever located, with adequate lead time and with incentives for those who have the will to arrive ahead of schedule. Or we can regulate toxic pollution . . . forever.

Consideration of alternatives encourages more appropriate relationships to uncertainty than the harm-driven approach. Because a harm-driven decision process has to judge whether a threshold of harm has been crossed, it creates incentive for proponents of the activity or activities in question to manipulate scientific information to concoct a quantitative point estimate of the activity's "safety" or "insignificant harm" that justifies the activity or fights off a call for precautionary action. Uncertainties need to "disappear" in order to arrive at this point estimate. This incentive to sweep uncertainties under the rug via manipulation of numbers is reduced in a comparative analysis of alternatives for reaching a positive goal. Such an analysis focuses on the performance potential of each alternative, that is, the likelihood that it will contribute to or at least not retard achievement of the goal.

Of course, analysis of alternatives being considered in a goal-driven process does require assessment of their potential harms as well as their potential to bring benefits, but the harms to be considered are not solely biological, for example, death, disease, or extinction. Both the harms and benefits of each alternative may be social, cultural, political (e.g., effects on democratic processes), economic, aesthetic, distributional (i.e., who bears the harms, who partakes of the benefits), or of some other category. This reduces incentive to hide uncertainties surrounding any one type of "harm" or "benefit."

If the harms of all alternatives are candidly presented, then it is likely that *every one* of the alternatives presents some threats of harm. These potential harms can be compared for their relative severity, irreversibility, and uncertainties. The harms of one alternative can also be compared with those of other alternatives for their distributional dimension, that is, who will be harmed in what ways. The incentive to come up with a "no-harm" prediction, then, is largely gone. The issue isn't (as with the harm-driven process) whether any given alternative is biologically "safe" or of "no significant harm." The issue is how one alternative stacks up against others on many types of harms and benefits.

Additionally, information that is missing regarding harms and benefits is weighed into the picture of each alternative. Thus, missing information may play an active role in whether a particular alternative is considered worth pursuing in contrast to an alternative for which the harms and benefits are better defined.

Ultimately, a "harmful" alternative (and what human activity, given our population numbers, does *not* cause some type of environmental stress or harm?) may thus be selected, because overall the nature of harms it poses and the benefits it brings make it the most promising candidate for reaching the positive goal.

Choices among alternatives will entail more precaution if a given goal has numerous targets, because any one target may slip into or depend upon a more conventional harm-driven risk assessment. For instance, one target for reaching the Swedish environmental quality goal of "no eutrophication" states, "Releases of nutrients to coastal waters, lakes, and watercourses and to groundwater should in the long term *not exceed levels that cause an adverse impact* on health, biological diversity or the possibility of versatile use" (Sweden Ministry of Environment 1997/98; emphasis added).

This is basically a harm-driven target, throwing analysis of alternatives into the usual risk-based mode of questioning. For instance, when analyzing nutrient-releasing activities (e.g., agricultural practices), what will be considered "an adverse impact"? Any observed departure from "natural" eutrophication? Slight diminishment of prey capture rates of aquatic predators or slight compromise of the immune system functioning of predators and prey? Defining and setting the "safe" level of a potentially harmful activity involve conventional risk analysis.

On the other hand, an accompanying target Sweden has established for reaching the "no-eutrophication" goal puts measurable limits on nutrient-releasing activities: "Nutrient levels in coastal and sea areas should be broadly similar to those that existed in the 1940s" (Sweden Ministry of Environment 1997/98). This, then, limits nutrient releases even if someone has produced a risk assessment declaring that such releases don't "exceed levels that cause an adverse impact." Of course, what is "broadly similar" leaves some room for defense of conventional practices. If positive targets are vague, movement toward them can be illusory.

When numerous alternatives are being considered, all parties have a stake in examining potential harms of alternatives they oppose and in pointing out the positive consequences of the alternative they support. This provides a rich information base for decision making. Harms are then considered in a context of comparative harms (and benefits), and if the process is broadly participatory, as it should be, a diversity of perspectives will determine the range of types of harms that will be considered.

Since most environmental and public health goals cannot be achieved without changes in behaviors of multiple sectors of society, goal-driven alternatives are likely to be examined for their comparative "fairness" regarding who bears the burden of change. Those alternatives that involve broadly distributed burdens (where appropriate) and provide broadly distributed benefits are most likely to receive broad public support.

Role of Science

The science involved in goal-driven analyses of alternatives is more likely than harm-driven analyses to include social sciences in addition to biological sciences. For example, an analysis may examine how consumers will accept a particular change or how labor will be affected under two different alternatives.

The science is also necessarily more complex, because a combination of activities in multiple sectors is generally involved in accomplishing positive public and environmental health goals. A goal of attaining "drinkable, fishable" water, for instance, requires rethinking industrial discharges of wastes to municipal water treatment facilities, human consumption of birth control substances and pharmaceuticals, transportation runoff, waste incineration and landfills, the production and use of industrial waste-laced agricultural fertilizers, and even the practice of combining industrial and human wastes in a single sewer system. Options to be considered necessarily include such measures as biological toilet systems, closed-cycle manufacturing processes, reduction of the health industry's dependency on drugs, increased use of mass transit and walkable urban nodal developments, population goals, and massive public education about links between various activities and pollution. The science of estimating the benefits of numerous options as well as the science of measuring progress and unanticipated harms is necessarily complex.

A goal-driven process reduces the drive to fiercely debate the fine points of whether a given piece of scientific evidence will be labeled "credible," because detailed scientific information is not, as in the harm-driven approach, a pivotal point by which change can be resisted. Given that a positive goal is being aspired to, change, innovation and new activities are *inevitable*. Of course, if some tout particular harmful industries or technologies as compatible with reaching the goal, while others provide evidence that those activities are incompatible with reaching the goal, there are bound to be debates.

When ambitious, public-interest goals such as health, recovery, restoration, or natural abundance are being pursued, the science of restoration and health must be as prominent as the science of potential harm. We know a lot, for instance, about how to kill weeds in overgrazed, degraded, arid, native grasslands. We know far less about how to restore healthy, native grassland ecosystems that will resist reinvasion by weeds, so the spraying is not endless. It does not take much science (or awareness of how little we know) to dismantle functioning ecosystems or to diminish sperm counts worldwide, but trying to restore them forces us to realize how much we don't know and how much we need to know. For instance, if we are attempting to revegetate an area with native grass seed, how local must the seed be to avoid weakening locally adapted remnant grasses via genetic contamination? If we are trying to restore sperm counts, for which toxics do we pursue reduction? Given that it is essentially impossible to attribute reduced sperm counts to individual toxics because of the difficulties of understanding interactive effects of immune suppressants, direct toxics, and endocrine-disrupting chemicals, does restoration of sperm counts require a fundamental, general switch from permitting toxics releases that "won't cause adverse effects" to prohibiting release of toxics in general unless there is no feasible alternative?

Pursuing meaningful, positive goals leads us to necessary collaborations among biological and social sciences, engineering, and political courage.

Involving Precaution

The precautionary principle is used throughout the process of selecting and achieving positive public goals (figure 18.2). Pursuit of a positive environmental health goal is more likely to raise questions about the *necessity* of a particular activity that retards or endangers achievement of the goal, particularly when sacrifices attendant to changing behaviors are being borne by other sectors of the society in order to achieve the goal. Harm-driven processes rarely question the necessity of harmful activities.

Involving the Public

A goal-driven precautionary process could theoretically be carried out by a government agency or any small group of people. However, *public* environmental and health goals, by their nature, are legitimately claimed as public goals only if broad participation by the public is included in setting the goals.

Diverse options, components, and systems for reaching goals are best brought forth by people of diverse experience, knowledge, skills, and perspective. Without broad participation, the process, and thus the results of the process, will likely be constricted and unambitious.

Analyzing alternatives is likewise best accomplished by considering information brought by scientists; people with local and indigenous knowledge; people who potentially will benefit and/or be harmed by any of the alternatives; and people who have cultural, ethical, political, and personal concerns or hopes.

Looping Back

Following implementation of actions intended to reach the goal, it is affected citizens who will most reliably track progress, unintended effects, and follow-through on commitments.

When alternatives that have been implemented are failing to bring hoped-for or promised progress toward the goal, new alternatives must be considered, analyzed, and selected (figure 18.2). As U.S. Forest Service research hydrologist Gordon Grant states (FSEEE 1996), "It's the kind of learning mentality that comes easily if you're in a research setting, but it's not unique to research. And in fact the distinction is less clear than a lot of people like to make it. If we're really serious about adaptive management, then you have to have some process to capture what you know, learn from your mistakes, and come back and do things differently. That works in science and it works in management."

Attaining ambitious positive goals necessarily entails "doing things differently" throughout a community, whether local, national, or global. For instance, the Swedish environmental quality legislation writes, "The Government proposes, within the framework of a fully developed *ecocycle strategy* aiming at better resource management, that all those concerned adjust their behavior with respect to a specified range of products so that the declared objectives can be achieved within one or two generations" (Sweden Ministry of Environment 1997/98; emphasis in original).

It takes a whole community—whether local or international—and transparent, participatory political processes to attain meaningful environmental or public health goals.

Conclusion

Positive, socially debated, goal-driven processes are essential for setting in motion the fundamental social, economic, legal, scientific, and political changes that are necessary locally, nationally, and globally to reorient human activities toward cooperation with nature and broadly shared health.

As local and national communities and a global community, we will always engage in major decision-making processes with regard to potential and documented harms posed by particular human activities and technologies, both ongoing and novel. In these processes, scientific information about potential harms of the activities being scrutinized is crucial.

When communities agree on positive public environmental health goals, however, scientific information about both potential harms and potential benefits of various options for realizing the goal is essential—but not sufficient. The attainment of positive environmental health goals requires forecaring, precaution, science, economics, politics, public involvement, and spirit.

Science in the service of reaching environmental health goals—science that helps humans understand both the need to avoid particular activities and the benefits of embracing activities that promote restoration or maintain environmental and public health—is science in the service of good.

References Cited

Bergen Ministerial Declaration on Sustainable Development in the ECE Region. 1990. UN Doc. A/CONF.151/PC/10 1 YB International Environmental Law 429, 4312.

Committee on Endangered and Threatened Fishes in the Klamath River Basin, National Research Council. 2002. *Interim Report from the Committee on Endangered and Threatened Fishes in the Klamath River Basin: Scientific Evaluation of Biological Opinions on Endangered and Threatened Fishes in the Klamath River Basin.* Washington, D.C.: National Academy Press.

Endangered Species Act. 1982. 16 U.S.C.

Federal Water Pollution Control Act (Clean Water Act). 1948, as amended 1972. Congressional Declaration of Goals and Policy. Section 101, U.S.C. Section 1251.

Forest Service Employees for Environmental Ethics (FSEEE). 1996. *Torrents of Change.* Videotape. Eugene, Ore.

Framework Convention on Climate Change. 1992. 9 May, 31 ILM 849.

Graham, J. 1999. "The Precautionary Principle: Refine It or Replace It?" *Risk in Perspective* 7, no. 3: 1–6.

Hitt, S. 2002. "A Duty to Conserve: The Moral Meaning of the Endangered Species Act." *Wild Earth* 12, no. 1: 65–68.

Houck, O. 1991. "The Regulation of Toxic Pollutants under the Clean Water Act." *Environmental Law Reporter* 21: 10528–60.

Johnson, K. N., S. Crim, K. Barber, M. Howell, and C. Cadwell. 1993. *Sustainable Harvest Levels and Short-Term Timber Sales for Options Considered in the Forest Ecosystem Management Assessment Team: Methods, Results, and Interpretations.* Unpublished manuscript. Corvallis: Oregon State University, Department for Forest Resources.

Kann, J., and L. Dunsmoor. 2002. *Comments on Scientific Evaluation of Biological Opinions on Endangered and Threatened Fishes in the Klamath River Basin, Interim Report from the Committee on Endangered and Threatened Fishes in the Klamath River Basin, February 6, 2002.* Unpublished paper prepared for the Klamath Tribes.

Ludwig, D., R. Hilborn, and C. Walters. 1993. "Uncertainty, Resource Exploitation, and Conservation: Lessons from History." *Science* 260 (2 April): 17, 36.

Ministerial Declaration Calling for Reduction of Pollution, 25 November 1987, 27 ILM 835.

O'Brien, M. 2000. *Making Better Environmental Decisions: An Alternative to Risk Assessment.* Cambridge, Mass.: MIT Press.

Oregon Natural Resources Council (ONRC) and others [not stated on document]. 2000. "A Citizens' Alternative for the Northwest Forest Plan." Eugene, Ore. http://www.onrc.org/programs/wforest/citizens.htm.

Oregon Natural Resource Council Action v. United States Forest Service, 59 F.Supp.2d 1085; 1999 WL 588253 (W.D.Wash. 1999).

Pollan, M. 2001. "Precautionary Principle." *The New York Times Magazine,* 9 December, 92, 94.

Sweden Ministry of Environment. 1997/98. "Swedish Environmental Quality Goals: An Environmental Policy for a Sustainable Sweden." Government Bill 1997/908:145. http://www.miljo.regeringen.se/english/english_index.htm.

U.S. Department of Agriculture (USDA), Forest Service (USFS), and U.S. Department of Interior (USDI), Bureau of Land Management. 2001. *Record of Decision and Standards and Guidelines for Amendments to the Survey and Manage, Protection Buffer, and Other Mitigation Measures Standards and Guidelines.* Portland, Ore.

CHAPTER 19

Toward Sustainable Chemistry*

Terry Collins

Chemistry has an important role to play in achieving a sustainable civilization on Earth. The present economy remains utterly dependent on a massive inward flow of natural resources that includes vast amounts of nonrenewables. This is followed by a reverse flow of economically spent matter back to the ecosphere. Chemical sustainability problems are determined largely by these economy-ecosphere material flows, which current chemistry education essentially ignores. It has become an imperative (Jonas 1984) that chemists lead in developing the technological dimension of a sustainable civilization.

When chemists teach their students about the compositions, outcomes, mechanisms, controlling forces, and economic value of chemical processes, the attendant dangers to human health and to the ecosphere must be emphasized across all courses. In dedicated advanced courses, we must challenge students to conceive of sustainable processes and orient them by emphasizing through concept and example how safe processes can be developed that are also profitable.

Green or sustainable chemistry (Anastas 1998) can contribute to achieving sustainability in three key areas. First, renewable energy technologies will be the central pillar of a sustainable high-technology civilization. Chemists can contribute to the development of the economically feasible conversion of solar

---

*Reprinted with modifications with permission from T. Collins, "Towards Sustainable Chemistry," *Science* 291, no. 48 (2001). Copyright 2001 American Association for the Advancement of Science.

into chemical energy and the improvement of solar to electrical energy conversion. Second, the reagents used by the chemical industry, today mostly derived from oil, must increasingly be obtained from renewable sources to reduce our dependence on fossilized carbon. This important area is beginning to flourish, but is not the subject of this essay. Third, polluting technologies must be replaced by benign alternatives. This field is receiving considerable attention, but the dedicated research community is small and is merely scratching the surface of an immense problem that I will now sketch.

Many forces give rise to chemical pollution, but there is one overarching scientific reason chemical technology pollutes. Chemists developing new processes strive principally to achieve reactions that only produce the desired product. This selectivity is achieved by using relatively simple reagent designs and employing almost the entire periodic table to attain diverse reactivity.

In contrast, nature accomplishes a huge range of selective biochemical processes mostly with just a handful of environmentally common elements. Selectivity is achieved through a reagent design that is much more elaborate than the synthetic one. For example, electric eels can store charge via concentration gradients of biochemically common alkali metal ions across the membranes of electroplaque cells. In contrast, most batteries used for storing charge require biochemically foreign, toxic elements, such as lead and cadmium. Because of this strategic difference, human–made technologies often distribute throughout the environment persistent pollutants that are toxic because they contain elements that are used sparingly or not at all in biochemistry.

Persistent bioaccumulative pollutants pose the greatest chemical threat to sustainability. They can be grouped into two classes. Toxic elements are the prototypical persistent pollutants; long–lived radioactive elements are especially dangerous examples. New toxicities continue to be discovered for biologically uncommon elements. The second class consists of degradation-resistant molecules. Many characterized examples originate from the chlorine industry (Thornton 2000) and are also potently bioaccumulative. For example, polychlorinated dibenzo-dioxins and -furans (PCDDs and PCDFs) are deadly, persistent organic pollutants. They can form in the bleaching of wood pulp with chlorine-based oxidants, the incineration of chlorine-containing compounds and organic matter, and the recycling of metals. The United Nations Environmental Program (UNEP) International Agreement on persistent organic pollutants lists twelve "priority" pollutant compounds and classes of compounds for global phaseout. All are organochlorines.

Imagine all of Earth's chemistry as a mail sorter's wall of letter slots in a post office, with the network of compartments extending toward infinity. Each compartment represents a separate chemistry so that, for example, thousands of compartments are associated with stratospheric chemistry or with a human cell. An environmentally mobile persistent pollutant can move from compartment to compartment, sampling a large number and finding those compartments that it can perturb. Many perturbations may be inconsequential, but others can cause unforeseen catastrophes, such as the ozone hole or some of the manifestations of endocrine disruption (Katz 1999). Most compartments remain unidentified and even for known compartments, the interactions of the pollutant with the compartment's contents can usually not be foreseen, giving ample reason for scientific humility when considering the safety of persistent mobile compounds. We should heed the historical lesson that persistent pollutants are capable of environmental mayhem and treat them with extreme caution. In cases where the use of a persistent pollutant is based on a compelling benefit, as with DDT in malaria-infested regions, chemists must face the challenge of finding safe alternatives.

Consider, for instance, the alarming reproductive damage that can be inflicted by minute quantities of endocrine-disrupting chemicals (EDCs), such as PCDDs, polychlorobiphenyls (PCBs), and the pesticides endosulfan and atrazine (Thornton 2000). EDCs disrupt the body's natural control over the reproductive system by mimicking or blocking the regulatory functions of the steroid hormones or altering the amounts of hormones in the body. Uncertainty still clouds our understanding of their full impact, but mass sterilization is one limiting conceivable outcome of ignoring the demonstrated dangers of EDCs. Our present knowledge strongly suggests that anthropogenic EDCs should be identified and eliminated altogether.

Stringent regulations based on the precautionary principle and the principle of "reversed onus" (Thornton 2000) should be developed to guard against the release of new environmentally mobile persistent compounds; a precise definition of persistence also needs to be developed. This would provide a regulatory foundation for weeding persistent bioaccumulative compounds out of all technology and highlight where research is needed to find safe alternatives. Groundbreaking legislative proposals toward this goal are about to be considered in the Swedish Parliament.[1]

In their current formal training, all chemistry students will learn that the chlorination of phenol proceeds by a mechanism known as electrophilic aromatic substitution. But very few will learn of EDCs and their dangers or come to know that prime examples of EDCs, namely PCDDs, are produced in trace quantities whenever phenol is chlorinated. This hazardous omission illustrates one important type of content that is simply missing from the conventional curriculum.

Green chemistry can dramatically reduce environmental burdens of both classes of persistent pollutants by moving the elemental balance of technology closer to that of biochemistry. Significant reductions in the dispersal of many persistent pollutants have already been achieved. By the late seventeenth century, the use of lead oxide as a correcting agent for acidic wine was banned on pain of death in Ulm in the duchy of Wurtemburg (Eisinger 1996). More recently, large reductions in lead pollution have been achieved in what are recognizable examples of green chemistry, for instance, by replacement of lead additives in paint with safe alternatives, by the development of cleaner batteries, and by the as-yet unfinished and sometimes flawed progression away from tetraethyl lead toward safer combustion promoters in fuels. PCDDs and PCDFs have been greatly reduced in the pulp and paper industry by the replacement of chlorine with chlorine dioxide as the principal bleaching agent.

Nevertheless, much more can and must be done. For example, chlorine-based oxidations such as pulp bleaching, water disinfection, household and institutional cleaning, and clothing care continue to produce huge volumes of organochlorine-containing effluent. Despite industry efforts to reduce pollutant concentrations, some of the inescapable trace contaminants are persistent, bioaccumulative carcinogens and/or EDCs. Chlorine-based oxidation technologies could be replaced with alternatives based on catalytic activation of nature's principal oxidizing agents, oxygen and hydrogen peroxide. My research group has patented tetraamido-macrocyclic ligand activators that are potent but selective peroxide-activating catalysts comprised of biochemically common elements for these and other fields of use. Environmental considerations also underpin the worldwide investigation and development of supercritical and near-critical carbon dioxide as a clean solvent. The present search for safer solvents in the green chemistry community is distinguished by a remarkable burst of creativity that perhaps reaches its zenith in ionic liquids. These solvents have unique properties such as the absence of any vapor pressure under standard conditions.

Pollutant production can also be reduced by improving process selectivity, reducing energy intensity, and minimizing the flow of matter to and from the ecosphere via atom economic processes, that is, processes optimized to reduce per unit of product the quantities of chemicals employed in the reactions as solvents and reagents or produced as by-products.

To achieve such sustainable chemistry requires a sea change in the chemical community. The principles of green or sustainable chemistry must become an integral part of chemical education and practice. However, there are several obstacles to overcome. First, chemists need to comprehensively incorporate environmental considerations into their decisions concerning the reactions and technologies to be developed in the laboratory. These questions need to become as important as those associated with the selectivity of the technology and how it works. Principles upon which to base these decisions have already been developed (Anastas 1998).

Second, it is critical that chemistry that is not really green does not get sold as such and that the public is not misled with false or insufficient safety information. For example, certain chlorine industry companies have sought to protect their profits by distorting scientific data to make dioxins appear to be less harmful to humans than they actually are (Thornton 2000). The general trust that chemical risk is treated in a fair and reasonable manner must be strengthened. Third, since many chemical sustainability goals such as those associated with solar energy conversion call for ambitious, highly creative research approaches, short-term and myopic thinking must be avoided. Government, universities, and industry must learn to value and support research programs that do not rapidly produce publications, but instead present reasonable promise of promoting sustainability. Fourth, chemistry exerts a near boundless influence on human action and is thus inextricably intertwined with ethics. An understanding of sustainability ethics (Jonas 1984) is therefore an essential component of a healthy chemical education.

The all-encompassing challenge lying before green chemists is to understand the ethical forces, chemical-ecosphere relations, educational needs, and research imperatives that sustainability brings center stage and to reconcile this understanding as much as possible with economic maxims. If chemists increasingly direct their strengths to contributing to a sustainable civilization, chemistry will become more interesting and compelling to people and may lose its "toxic" image. It will become more worthy of public support and spawn exciting economic enterprises that nurture sustainability.

Endnotes

1. Editor's Note: These deliberations resulted in Swedish Government Bill 2000/01:65, which would phase out persistent and bioaccumulative chemicals in products and processes over a ten-year period. This policy, however, is partly dependent on the outcome of upcoming European Union chemicals legislation.

References Cited

Anastas, P. T., and J. C. Warner. 1998. *Green Chemistry: Theory and Practice.* Oxford: Oxford University Press.

Eisinger, J. 1996. *Natural History* 105: 48.

Jonas, H. 1984. *The Imperative of Responsibility: In Search of an Ethics for the Technological Age.* Chicago: University of Chicago Press.

Katz, R. K., ed. 1999. *Endocrine Disruptors: Effects on Male and Female Reproductive Systems.* Boca Raton, Fla.: CRC Press.

Thornton, J. 2000. *Pandora's Poison: Chlorine, Health, and a New Environmental Strategy.* Cambridge: MIT Press.

PART VI

Science to Support Precautionary Decision Making

The chapters in this part outline an approach to environmental science to more effectively address uncertainty and the complexity of systems and support precautionary decisions. In essence, they build a vision for science and precaution and for the methods and techniques in science that would lead towards that vision. The ideas presented build from many of the recommendations and analyses presented throughout this volume.

Carl F. Cranor asks in chapter 20 what research strategies would further the prevention of serious threats to human health and the environment, prevent threats from materializing into harm, and, ultimately, contribute to a better future for humans and the environment. He challenges scientists to develop methods for identifying early warnings of hazards—to prevent risks before they occur.

Elizabeth A. Guillette presents her research on the effects of pesticides on the Yaqui Indian children of Mexico as a case study on innovative methods to assess subtle risks. In chapter 21, she argues that the range of possible chemical-induced human harm is often concealed from our present scientific understanding and outlines the need for new methods and approaches to environmental science to protect those most vulnerable in the population. These methods must include close interaction with affected communities.

Based on her vast research and policy work on the risks of tributyltin to marine fauna, Cato C. ten Hallers-Tjabbes argues in chapter 22 for the need for a more effective interface between scientists and policy makers. Scientists must be trained to communicate results to policy makers in a dynamic fashion. This could improve the information basis of decisions, characterize uncertainties, and refine the types of questions asked and methods used as well as the expediency of preventive actions.

In chapter 23, Richard Levins reflects on his decades of scientific research in ecology and health. He argues that there are many scientific methods, depending on the nature of the problem and the stage of the investigation. Thus, methods must be appropriate to the problem being studied. He suggests that we use science to understand how natural systems deal with disturbances and cope with uncertainties to model our own ways of studying and preventing damage to health and ecosystems.

CHAPTER 20

What Could Precautionary Science Be?
Research for Early Warnings and a Better Future

Carl F. Cranor

One approach to developing a science that more effectively addresses uncertainty and complexity is to examine how scientific research might be modified to serve more precautionary aims and thus better protect human health and the environment. That is, are there more precautionary and preventive approaches to using science to help guide social decisions? Can certain types of research help produce decisions that are consistent with good science and the presuppositions of the precautionary principle? In this chapter, I outline such an approach.

This chapter is complementary to other considerations related to science and the precautionary principle both in this book and in my previous writing (Cranor 1999, 2001), including attenuating stringent demands for full scientific documentation of threats of harm before regulatory action, helping scientists and agency researchers reduce the number and frequency of false negatives in environmental decisions, taking precautionary actions based on plausible warnings, and shifting burdens of proof to provide incentives for prevention.

The approach in this chapter, however, focuses on the research agenda through the *Gedankenspiel* (thought play) question, "Given what we know at present, what could precautionary science be?" What research strategies would further the prevention of serious threats to human health and the environment, prevent threats from materializing into harm, and, ultimately, contribute to a better future for humans and the environment? I begin to answer this by

formulating some questions and issues that follow from an implicit understanding of the precautionary principle. I then outline a precautionary research
strategy that builds, among other things, on inventories and evaluations of the
state of the environment and public health, anticipation of potential harms,
and generalizations of some of these strategies. I offer some specific suggestions
on directions for toxics research in general and carcinogens in particular.

Background: The Case for Precautionary Science

Central to the idea or concept of precaution is that something of significance
or value to us is threatened and merits some kind of precautionary and protective action. We take precautions toward highly valued people and things
around us such as children or other loved ones, or precious art objects, such as
Michelangelo's *Last Judgment*. Because we value them so highly, we plan and
organize our activities in order to prevent threats to their integrity, well-being,
or current condition. For example, the curators of the Vatican Museum do not
wait to see whether smoking in the Sistine Chapel will pose a threat of serious or irreversible damage to the *Last Judgment* before they decide that smoking will not be permitted. They may even decide that the cumulative effects
of permitting thousands of people per day to view the *Last Judgment* poses too
great a threat to the painting. We also take precautionary action toward dam-
aged things about which we care in order to prevent further harm, to rectify
existing damage, or to restore them to some desirable condition.

The precautionary principle, however, applies specifically to the environment and human health and implies that we place a high value on these things.
Some generally accepted facts about the current state of the environment indicate that it is under considerable pressure, continual threat, and degraded from
healthier previous states, and that these conditions will only worsen because
of increasing human population pressures. Consider: (1) the world's forested
lands have shrunk substantially over the past fifty years (Rodgers 1992: 8);
(2) humans are overpumping aquifers in China, India, North Africa, Saudi Ara-
bia, and the United States by about 160 billion tons of water per year (Brown
et al. 2000: 6–7); and (3) about "two-thirds of major marine fisheries are fully
exploited, overexploited or depleted" (Lubchenco 1998: 492).

In the world of toxic substances, the creation of chemical substances
appears to be beyond the control of institutions whose job is to protect the
public and the environment. Businesses are producing and using many more
chemical compounds than our legal institutions can evaluate properly (see
chapter 9). The former Office of Technology Assessment (OTA) found that
federal regulatory agencies had addressed only one-half to one-third of known
carcinogens within their jurisdiction (U.S. Congress OTA 1987). Current
methods of assessing the substances one at a time, after they are in commerce,

will preclude understanding their toxicity and removing any significant number of substances if they pose problems.

In general, little is known about the universe of approximately 100,000 chemical substances or their derivatives used in commerce. Most have not been well assessed for health effects (Huff and Hoel 1992; U.S. Congress OTA 1995; Zeise 1999). In 1985, for 75 percent of the 3,000 top-volume chemicals in commerce, the most basic toxicity results could not be found in the public record, with little change in that percentage as recently as 1998 (National Research Council 1985; U.S. Congress OTA 1987; Environmental Health Newsletter 1998). An additional 800 to 1,000 new substances are added to the list each year with little or no toxicity testing (U.S. Congress OTA 1987: 127).

The knowledge gaps about toxic substances will be slow to close because of the "sparseness and uncertainty of the scientific knowledge of the health hazards addressed" (National Research Council 1984). Animal and human studies are costly and take years to conduct, interpret, and understand. These problems are exacerbated for substances with long latency periods or associated with erratic exposure patterns (Huff and Rall 1992).

Given the current size of the chemical universe and our current understanding of it, we are likely to remain forever ignorant of the properties of most of these substances. Thus, we need to find more precautionary approaches to addressing the universe of potentially toxic substances in order to prevent some risks from becoming real threats and threats from materializing into harm. Ultimately, such an approach must provide long-term sustainable conditions for human and ecosystem life.

Inventories

In order to know the current condition of the environment and public health, we need inventories and scientific evaluation of the status quo. We need to know what the current conditions are in order to know toward what things we should take a precautionary approach. What inventories of the current state of the environment and public health would we need in order to take a fully precautionary approach to preserve valuable environments, to remedy dreadful problems, and to develop the kind of world in which we want to live? This is not to suggest that we should provide an inventory of only those aspects of the environment and public health we currently care about, although we should probably begin with that. A broader inventory of things not currently of concern might reveal areas that we would care about if we knew their present condition.

Meanwhile, we are facing major environmental problems that need remediation—devastated ecosystems such as the Baltic or North Seas that are not only valued environments, but also important sources of food for several countries. We face human health threats that appear to be out of control, such as the presence of PCBs, dioxins, or other persistent organic pollutants (POPs) in humans and large mammals around the world, including near-toxic levels in animals and humans in the Arctic. We want to protect highly valued or pristine environmental resources—the Grand Canyon, the Everglades, or the Brazilian rainforest. In many areas scientists have performed the necessary inventory and evaluative tasks. However, in other areas, ignorance persists.

Health monitoring and disease registries are two kinds of inventories that further precaution. They provide a snapshot of the current condition of the environment or human health. Compiled over time, inventories provide evidence of trends—whether things are getting better or worse. They also provide important basic information as well as baselines from which to launch further research. Consider several examples from the area of toxic substances.

- One group of researchers has called for a childhood cancer registry. This would provide snapshots of current incidence and death from cancer, information on geographic distributions of disease, and other information valuable for communities and further epidemiological studies. Epidemiological studies would, in turn, foster further research to help protect this susceptible subpopulation through such avenues as improving methods, developing biomarkers for childhood cancer, and refining exposure assessment (Carroquina et al. 1998). Registries for other childhood diseases—neurological or reproductive—might also be called for in the spirit of precautionary principle. One example is the Health Track project (www.healthtrack.org)
- PCBs and other persistent organic substances have been known for some time to be problematic in the environment. Nevertheless, manufacturers have introduced substances that are quite similar to PCBs, polybrominated diphenylethers (PBDEs). Hooper and MacDonald (2000) have shown that concentrations of PBDEs are increasing in breast milk among European women. PBDEs are widely used in products in the United States and are likely to pose risks to adults and children alike. However, the United States has no breast milk monitoring program. We will not know whether this substance is a problem until we have data, but at present there is no systematic effort to gather it.
- We also need a registry of communities whose members are highly exposed to suspected toxic substances. Communities that experience both exposure to multiple toxicants and low socioeconomic conditions are vulnerable to adverse health effects (Levins forthcoming). The aims would be to identify them, to take steps toward reducing exposure, to provide the basis for health studies, and ultimately to improve people's health and living conditions. A similar registry of communities with high disease rates, even if causes are

unknown, might reveal groups who are already vulnerable and in need of precautionary action to remedy existing problems and prevent future ones (Tickner 2001).

- Most recently, the National Academy of Sciences has called for general monitoring of our biological and agricultural resources, so that a record can be maintained over time and so we have background information against which to compare any effects from transgenic plants that might be produced in the future (NRC 2002).

Evaluations

Without such registries and monitoring, we cannot assess the condition of the environment or public health. Once such information is available, scientists and the larger public can evaluate the results revealed by the inventory and decide what other registries, monitoring, or inventories the scientific community should be developing.

The second group of preliminary questions for a more precautionary approach to science has to do with evaluation. Are the current states of the environment and public health desirable or undesirable? What are the threats to health? That is, we need to *evaluate or assess* the condition of the various environments and the public health (with attention to subpopulations) in the inventory. This is a quintessentially scientific task that ultimately helps to inform broader public discussion.

Scientists are uniquely qualified to evaluate these conditions, and their evaluation is needed to guide the social discussion. Ecologists are experts, for example, on whether an ecosystem is in a healthy state or whether it is in a degraded or seriously threatened one. Public health professionals can assess the condition of human health and threats to it as well as conditions likely to prove harmful. Such evaluation is a legitimate scientific activity. Scientists must judge whether the environment is in a better or worse state than it was at some earlier time, whether it is in anything like a healthy state at all, and whether, if current trends continue, it will be in a better or worse state in the future.

Some scientists may be discomfited by the idea of making such evaluations, since evaluative judgments are supposedly excluded from "objective" scientific study and inquiry. Making this separation is a significant part of graduate school training. Yet the kinds of evaluations I suggest are well within the purview of scientific research. For the most part, scientists have not been reticent to speak out in this way, but in implementing the precautionary principle, they should be even more outspoken in assessing and evaluating the current condition of the environment and the public health that may be at risk.

Education

The facts and evaluations listed above represent the consensus of a large portion of the scientific community, one that is shared by an informed portion of the public. Disappearing forests, declining aquifers, proliferation of persistent organic substances, and degraded farmland may be known to experts and, if the effects have been widely publicized, to some in the broader public. Many effects are likely to be beyond the awareness of U.S. citizens, however, because we draw on the rest of the world for many of our natural resources and for disposal of some of our wastes. The United States is the 10,000-pound gorilla in the world environment, using large percentages of the energy and vast quantities of timber, metals, and other natural resources from around the world. But few of the effects of our collective actions are manifested within our national boundaries. Other problems may not be evident because of fortuitous circumstances. The effects of global warming on the North American climate may not be as significant in the short run as elsewhere because of a climatological accident: for up to two decades the main effect may be longer growing seasons (McKibben 2001). Nor does the United States suffer some shortages of natural resources and all the adverse impacts of waste disposal, because this continent has been endowed with large amounts of land and good coal, natural gas, mineral, and water resources. But they will not last forever.

Education furthers precaution. By providing an inventory of the state of the environment and a scientific evaluation of it, scientists can serve an important public educational function. There is anecdotal evidence that citizens informed about environmental problems respond and take appropriate action. In addition, the more the general public cares about the environment and the state of public health, as the citizens of Northern European countries do, the greater will be the educational effect of providing information about the state of the environment and public health (Wahlstrom 1999).

Revealing Commons Problems

A review of existing environmental and public health problems would show that many have resulted from commons effects, a complex social, legal, and biological problem. A commons problem has the following characteristics (Ostrom and Ostrom 1977, 157):

- No one in particular owns the resource.
- Many users have access to the resource.
- No single user can control the activities of others; conversely, all must cooperate to make joint agreements effective.
- Demand upon the resource exceeds the supply.

At the macro level are such problems as global warming and CFC depletion of the ozone layer. A smaller commons effect might be loss of species because of development in California, adverse health effects from air pollution in the Los Angeles air basin, or depletion of the Ogallala aquifer.

Commons problems tend to hide the effects of human actions on the environment and public health:

• In many cases there is no obvious cause-and-effect relationship between the actions of an individual or a small number of individuals and visible adverse effects. It takes the cumulative effects of large numbers of human actions to produce untoward effects on the environment or public health. Thus, *the causal relationship between human action and environmental or human effect may be difficult to detect.* (Of course, it is easy to take advantage of this feature to frustrate public action on a problem.)
• Even cumulative effects are not easily observed. The subtle scientific connections between collective human actions and adverse effects must be revealed by patient, time-consuming, interdisciplinary research such as the investigations on global transport of POPs or the toxicological effects of air toxics.
• There may be long *latency periods* between human actions that ultimately contribute to an adverse effect and a clear appearance of the effects, for example, CFC ozone depletion, carbon dioxide global warming effects, and the migration of POPs through the environment.

Because many environmental effects are hidden, subtle, the result of latent processes or cumulative commons contributions, or all of the above, science has a further *revelatory* role to play. Scientists need to reveal and identify these effects on the environment and to report them to the public. This is an aspect of providing an inventory, but it is more. It aims to reveal environmental and public health problems that are difficult to discern.

The Connectedness Principle

Beyond inventories and revealing commons problems, however, the concept of precautionary actions suggests that research should *anticipate* effects and try to identify potential commons issues at very early stages.

One way we can anticipate potential harms and help organize research agendas is by attending to a presupposition of commons problems. We might call this the "connectedness principle" in honor of John Muir, who noted, "When we try to pick out anything by itself, we find it hitched to everything else in the universe" (Muir 1988: 110). Surely this is an overstatement, but it reminds us of our connections to the biological world and its importance to us. More significantly, it reminds us that we live in a closed world system in which we can no longer ignore the impacts of our activities and their cumulative and interactive impacts. Human beings can no longer dispose of wastes or extract resources at will.

This observation suggests that we might adopt the connectedness principle as a working presumption of research and public policy toward the environment and human health. By now we have had enough evidence in the form of "surprises" to presume that human industrial activities *will* affect—probably adversely—the environment and human welfare mediated through the environment, on many levels. As population pressures increase, such effects will only intensify. Anticipatory research would look for such effects.

Anticipatory Science

The next *Gedankenspiel* appropriate to the idea, implicit in the concept of precaution, of anticipating potential harms to the environment and public health might be to ask, "If we were gods, what would we want to know about the world and our effect on it in order to *anticipate* threats to human health and the environment? What research might help us pass on a better world to future generations?" The exact nature of such research must be left to experts informed by a wide range of societal stakeholders. But inventories and current assessments of the environment and human health will indicate where problems are likely to occur.

An important first step in such scientific activity is to *mine existing knowledge*—learn from it, extend it, and apply it to prevent new possible risks from becoming serious and real threats. Consider obvious cases of which we are aware.

- If certain substances have been identified as toxic threats because they persist or bioaccumulate, becoming more toxic as they do so; migrate through the food chain; and are transported by evaporation, condensation, and global atmospheric force, then, in absence of strong evidence to the contrary, we have good reason to believe that chemically and biologically similar substances are likely to pose similar threats (U.S. EPA 1996; Brunner et al. 1996; Aulerich et al. 1986; Hornshaw et al. 1983).
- If we know that there are appropriately deep biological analogies between one benzidine dye and other benzidine dyes, and the first is a human carcinogen, then we have reason to believe that the others are human carcinogens until there is compelling scientific and social reasons to treat them differently.

If we know or have good reason to believe that human activities have greatly increased the number of plant and animal species that have gone extinct or have been put on endangered species lists, we know enough to infer that continued patterns of human development will have similar effects. We should learn from them and then plan our activities using this assumption to prevent or minimize such effects. Scientists could look for *deep, biologically plausible, and persuasive analogies* (Thagard 1999) between human actions and known adverse ecological and human health impacts in order to anticipate and identify as yet unknown effects. The aim should be to ensure that we have extracted all that we can from the existing body of scientific knowledge in order to prevent future harm. Deep analogies would point us toward other adverse impacts that we could seek to avoid.

Thus, scientists can draw on their reservoir of knowledge to make educated guesses or predictions about new problems, especially to provide early warnings about new commons problems. For example, Molina and Rowland (1974) knew enough from their understanding of chemical structures and interactions to make reasonable predictions about the effects of CFCs on the ozone layer. Their predictions were subsequently verified by further empirical research. Others have made predictions about the adverse impact of proposed development of road networks in the Brazilian rainforest (Laurance et al. 2001).

Beyond mining the existing knowledge base, scientists could seek to *develop relevant scientific theories and data*. For example, ecologists might look for "canary conditions" as indicators of the health of ecosystems. As I discuss below, the success of a more precautionary approach to science over the longer run is likely to depend upon understanding various kinds of biological mechanisms that contribute to harm to ecological or human health and generalizing from such results.

Toxics

We can further illustrate the use of pertinent theories and data by focusing on toxic substances generally. Scientists could develop diagnostic tests or indicators to identify likely toxicants before they pose a threat of serious or irreversible damage to the environment or human health. (Once there is a threat, governmental agencies, because of existing laws and legal and political pressures, are typically required to produce considerable corroborative science to establish the actual harm or threat.) Then, as a matter of social policy, we could create a strong rebuttable presumption against their coming to market or a strong presumption toward removing those already in the market. The following are examples of this kind of research.

- The U.S. EPA has developed screening techniques to identify potentially toxic substances under the Premanufacture Notification provision of the Toxic Substances Control Act to allow action before exposure occurs. More scientific research would aid this effort and help to validate red flags for human health effects (U.S. Congress OTA 1995). Such information would be invaluable to companies in developing more environmentally benign products.

- A science advisory panel in California has considered about fifty neurotoxins that are cholinesterase inhibitors. They appear to act by essentially the same mechanism to produce adverse effects on the cholinesterase of insects and mammals. If the mechanisms are sufficiently similar, there should be no need to investigate and perform individual risk assessments of each cholinesterase inhibitor before judging that it may pose substantial risks to human and other animal health. Moreover, if vegetables and fruits are sprayed with several different substances that contain cholinesterase inhibitors, this should be assumed to be an additive risk rather than individual risks that are unrelated to each other (Byus 2001).

- There is much to be gained from understanding the basic toxicity mechanisms that are common to many different substances. As mechanistic understanding is expanded, scientists might develop an inventory of toxicity mechanisms common to different substances and to use these to identify other toxicants. Such information can also serve to identify molecular changes that serve as early warnings of potential impacts (e.g., is the chemical a cholinesterase inhibitor or does it bind to the estrogen receptor?).

- For many reasons, as a matter of social policy we should respect the value of laboratory animal studies as reasonable indicators of human toxicity risks. They are far from ideal because they are costly and time-consuming (Cranor and Eastmond 2001), but they should be seen as interim procedures until faster, sufficiently accurate short-term tests are developed (Cranor 1997b).

- In identifying and assessing potential carcinogens, scientists could do the following:

  - They could look for mutagens or strong mutagens. Perhaps we should cease being concerned with whether mutagens are sufficiently correlated with animal or human carcinogens in order to validate mutagenic tests.
  - They could look for alkylating and clastogenic agents. An alkylating agent induces genetic mutations. Clastogens are agents that cause chromosomal breakage.
  - They could look for substances that attach to Ah receptors, which activate biological processes that are involved in carcinogenicity (U.S. DHHS 2001). If substances have either effect, we may know enough not to permit them into commerce or to eliminate or lower exposures if they are in commerce.

Generalization

As generalizations from the above examples, scientists could begin to look for other modes of action that are critical steps in toxic pathways and that would permit reliable inferences across substances. It is a reasonable conjecture that the more scientists know about the biological processes at the "molecular, cellular, tissue, and organ" level, the more they will find them "strikingly similar from one mammalian species to another" (Huff and Rall 1992: 434). And, more fundamentally, "the more we know about the similarities of structure and function of higher organisms at the molecular level, the more we are convinced that mechanisms of chemical toxicity are, to a large extent, identical in animals and man" (Huff 1993).

They should, in short, develop cost-effective tests to identify toxicants before they become problems and before they become serious threats.

Both an advantage and a problem of this approach is that it requires fundamental scientific research. Thus, research to identify modes of toxic action fits well with a long-term scientific agenda. It also tends to be difficult and time consuming. However, once scientists have discovered a fundamental mode of action, it may be pertinent to a variety of substances with widespread application, serving well the aims of precaution. Beyond this, the next step would be to design a battery of tests to avoid developing or permitting into commerce substances that exhibit known toxic modes of action.

Consider two further ideas. First, there may be promise in genome-wide expression arrays to permit identification of toxic effects on the human genome, or on particular aspects of it, such as on onco-genes or tumor-suppressor genes, so that potential genetic threats from substances can be identified and eliminated before they become actual threats to humans.

Kenneth Olden, Director of National Institutes of Environmental Health Sciences and the National Toxicology Program, has suggested the possibility of identifying some two hundred "susceptibility genes" in the human genome. Those who are most at risk from various kinds of exposures could then be identified and public health policies developed to protect them. Such a view rests upon the scientific advances of the human genome project as well as the principle that every citizen has equal right to be protected against harm from exposures to toxic substances (Cranor 1997a). Moreover, one might create a susceptibility gene array as a kind of screening device for potentially toxic substances for chemicals or other compounds that would threaten significant adverse reactions in susceptible subpopulations. (Of course, while there is good reason to develop and utilize susceptibility genome arrays, there are downside risks that would also have to be evaluated in their use, e.g., the possibility of medical or workplace genetic discrimination.)

Another idea would be to consider identifying other "necessary, but not sufficient" steps in a toxicity process (like the Ah receptor), studying the relevant mode of action, and searching for other substances that exhibit the same or similar modes of action. Such an approach would be another generalization across substances, looking for a particular mode of action in order to screen out substances before they became real threats.

The examples just considered are known and some progress has been made in using a precautionary approach with some of them. They are suggested to indicate how we might begin to think about and conceive a precautionary scientific approach.

Conclusion

A plausible understanding of the precautionary principle suggests the following roles for scientific research. As a preliminary step, scientists should continue and accelerate their inventory of the state of the world and human health. They could seek to reveal the connections between human activities and environmental effects as well as environmentally mediated effects on human well-being. Because of the subtlety of many problems, especially those resulting from commons problems, it is important to anticipate commons problems. Here, scientists would look for effects that may be individually insignificant, but cumulatively harmful. They might ask the philosopher's generalization question:"What would happen to the environment or human health if a large number of people did this?" And, scientists should not shrink from evaluating the conditions of ecosystems and human health, according to their expertise, and reporting their evaluations to the wider public.

A research agenda that sought to identify potential threats to human health and the environment would best serve the aims of precaution. The alternative is to continue to rely upon time-consuming, corroborative science to confirm—case by case, in legally and politically difficult circumstances—adverse effects or threats to human health or the environment.

A more precautionary approach to science has a number of attractions. It builds on a basic research agenda. If successful, it will remove potential threats before they become serious and need to be addressed by corroborative science, helping to minimize false negatives. It will help avoid time-consuming and costly legal and scientific battles over the seriousness of a threat. It will arm companies with the knowledge to develop products and processes that support a more environmentally sustainable future and government entities with the information needed for wise planning. It will make this information available at an early stage, so choices on products and plans can be made in a cost-effective way. It will open the possibility of leaving the world and human health better than the current generation finds it.

References Cited

Aulerich, R. J., R. K. Ringer, and J. Safronoff. 1986. "Assessment of Primary and Secondary Toxicity of Aroclor 1254 to Mink." *Archives of Environmental Contamination and Toxicology* 15: 393–99.

Brown, L. R., C. Flavin, and H. French. 2000. *State of the World 2000.* New York: W. W. Norton, Co.

Brunner, M. J., T. M. Sullivan, A. W. Singer, M. J. Ryan, J. D. Toft II, R. S. Menton, S. W. Graves, and A. C. Peters. 1996. "An Assessment of the Chronic Toxicity and Oncogenicity of Arochlor-1242, Aroclor-1254, and Aroclor-1260 Administered in Diet to Rats," Batelle Study No. SC920192. Chronic Toxicity and Oncogenicity Report, Columbus, Ohio.

Byus, C. 2002. Department of Biochemistry and Biomedical Sciences Program, University of California, Riverside, and member of California's Air Resources Board Science Advisory Panel. Personal communication.

Carroquino, M. J., S. K. Galson, J. Licht, R. W. Amler, F. P. Perera, L. D. Claxton, and P. J. Landrigan. 1998. "The U.S. EPA Conference on Preventable Causes of Cancer in Children: A Research Agenda." *Environmental Health Perspectives* 106, suppl. 3: 867–73.

Cranor, C. F. 1993, 1997. *Regulating Toxic Substances.* New York: Oxford University Press.

———. 1997a. "Eggshell Skulls and Loss of Hair from Fright: Some Moral and Legal Principles That Protect Susceptible Subpopulations." *Environmental Toxicology and Pharmacology* 4: 239–45.

———. 1997b. "The Normative Nature of Risk Assessment: Features and Possibilities." *Risk: Health, Safety, and Environment* 8: 123–36.

———. 1999. "Asymmetric Information, the Precautionary Principle, and Burdens of Proof in Environmental Health Protections." In *Protecting Public Health and the Environment: Implementing the Precautionary Principle,* edited by C. Raffensperger and J. Tickner. Washington, D.C.: Island Press.

———. 2001. "Learning from the Law to Address Uncertainty in the Precautionary Principle." *Science and Engineering Ethics* 7: 313–26.

Cranor, C. F., and D. A. Eastmond. 2001. "Scientific Ignorance and Reliable Patterns of Evidence in Toxic Tort Causation: Is There a Need for Liability Reform?" *Law and Contemporary Problems* 64: 5–48.

Environmental Health Newsletter. 1998. "EPA, EDF, CMA Agree on Testing Program Targeting 2,800 Chemicals," vol. 37: 193. Silver Spring, Md.: Business Publishers, Inc.

Hornshaw, T. C., H. E. Aulerich, and Johnson. 1983. "Feeding Great Lakes Fish to Mink: Effects on Mink and Accumulation and Elimination of PCBs by Mink." *Journal of Toxicology and Environmental Health* 11: 933–46.

Huff, J. 1993. "Chemicals and Cancer in Humans: First Evidence in Experimental Animals." *Environmental Health Perspectives* 100: 201–10.

Huff, J., and D. Hoel. 1992. "Perspective and Overview of the Concepts and Value of Hazard Identification as the Initial Phase of Risk Assessment for Cancer and Human Health." *Scandanavian Journal of Environmental Health* 18: 83.

Huff, J., and D. P. Rall. 1992. "Relevance to Humans of Carcinogenesis Results from Laboratory Animal Toxicology Studies." In *Public Health and Preventive Medicine,* 13th ed., edited by J. M. Last and B. Wallace, 433–44. Norwalk, Conn.: Appleton & Lange.

International Agency for Research on Cancer. Last modified 9 August 2000. "Overall Evaluations of Carcinogenicity to Humans: Probably Carcinogenic to Humans." http://193.51.164.11/monoeval/crthgr02a.html.

Laurance, W. F., M. A. Cochrane, S. Bergen, P. M. Fearnside, P. Delamonica, C. Barber, S. D'Angelo, and T. Fernandes. 2001 "The Future of the Brazilian Amazon." *Science* 291: 438–39.

Levins, R. "An Ecologist Looks at Health." Unpublished manuscript, on file with author.

Lubchenco, J. 1998. "Entering the Century of the Environment: A New Social Contract for Science." *Science* 279: 491–97.

McDonald, T. A., and K. Hooper. 2000. "The PBDEs: An Emerging Environmental Challenge and Another Reason for Breast-Milk Monitoring Programs." *Environmental Health Perspectives* 108: 387–92.

McKibben, B. 2001. "Some Like it Hot." Review of *Climate Change 2001: Third Assessment Report, The Scientific Basis, Impacts, Adaptation, and Vulnerability, Mitigation and the Bush Administrations' National Energy Policy: Report of the National Energy Policy Development Group. The New York Review of Books.*

Molina, M. J., and F. S. Rowland. 1974. "Stratospheric Sink for Chlorofluoromethanes: Chlorine Atom Catalyzed Destruction of Ozone." *Nature* 249, no. 5460: 810–12.

Muir, J. 1988. *My First Summer in the Sierras.* San Francisco: Sierra Club Books.

National Research Council. 1984. *Risk Assessment in the Federal Government: Managing the Process.* Washington, D.C.: U.S. Government Printing Office.

———. 1985. *Screening and Testing Chemicals.* Washington, D.C.: U.S. Government Printing Office.

———. 2002. *The Environmental Effects of Transgenic Plants.* Washington, D.C.: U.S. Government Printing Office.

Ostrom, V., and E. Ostrom. 1977. "Theory of Institutional Analysis of Common Pool Problems." In *Managing the Commons,* edited by G. Hardin and J. Baden, 157–62. San Francisco: W.H. Freeman and Company.

Rodgers, W. H., Jr. 1994. *Environmental Law.* St. Paul, Minn.: West Publishing Co.

Thagard, P. 1999. *How Scientists Explain Disease.* Princeton: Princeton University Press.

Tickner, J. 2001. Personal communication.

Travis, C. C., and S. T. Hester. 1991. "Global Chemical Pollution." *Environmental Science and Technology* 25: 814–19.

United Nations. 1992. *Agenda 21: The Rio Declaration on Environment and Development.*

U.S. Congress, Office of Technology Assessment (OTA). 1987. *Identifying and Regulating Carcinogens.* Washington, D.C.: U.S. Government Printing Office.

———. 1995. *Screening and Testing Chemicals in Commerce.* Washington, D.C.: U.S. Government Printing Office.

U.S. Department of Health and Human Services (U.S. DHHS). 2001. *Ninth Annual Report on Carcinogens.*

U.S. Envrionmental Protection Agency (U.S. EPA). 1996. "PCBs: Cancer Dose-Response Assessment and Application to Environmental Mixtures." APA/600/P-96/001F September 1996.

Wahlstrom, B. 1999. "The Precautionary Approach to Chemical Management—Swedish Views and Experiences." In *Protecting Public Health and the Environment: Implementing the Precautionary Principle,* edited by C. Raffensperger and J. Tickner. Washington, D.C.: Island Press.

Zeise, Lauren. 1999. Chief, Reproductive and Cancer Hazard Assessment Section, California Environmental Protection Agency. December 1999. Personal communication.

CHAPTER 21

The Children of the Yaqui Valley: Precautionary Science and Communities

Elizabeth A. Guillette

The precautionary methods of public health tend to be based on the prevention of disease. In most parts of the world children are vaccinated, and infectious diseases tend to be rapidly confined with treatment. The conventional environmental factors considered by public health agencies are sanitation, personal habits, and access to sufficient food. Such factors are relatively easy to observe and measure, as are their effects. Factors where unknown outcomes may be involved, such as exposure to synthetic chemicals as a contributing cause to ill health, tend to be overlooked. The ambiguous nature of contaminant exposure contributes to the difficulty of observing clear correlations. In addition, epidemiological studies involving contamination and birth defects and cancer in young children often fail to achieve statistical significance because occurrences are rare (Axelrod et al. 2001). Documentation of mental and health changes is difficult to obtain and analyze, as many countries, including the United States, fail to keep records of birth defects and various chronic diseases.

The problem becomes more complex with the detection of childhood developmental disorders, because symptoms may fall within ranges of behavior that are considered normal by current standards. True baseline data on standards for normal development before the industrial, high-technology era are not available for comparison with today's data. The innate potential abilities of children will never be known, as every person is now exposed, to some degree, to mixed toxicants that can affect development (DeRosa et al. 1998).

Diagnosed disease is the focus of many human studies, including those involving endocrine-disrupting contaminants. In this chapter I discuss a study that focused, instead, on hidden effects of toxic exposures on children's health and development. Many of these subtle effects are easily overlooked in typical childhood health examinations. I argue that to protect children's health from environmental risk more effectively, conventional science must be combined with community involvement. Research findings, especially if they imply environmental risks for children, have direct meaning to families. The spread of knowledge stimulates action for change. Scientists must work with communities to promote precautionary action.

Subtle Health Changes: The Yaqui Valley of Mexico

The Yaqui Valley study (Guillette et al. 1998) serves as an example of how scientific methods can be adapted and joined with community observations and action in order to support precautionary policies. The indigenous people of the Yaqui Valley, located in the state of Sonora, Mexico, underwent a philosophical division when modern farming techniques were introduced in the early 1950s. Some individuals embraced the use of equipment and pesticides, and large areas were planted in single crops such as chilies and cotton. New towns arose in the valley to provide the necessary infrastructure for farming, family residency, and support services.

At the same time, other indigenous families preferred to retain traditional farming and ranching, which was possible to do in the nearby foothills. People from the lower section of the valley migrated to the one existing town. Adding to the population were individuals from the higher elevations seeking the benefits of modernization without modern-style agriculture. The result was a population with close relatives residing in both the foothills and the valley itself.

Today, all the towns in both the valley, where modern agriculture is carried on, and the foothills, where traditional farming continues, have been modernized to similar degrees. All inhabitants retain customary dietary practices and cultural values, including the avoidance of drugs and alcohol by women. Researchers say the valley and foothill towns offer a worthy comparison, comparable to a controlled laboratory experiment, due to similar genetic background, ethnicity and culture, and infrastructure. Modern agriculture did not bring greater prosperity to the valley, resulting in a similar socioeconomic status in all towns. The only difference is in levels of chemical exposure (Weiss 2001).

Chemists at the Instituto Tecnologico de Sonora (ITSON) had an interest in the pesticide levels found in newborn cord blood and breast milk, indicating persistent pesticides in mother and child (Banuelos and Montenegro

1991). Emphasis was placed on the high levels of contaminants carried by the mothers. The school expressed a desire for cancer studies, particularly breast cancer, based on reports from fellow employees with relatives in the agricultural region. My initial contact with these chemists was initiated by a fellow anthropologist familiar with the area and my own interests in contaminant research. The Yaqui study grew out of attention to concerns that came from the valley's pesticide use–agricultural communities. That is, the science was inspired by, and drew its direction from, community questioning.

A main principle of anthropology is an understanding of beliefs systems and the level of knowledge regarding the question at hand. In 1995, focus groups, in which a small group of individuals with a common interest discuss a given topic, were organized. Separate focus groups were held with parents and with community leaders. I asked open-ended questions about the community's perceptions on pesticides and their safety, the use of pesticides in homes, sources of community pesticide exposures, and perceptions of the occurrence of disease. The sessions also introduced ITSON students and staff to a new method of research.

Groups in the less-exposed foothills expressed a strong belief that pesticides were harmful to human health. They claimed they did not use them in gardens, the traditional method of agriculture, or in the home. These people said birth defects, mental retardation, and cancers were either very rare or absent in the area. Men and women gave similar replies to all questions.

The replies of men and women living in the valley varied. The men of the valley knew that massive direct contact with pesticides would result in death, and they acknowledged that a pesticide spill in the area had caused large numbers of fish to die. At the same time, they expressed the belief that pesticides were relatively safe for humans. Both fathers and political leaders said they had experienced multiple exposures without obvious harm. Men were also hesitant to discuss disease incidence, although this may have been a realistic reflection of culture.

Valley women felt pesticide exposure was harmful to health. Short-term consequences included headaches, nausea, and vomiting. They spoke openly about breast cancer and various blood diseases, noting that rates appeared high among young adults. Other than an apparent increase in birth disorders over the years, no person could identify childhood health changes. But the mothers in the valley all insisted that something was different about their children, although they could not identify specifics. A comparison of their own childhood with the present yielded no clues, indicating changes occurred very slowly or were very subtle.

At first glance, the valley children appeared normal and typical to me, yet I knew that mothers are usually correct in recognizing subtle problems. But I did not know what type of changes might be occurring or whether the changes would be evident on the community level. One conversational question, "Do your children have trouble remembering things?" provided a hint that subtle changes were occurring. Although all mothers answered "yes," the reasoning behind the answer varied markedly between the foothill and valley mothers. Mothers in the valley said if they told their child about an upcoming event, the child would not remember being told. Foothill mothers said forgetfulness was a problem because their child had to be told several times to perform a duty, such as cleaning their room. Such answers indicated the traditional survey type of research would not be adequate in this situation. The situation demanded a broad-based evaluation of physical and mental developmental abilities of the children themselves.

A double blind study was impossible. The poverty of the area, the limited transportation for group testing, and the scarcity of modern facilities meant the research would have to be performed in individual homes under varying circumstances. This called for a research protocol different from the conventional scientific approach to study. We allowed more room for variation in test settings, for example, than most scientific studies allow. This adaptability proved to be advantageous, providing information that would have been impossible to observe otherwise.

The initial research involved four- and five-year-old children. This eliminated any differential effects of schooling. We designed the testing of physical and mental abilities to appear as play. We chose standard developmental-scale activities, such as catching a ball and jumping, to test neurophysical abilities such as eye-hand coordination and stamina. We used items from various IQ tests, such as drawing a person and memory tests, to represent neuromental abilities.

No individual child in any town did extremely poorly or extremely well on all exercises. The nature of the research format prevented a comparison of total abilities on an individual level. The itemized scores of children within a given community were statistically compared to other communities. Thus, rather than determining risk factors for individual children, we designated as living in a given community as the risk factor. Towns varied between populations of 800 to 1,000, limiting the number of four- and five-year-olds. Ten to twelve children were tested in each of three valley towns (total of thirty-three) and seventeen in the foothill town. Males and females were evenly divided. Various growth statistics, such as height, weight, and arm circumference, showed no significance difference between sites.

The four- and five-year-old children of the valley community scored statistically lower in all investigated areas, including gross and fine eye-hand coordination, stamina, memory, and perceptual abilities, than did the children of the less-exposed foothill community (Guillette et al. 1998).

The living situations of the families inspired unconventional testing methods. Although we tested all four- and five-year-olds within single households, neighboring children and siblings out of the testing age range were invited to join in the play. The companionship seemed to prompt each child to do his or her best and forget that this was a test. (Only the child[ren] falling in the study group were scored on abilities.) During the drawing exercise, the child or children undergoing testing sat with the field staff to prevent input from others, although all other children were provided with paper and pencil. The group involvement allowed for observations of various interactions. Overall, the valley children who were not undergoing a specific test tended to be less active in play. Even though they were handed a ball, they tended to stand separately without talking or playing. The foothill children, by contrast, immediately started their own games with much cooperative interaction between individuals of all ages.

Many of the same children, along with age cohorts, were tested again two years later. Developmental activities were adapted to be representative of six- and seven-year-olds. The pesticide-exposed children, when compared to those less exposed, continued to show deficits in coordination and stamina. We had learned in our first round that the original test for balance, standing on one foot, was inappropriate for preschool children, because they were taught that standing on one foot was dangerous and could lead to injury. For this older group, we devised a new test for balance involving walking on a 5-cm-wide plank on the ground. The children enjoyed this activity, but the valley children showed a decreased sense of balance, compared with the foothills children, especially when combined with the coordination involved with having to turn around and walk back on the board.

Children, now having a year or two of schooling, were asked to draw a person under a tree and then add a dog. The valley children's drawings of people were similar to those drawn by the less-exposed four-year-olds two years previously. The dogs tended to have incomplete body parts (see figure 21.1). Mental functioning, including problem solving and memory, remained compromised. General behaviors when under the stress of mental exercises indicated frustration and short attention spans. In addition, the exposed children usually did not recognize their errors when given opportunity to correct a response, in contrast to the less-exposed group.

We interviewed valley schoolteachers during this second visit, when the children under study were of school age. Teachers complained that they needed to repeat facts more often than before and that children daydreamed more. Teachers also claimed an increase in classroom disruptive behaviors. In addition, a health survey of these children revealed a significantly higher incidence of overall illness, with infections occurring at almost twice the rate of those less exposed (Guillette 2000a).

The previous pesticide study performed by ITSON showed that the studied

Foothills   Valley

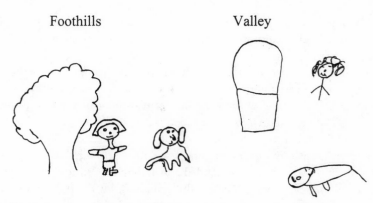

Figure 21.1. Drawings of a person, tree, and dog done by seven-year-old foothill and valley children of the Yaqui valley region of Mexico after one year of formal schooling.

valley children who were exposed to various pesticides in utero and during breast-feeding. Using equipment that was able to test for seventeen pesticides, assorted organochlorine compounds, including DDT, lindane, and endrin, were found in cord bloods and breast milk one month post partum (Banuelos and Montengegro 1991). No further studies have been adequately performed. According to the evidence of empty cartons and barrels found in the region, newer pesticides, reflective of development and use in the United States, have been introduced into the agricultural region since this initial study. The assumption is that the valley children have been exposed to multiple types and classes of pesticides, both in utero and continuing during their lifetime. Cell and system damage occurring from in utero exposure is thought to be permanent with lasting effects of attention span, intelligence, and behavior (Schettler et al. 2000). Therefore, the noticed decreased functional and mental abilities of the studied children are likely to have begun in utero and are possibly aggravated by continuing exposures. All parents of the tested children were given immediate feedback regarding the strengths and weaknesses of their individual children in both the initial and second study. Community meetings were held in each town following analysis of data. The mothers expressed gratitude that someone cared enough to discover that their suspicions were correct. At the same time, the families were faced with the dilemma of whether to leave the area or remain. Generally, they saw the risk of leaving the area and not finding suitable income as greater than remaining.

Researchers provided education on avoiding pesticides at home and dispelled various myths about pesticides. In the following years, we noted that mothers decreased the use of home pesticides but the many ambient exposures continued. While the Mexican government is making headway in eliminating

DDT and other persistent pesticides, these poor farmers are powerless to change the terms of the bank loans that require them to use pesticides in order to meet the U.S. demand for fruit and vegetables that look perfect (Villa 2001).

Community Recognition of the Dangers of Contamination

The Yaqui Valley study demonstrates the value of recognizing community concerns and delving into the unknown with innovative scientific methods. The valley towns are not unique in perceiving health and behavior change associated with contamination. As an anthropologist involved with communities, I have heard concerns from people around the world. Local midwives in agricultural sections of southern Africa reported a startling increase in children born with genital malformation (Ranots 1994). Premature breast development (thelarchy) has become epidemic in Puerto Rico, along with an increase of breast tissue formation in young boys (Bourdony 2001). Families living near the closed air force bases in the Philippines recognize health and behavioral change in their children, with mental and physical retardation at the forefront (Baldonado 1999). Respiratory disorders and growth changes are found in the Bhopal region of India (Sarangi 2002). Increased illness rates and learning disorders occur with those dependent on the Mississippi Delta for subsistence (Frate 1995). Concerned citizens have begun their own community assessments of the impacts of contamination from various sources, including pulp mill effluent in the Far East and arsenic in India.

In all instances, the lack of formal education has not stopped citizens in these communities from recognizing the presence of deviations from normal health and development (Guillette 2002). I have provided the methodology used in the Yaqui Valley to these communities, as their main concerns have been the hidden or subtle deficits in children.

Researchers must diverge from the traditional survey interview and hospital record reviews and explore new avenues for revealing the possible effects of various contaminants. For instance, young children might be evaluated with a comprehensive neuromuscular, neuromental developmental approach, similar to the one we used. Then the families of children scoring low and high might be investigated with a detailed exposure assessment (Guillette 2000c).

Community Action

The hidden deficits found in the Yaqui Valley may be regarded as relatively small when compared with gross birth defects or childhood cancers. However, those small deficits may have enormous implications for public health. A slight

downward shift of intelligence was found to be associated with in utero exposure to polychlorinated biphenyls (PCBs) (Jacobson et al. 1996). For an individual of normal intelligence, a loss of five IQ points may be inconsequential.
But on a societal level, the consequences are great, as the number of individuals in the highly intelligent group diminishes and the number of individuals
requiring special education and training programs rises substantially (Weiss
1997). The Yaqui Valley children, although not directly tested for IQ, demonstrated a lack of perceptual and mental-processing abilities. When such neuromental changes are combined with developmental changes such as decreased
coordination, stamina, and balance, the societal consequences may extend
beyond the need for special education. Everyday household duties and job
performance are affected.

Community action to avoid unnecessary exposure to contaminants is
growing. A basic realization that pesticides were likely to contribute to social
and medical problems led the town of Hudson, Canada, to pass a ban on the
cosmetic use of pesticides on lawns and public places in 1992 (Epstein 2001).
Other towns followed with similar restrictions on pesticide use. Halifax, in
2000, became the first major city to pass a bylaw banning cosmetic pesticide
applications (Pesticide By-law Advisory Committee 2000).

I participated in a presentation to the Halifax Regional Municipal Council. The council took an interdisciplinary approach, hearing testimony from
research scientists, physicians, and individuals who suffered acute reactions to
pesticides or multiple chemical sensitivities attributed to childhood exposure.
A progressive lawn-care company presented evidence that integrated pest
management was possible in the area.

Soon after Halifax adopted its restriction, protests by lawn-care companies
in the township of Hudson, the place of the earlier ban, resulted in a case being
brought to the Supreme Court of Canada. Under question was the right of
townships to pass bylaws not reflecting national laws. Defending attorneys presented the contextual approach implied by the precautionary principle—a
community's right to take action to protect itself in advance of scientific
proof—as consistent with principles of international law and policy (Epstein
2001). The court, in its ruling, referred to the precautionary principle and
affirmed that the pesticide control by-laws were legal preventive actions
(Epstein 2001).

The Need to Unify Communities, Science, and Policy for Good Health

My participation in many Canadian communities involved presenting the
methodology and findings of my Yaqui Valley research. In such instances, I believe
that scientists should limit themselves to their own area of contaminant research

and findings. Advocacy of policies, by explanation of how research findings in one locale may apply in another, is a necessity, bringing out that human contaminant research is in its infancy. For instance, I usually stress that the Mexican children may have been exposed to higher amounts of pesticides than other children, but with the unknowns of the dangers of small exposures on the developing fetus, limiting the unnecessary use of chemicals is a sound policy for public interest. Advocacy differs from activism, or the demanding of change, which has the potential danger of compromising interpretation of research value and honesty.

Scientists must not forget the damage that occurs from exposures to multiple compounds. Research involving various contaminants and their effects on humans suggests that effects of endocrine disruption are similar, although mechanisms of action may vary (Guillette 2000b; Gnu et al. 1994; Jacobson and Jacobson 1996). Damage to the neuromuscular, immune, thyroid, and reproductive systems could have greater social and medical impacts than a small decrease in intelligence. As the quality of overall health decreases, so will the qualities of productivity, innovation, and leadership.

The challenge is to link physical, social, and medical science to what is happening within communities. Precautionary investigation must extend beyond the suspected into the unknown, generating new hypotheses from community knowledge. Interdisciplinary models for the methodology of research will help to clarify some of the uncertainties involved with contamination. This unification of multiple scientific approaches and disciplines, demonstrating likely correlations between contamination and undesirable human outcomes, will provide the public with sound knowledge needed to create change in policy and in their individual lives.

But the precautionary principle is about taking protective action even when such correlations cannot be proved scientifically. Science must investigate life and living conditions in the real world. The real world is made up of people with various genetic makeups, daily lifestyle practices, and cultural beliefs—and various exposures to multiple contaminants. Each of these can affect people's health, and the role of each factor remains uncertain. Individuals apply precaution in daily life by taking prudent steps without full knowledge. The same is true with communities as they recognize health changes occurring with contamination (Sclove and Scammell 1999). The precautionary principle must be applied on all levels and on a continuum involving people, policy, and science.

References Cited

Axelrod, D., D. Davis, R. A. Hajek, and L. A. Jones. 2001. "It's Time to Rethink Dose: The Case for Combining Cancer and Birth and Developmental Defects." *Environmental Health Perspectives* 109: A246–49.

Baldonado, M. Peoples Committee for Base Cleanup. Quexon City, Philippines. 8 April 1999. Personal communication.

Banuelos, M. G. and M. M. Montenegro. 1991. "Principles vias de contaminacion por plaquicidas en neonatos-lactanes residents en Pueblo Yaqui, Sonora Mexico." *Instituto Tecologico de Sonora* 1, no. 2: 33–42.

Bourdony, C. University of Puerto Rico School of Medicine, San Juan, P.R. 19 October 2001. Personal communication.

Canada v. Hudson, Case Report. 2001. *Municipal and Planning Law Reports, Third Series* 9: 1–33.

DeRoas, C. P. R., H. Pohl, and D. Jones. 1998. "Environmental Exposures that Affect the Endocrine System, Public Health Implications." *Journal of Toxicology and Environmental Health* B8, Crit. Rev. 1: 3–28.

Epstein, H. 2001. "Case Comment: Sprayteck v. Town of Hudson." *Municipal and Planning Law Reports, Third Series* 19: 56–66.

Frate, D. 1995. "Agricultural Pesticide Exposure through Subsistence Fishing in the Mississippi Delta: An Anthropological Perspective." Presented at the Society for Applied Anthropology 1995 Annual Meeting. Albuquerque.

Guillette, E. 2000a. "A Comparison of Illness Rates between Children Exposed to Agricultural Pesticides and Non-Agricultural Children in Sonora, Mexico." In *E-Hormone: The Cutting Edge of Endocrine Disrupter Research*. New Orleans: Tulane/Xavier Center for Bioenvironmental Research.

———. 2000b. "An Anthropological Interpretation of Endocrine Disruption in Children." In *Environmental Endocrine Disrupters: An Evolutionary and Comparative Perspective*, edited by L. J. Guillette and D. A. Crain, 322–34. New York: Taylor and Francis Publishers.

———. 2000c. "Examining Childhood Development in Contaminated Urban Settings." *Environmental Health Perspectives* 108, suppl. 3: 389–93.

———. 2002. "Hidden Outcomes from Contamination, The Need for Community Assessment." Presented at the Health Canada Environmental and Occupational Toxicology Seminar Series. Ottawa: Health Canada.

———. In press. *Communities at Risk: The Assessment of Children*. Kagaku (Science) Japan.

Guillette, E., M. Meza, M. Aquilar, A. Sota, and I. Garcia. 1998. "An Anthropological Approach to the Evaluation of Preschool Children Exposed to Pesticides in Mexico." *Environmental Health Perspectives* 106: 347–52.

Gnu, Y. L., C. Lin, W. Ryan, and C. Hsu. 1994. "Musculoskeletal Changes in Children Prenatally Exposed to Polychlorinated Biphenyls and Related Compounds." *Journal of Toxicology, Environment, and Health* 4: 83–93.

Jacobson, J., and S. Jacobson. 1996. "Intellectual Impairment in Children Exposed to Polychlorinated Biphenyls in Utero." *New England Journal of Medicine* 335: 783–89.

Pesticide By-law Advisory Committee. 2000. *Report of the Majority of the Pesticide By-law Advisory Committee to Halifax Regional Municipality: Putting People and Ecosystem Health First.* 1 October. Halifax, Canada.

Ranots, A. Maluti Nursing School, Ficksburges, South Africa. 1 July 1994. Personal communication.

Sarangi, S., Sambhavna Trust, Bhopal, India. 20 February 2002. Personal communication.

Schettler, T. G. Solomon, M. Valenti, and A. Huddle. 1999. *Generations at Risk: Reproductive Health and the Environment.* Cambridge, Mass.: MIT Press.

Sclove, R., and M. Scammell. 1999. "Practicing the Principle." In *Protecting Public Health and the Environment: Implementing the Precautionary Principle,* edited by C. Raffensperger and J. Tickner, 252–65. Washington, D.C.: Island Press.

Villa, F. L. 2001. *Instituto Tecnologico de Sonora, Direccion de Investigacion y Estudios de Postgrado.* Obregon, Mexico.

Weiss, B. 1997. "Endocrine Disruptors and Sexually Dimorphic Behaviors: A Question of Heads and Tails." *NeuroToxicology* 581, no. 2: 581–86.

———. 2000. "Vulnerability of Children and the Developing Brain to Neurotoxic Hazards." *Environmental Health Perspectives* 108: 375–81.

———. 2001. Interview in *Toxic Legacies,* a television video co-produced by The Nature of Things, Force Four Entertainment, and Discovery Health U.S.A.

CHAPTER 22

Science Communication and Precautionary Policy: A Marine Case Study*

Cato C. ten Hallers-Tjabbes

The concept of precaution has had a place in international decisions on the protection of the marine environment since 1987, when participants at the Second International North Sea Conference on the Protection of the North Sea declared:

> In order to protect the North Sea from possible damaging effects of the most dangerous substances, a precautionary approach is necessary which may require action to control inputs of such substances even before a causal link has been established by absolute clear evidence. (London Declaration, 1987)

The wording of the London Declaration indicates a distinct role for science: to pass on signals of potential harmful effects or damage. It deviates from the former assumption that action must be based on clear, dose-effect-related proof of harm. This precautionary approach calls for a different approach to scientific communication as well, one that issues timely warnings and allows decision makers to grasp the reasons underlying scientists' concerns. It involves interactive communication rather than mere reporting on findings in the form of numbers and written reports. It challenges scientists to understand how decision makers see the world and to recognize which information will help

---

*Netherlands Institute for Sea Research publication 3695.

clarify the message that damage and harmful effects are likely to occur. In order to frame that message, scientists need to understand not only their own specialized studies but also how their knowledge links to broader scientific information. They also need to gain a basic understanding of the world of decision making and its most relevant processes. A similar approach is needed to communicate to the general public and to those involved in activities that are foreseen to cause environmental damage or harm.

A group of marine scientists of which I am a part has learned much about what this kind of scientific communication entails. During the past decade, we have been working both to study a particular problem and to make sure that our findings are translated into appropriate policy. What we have learned may be helpful for other scientists who want to ensure that their research supports wise policies.

Precaution in the Marine Environment

The very first reference to the concept of precaution in connection with environmental protection occurred in a marine context. Perhaps this can be explained by the character of the marine environment, which is usually submerged and turbid, so that signals of deterioration do not leap to the eye. And not only signals of deterioration remain hidden; there is much more going on than we know of. Recently, whole new communities that do not need oxygen for fueling their energy have been detected in several deep oceans. At terrestrial locations, early signals of environmental decay are likely to leave some mark in the vicinity of someone's backyard, whereas at sea, environmental disasters may take place without anyone noticing (Jensen 1989; Sperling 1986, 1988; Hey 1992). The marine environment is a unified system, where everything is connected, if only by the flow of water masses. It is impossible to see what is happening in it, let alone predict what might happen, without concerted, long-term study and observation. Therefore, it is critical for scientists who are aware of the presence of such hidden conditions to pass the information on, along with explanations of why it might be happening.

In the marine environment, the limits of the theory of "environmental capacity" (Cairns 1977; Pravdic 1985; Stebbing 1992; Thorne-Miller 1994) were becoming apparent in the latter decades of the twentieth century, following reports on sudden and unforeseen collapses of marine systems. In Denmark oxygen depletion due to eutrophication (excess of nutrients) caused a massive dying of organisms, while the measured eutrophication levels had appeared to remain within critical limits (Jensen 1989). Reports of local extinction of fish stocks that none had foreseen began to accumulate. One had assumed that the marine environment itself would signal when harm was about to become irreversible or deleterious on a larger scale. The belief that activities and inputs could be banned *after* harmful effect was fully proved led to halfway solutions. Ecosystems were calculated, on the basis of a dose-effect relationship, to be able to cope with considerable levels of harmful input. However, time and again, no matter what "safe" levels were maintained, ecosystems deteriorated.

The adoption of the precautionary concept, based on these concerns, represented a major shift in perception of the human responsibility for the marine environment. The 1987 London Declaration, adopted by the North Sea States, invoked precaution not only as a general principle but also, specifically, in regard to "reducing emissions of substances that are persistent, toxic and liable to bioaccumulate at source by the use of the best available technology and other appropriate measures." The Declaration continued: "This applies especially when there is reason to assume that certain damage or harmful effects on the living resources of the sea are likely to be caused by such substances, even when there is no scientific evidence to prove a causal link between emissions and effects ('the principle of precautionary action')" (London Declaration, 1987).

In 1990, the Third International North Sea Conference on the Protection of the North Sea reiterated the precautionary principle as a goal-setting instrument for future work. Precaution was also incorporated in the Convention for the Protection of the Environment of the North-East Atlantic (OSPAR 1992).

Potentially harmful substances enter the marine environment through specific routes, either from objects at sea (shipping, mineral extraction, dumping activities at sea), through aquatic sources entering the sea (rivers, estuaries, coastal runoff), or through the air via the surface microlayer, picked up through spray and wave action. The environmental supervision of such routes is subject to specific international and regional regulations in which a precautionary approach is widely applied (Freestone and Hey 1996). Some of the most important of these are mentioned below.

The International Maritime Organization's Marine Environment Protection Committee is the main platform for global decision making and consultation on the environmental aspects of marine shipping and for the MARPOL Convention (1973/1978). State representatives consult on a regular basis in MEPC meetings.

The 1972 London Convention (LC 1972) and its 1996 Protocol govern dumping at sea worldwide. The convention incorporates what has been termed a "reverse listing" concept, which means that only those substances specifically permitted may be released into the sea—all others are banned.

Further application of precaution in marine environmental protection is set out as one of the objectives of the United Nations Conference on Environment and Development (UNCED 1992: 92). UNCED agenda 21, section 17.22, proclaims that states commit themselves to "apply preventative, precautionary and anticipatory approaches so as to avoid degradation of the marine environment as well as to reduce the risk of long-term or irreversible effects upon it."

Tributyltin: An Endocrine Disruptor in the Sea

As the case study below will demonstrate, declarations, policies, and even laws are not always enough to change harmful practices. The "assimilative capacity" assumption still prevails in many quarters of the marine community. However, the history of decision makers' responses to harmful impacts of one particular substance, tributyltin (TBT), is the record of change from an assimilative capacity approach to a precautionary one. This has been the focus of my work for the past decade.

I will describe in some detail a project we performed from 1998 to 2001, called "Action to Demonstrate the Harmful Impact of TBT: Effective Communication Strategies between Policy Makers and Scientists in Support of Policy Development" (the TBT communication project). But the roots of this project go back much further.

TBT is an ingredient that has been used in "antifouling" paints to protect the hulls of ships from barnacles and other organisms that attach to the hull. TBT was first used as antifouling on marine ships in the early 1970s and slowly became the most commonly used antifouling agent. However, alternatives remained being used and new ones kept being developed, which found their way to content users. TBT is one of the group of organotins, industrial products that are also used as pesticides, as wood preservatives, and as stabilizers in plastics such as PVC. Apart from TBT, triphenyltin (TPT, also used as a pesticide) was sometimes also included in antifouling paints. All organotins are highly toxic to several biological processes, including those of immune and nerve systems, and TBT is among the most toxic of them all (Snoeij et al. 1987; Fent 1996). Effects occur throughout the animal kingdom, including in humans (Whalen et al. 1999). TBT is also the only organotin that is directly and purposefully released into the water, where it can freely enter all that lives there. By its nature TBT is even more toxic in the sea than in fresh water, and it behaves more unpredictably than most other toxic compounds. TBT is toxic to very basic biological mechanisms, and it has androgenic endocrine disruption effects leading to higher testosterone (a male sex hormone) levels in females. This androgenic effect has first been found in coastal marine snails; recently studies indicate that this effect may occur more widely in the animal kingdom, including in humans (Heidrich et al 1999).

TBT had been suspect of deleterious effects since the 1970s (Dundee

1979), but only after oyster cultures suffered (economic) loss from TBT in a yachting area in France did local policy respond (Alzieu et al. 1991). Soon after, the decline of a coastal snail species due to TBT contamination from yachting and marinas caused a stir in the decision world (Mee 1996), although only in relation to small craft in coastal environments.

The decline was due to what is now known as the most sensitive effect from TBT, imposex, the growth of male parts, such as a penis, in female snails. Imposex occurs at concentrations of less than 1 nanogram TBT per liter of seawater (a concentration of 10^{-9}) and can lead to aborted eggs and sterile female snails, so the population disappears, unable to reproduce (Bryan et al. 1987; Gibbs and Bryan 1986). No other compound has such effects. TBT is one of the toxic compounds for which quantitative structure activity relationships (QSARs) cannot adequately predict environmental behavior (Gray 1996). QSARs predict toxicity of compounds on the basis of similarities in their chemical structures and their ability to pass through biological membranes. A report from the late 1980s expressed serious concern about TBT, calling it "the most toxic compound ever brought into the marine environment" (Ward 1988).

After the European Union and the United States took action to restrict TBT, the International Maritime Organization (IMO, the UN Agency for global marine shipping, regulations for shipping, and a platform for consultation of member States) and its Marine Environment Protection Committee (MEPC) recognized the harmful impact of TBT in *coastal* environments from abundant small shipping and yachting. The IMO recommended a global ban on TBT-containing paints on ships smaller than 25 meters (Resolution MEPC 46[30], 1990). A recommendation by the IMO in itself does not have a mandatory status. Several states responded by a mandatory ban in their state legislation; many others did not. At the same time, IMO-MEPC declared that "concentrations of TBT in open sea are too low to cause harmful effects" (IMO 1990). This was based on conjecture, not evidence, and on assumptions of the marine environment's assimilative capacity.

The London Convention 1972 (LC 72), regulating dumping of matter at sea, has long ignored the fact that harbor-dredged material acted as a channel for secondary input of TBT into the sea (ten Hallers-Tjabbes 1995a). To date, this compounding of materials is not taken into account when evaluating whether dredged materials may be dumped at sea.

The slow recognition of the serious impact of TBT may also bear on the androgenic, not estrogenic, character of the toxic action of TBT. Estrogenic environmental effects have received ample attention in the scientific and public debate, often without being clearly linked to responsible compounds; androgenic effects, even although proven in the case of TBT, have only marginally been on the public and political agenda.

Scientists Meet Policy

My colleagues and I at the Netherlands Institute for Sea Research entered the picture in 1991 as scientists wondering why time and again, TBT was identified as a coastal problem and caused by small shipping only, while nobody knew what happened in deeper waters, where the bigger ships sail. So far, both the scientific community and the IMO had ignored the possibility that the substance could be harmful elsewhere in the marine environment, although impacts might be difficult to detect and characterize (IMO 1990, 1994). Knowing that TBT could cause harm at extremely low concentrations, we decided to investigate possible effects in offshore waters.

Our first investigations, completed in the North Sea in 1991, revealed imposex in the offshore snail species *Buccinum undatum* ("common whelk" or "whelk"). Imposex incidence was clearly correlated with the number of ships passing the research locations (ten Hallers-Tjabbes et al. 1994). This phenomenon had not been observed in studies on the same species in the early 1970s, when TBT was not yet used in marine shipping (Ten Hallers-Tjabbes 1979). Furthermore, whelk populations had disappeared from several densely shipped areas since then, in which decline TBT might have played a major role (ten Hallers-Tjabbes et al. 1996).

Recognizing the relevance of these findings for the international decision process at IMO and for the countries bordering the North Sea, we took steps to enter these findings into the decision process of the International Conferences on the North Sea. At an intermediate Conference in 1993, the North Sea States determined to ask the IMO to regulate TBT more strictly (Copenhagen Statement of Conclusions 1993). However, at its 35th Meeting (IMO 1994), IMO-MEPC stated that further regulation of TBT was not justified, due to the adequacy of Resolution 46(30), which recommends banning TBT from small ships. In 1995, the North Sea States decided to take concerted action to ask IMO for stricter control of TBT, including a possible ban (Esbjerg Decalaration 95).

In the course of our research to gain a better understanding of the extent of the impact of TBT on the common whelk in open sea, we investigated the whelks' burden of TBT and metabolites. We found that TBT levels were correlated to shipping density and to imposex incidence, showing that TBT did play a role (ten Hallers-Tjabbes et al. 1996; Mensink et al. 1996b). The cause-and-effect relationship between imposex in juvenile *B. undatum* and TBT exposure was subsequently shown in experimental studies (Mensink et al. 1996b, 1996a, 1999, 2002).

After the first failures to convince the IMO to take action, we recognized the need to understand the impact of TBT in relation to shipping elsewhere.

We sought support to study imposex in offshore snail species in shipping areas in Southeast Asia, a region where we had done research before. We did this for two reasons.

First, we understood that environmental messages from Europe and other northern hemisphere countries are often perceived as irrelevant in other areas of the world. This can cause very real environmental problems in those areas to be ignored or, worse, environmental problems to be transferred from one region to another, by dumping the compound that is no longer allowed by Northern States in other areas of the world (Ward 1988). In that sense TBT may pose a hidden threat in the marine environment (Stigliani et al. 1991). Second, we recognized that the IMO decision-making process gave states with larger fleets a greater voice. Singapore and Malaysia have extensive commercial fleets.

Our research in the region found a clear relationship between imposex incidence and the proximity of shipping routes in twenty-two snail species in the seas off Singapore, Malaysia, and Thailand (Swennen et al. 1997). We communicated the results of the field study to regional decision makers, scientists, and public institutes. Formal support from the Dutch government throughout the research process ensured that the results were summarized and submitted to IMO-MEPC (IMO 1996).

This time, the strategy was effective. Our information was bolstered by a message from the North Sea States and a statement of concern by Japan. The result was a decision by IMO-MEPC to move toward stricter control and a possible ban of TBT (ten Hallers-Tjabbes 1997). Following a two-year period of investigations by a correspondence group, in 1998 IMO-MEPC decided on a global phaseout of all organotins used in antifouling systems for ships within ten years (IMO, 1998a, 1998b).

During this policy process, critics frequently claimed that clear proof was lacking of ecological damage from TBT (CEFIC 1996; Evans et al. 1995, 1996, 2000; Green et al., 1997). This foreshadowed the vigorous opposition we would encounter in the next phase of our work. The tendency to present the ban on TBT for small ships as a universal measure often leads to confusion (Ten Hallers-Tjabbes and Boon 1995).

The above shows that findings as communicated by scientists inspired policy makers to take action; however, the policy action met with limitations related to the nature of the global policy process (only northern hemisphere or "developed" States had expressed concerns). We scientists then combined our understanding of the policy process with our scientific expertise and proposed follow-up research that had potential to raise awareness in other parts of the world, in particular in states that had an important role in the relevant policy process.

The TBT Communication Project

A general policy, such as an outright ban, can only be effective when it is implemented regionally and supported by policy makers, the public, and the users of the material. In the late 1990s in the European Union (EU), such support was present in northern Europe, but in southern Europe the problem of TBT was less clearly present on the environmental agenda. Experience with a forerunner, the 1989 partial ban on TBT from small vessels by the EU and other countries, had shown that implementation of such measures was not automatic. Several studies showed continued illegal use of TBT on small vessels in areas where TBT was legally banned (Smedes 1994; Kettle 2000).

We decided to use our experience with both the science and politics of TBT to explore how, in a precautionary framework, scientists can step beyond their mode of perceiving the world and bring their science more effectively to the policy processes. This was the basis for the TBT communication project we launched in 1998, with the support of a grant from the European Commission environmental program (EU Life Programme).

We began to explore the benefits of our approach to TBT science and policy and the potential to expand it to support implementation of the TBT ban. A major objective was to prevent a transfer of this environmental problem from regions where the problem had been recognized and support for implementation was present to areas where such (illegal) transfer could go unnoticed. We knew that there remained strong pressure to go on using TBT, and the less awareness in a community of the risks of TBT, the easier it would be to ship TBT there from other communities where a ban had been implemented and market it. Neither users of antifouling paints nor local administrators would get wind of a threat, which would also mean a delay in effective regulation to prevent the use of TBT. In addition to the documented illegal use of TBT on small ships in several countries, there are other channels for exposure, such as the dumping of abandoned ships or the stripping of abandoned ships in tropical countries, where TBT-holding paints are just scraped off at beaches, while nobody seems to be aware of the risks. Instead of transferring a problem, we wanted to try transferring a solution.

We had reason to believe that the problems associated with TBT did indeed exist throughout the seas around southern as well as northern Europe. In southwestern Spain, previous studies had shown high levels of TBT in water, sediment, and animals near harbors (Gomez-Ariza et al. 1999); yet the continental shelf sea had never been investigated. We planned, therefore, to engage scientists from several countries both in investigating TBT-related effects and communicating their present and previous findings to users, policy makers, and the public. We acted on the belief that scientists have a social responsibility to explain their work to the public and that if scientists and policy makers communicate with one another, the marine environment will benefit.

Moreover, we hoped this project would develop useful models for communicating scientific reasons for concern, issuing early warnings, and assisting in the development of preventive environmental policies. We aimed not only to demonstrate the adverse impact of TBT but also to show the benefits of direct and flexible communication based on mutual understanding between scientists and policy makers. The case study explored the potential of a precautionary approach as a frame for scientists to communicate underlying reasons for recognizing environmental threats, as emerging from scientific study and expertise, and so contribute to policy planning. This is a different process from aiming for proof in numbers. *We are convinced that scientists who study complex and uncertain systems need to be aware that they have an important role in clearly communicating results and judgment. This transfer of insight is ultimately central to achieving precautionary decisions—rather than aiming for perfect numbers.*

Project Design

My colleagues and I at the Netherlands Institute for Sea Research (NIOZ), who initiated the project, set out to share with colleagues in southern Europe our experience in investigating the impact of TBT in the open sea and in communicating with decision makers. We chose as partners marine scientists in Spain, Italy, and the Netherlands who had specific expertise in this field of research and were willing to seek cooperation with regional and national policy makers and pass on their findings to users, such as shippers, fishermen, boat manufacturers, and paint sellers as well as environmental organizations, the public, and the press. In addition, we obtained the cooperation of marine scientists in Portugal and the United Kingdom to conduct open-sea research. The project involved ten core scientists, while several others came in at different stages.

Key to the project was an additional network of policy makers and experts in international law and policy issues that supported the scientists in the communication process throughout the project. The network consisted of policy makers who were key persons in the IMO process and staff of the Netherlands government, the EU, and the IMO; a policy maker in environmental planning and a university professor in international natural resources law. Their guidance was instrumental in facilitating communication and set the stage for direct input into the global policy channel of IMO-MEPC. Network and project members met together on various occasions, during which time was reserved for informal exchange of ideas.

Throughout the project, an iterative, formal, and informal approach to communication played a pivotal role. At project meetings the communication strategy was discussed jointly between project and network members. Informal meetings and ongoing contacts aimed to stimulate mutual understanding between scientists and policy makers and between scientists of different nationalities and different disciplines. We developed and performed a scientific program and explored channels for communication with the world outside our scientific community. We designed project and public meetings with communication as a primary goal and drafted press messages. For evaluating our joint performances in both science and in communication we asked ourselves what had gone well, what could be improved, and what we had learned and then went from there. At NIOZ we kept track of progress in communication and in establishing contacts and of received responses and provided feedback. Essential in the project process was understanding of the ways in which individuals perceive the world around them, how we make our own "personal map." (A personal map refers to the way people represent the outside world in their mind. Such a map is based on the reflection and interpretation of the information that they receive from the outside world, combined with their own internal information. The map reflects people's backgrounds, such as the community to which they belong and its rules, its structure and the processes occurring, the values of the individual human and of relevant peer groups, and the individual character of each person.)

Doing the Science

We scientists investigated relationships between offshore shipping and levels of organotins and presence of imposex in the Atlantic Ocean off the Iberian Peninsula and in the Tyrrhenian Sea off Sicily. The North Sea was revisited to investigate changes in earlier trends.

For a proper assessment of the impact of TBT we used shipping densities (low = fewer than 5 ships per day, medium = between 5 and 10 ships, and high = more than 10 ships), within a given distance of a sampling station, as a measure for the input of TBT. Given the different conditions in the marine systems investigated, that area was larger in the North Sea and the Bay of Cadiz than in the Tyrrhenian Sea, where tidal streams have a much smaller effect on dispersing water laterally.

In all areas investigated, the action of TBT was reflected both in high levels of imposex and in elevated burdens of TBT in snails and sediment in areas with high and with intermediate shipping density, along shipping routes. At several sites off southwestern Spain, off Portugal, and in the North Sea, the rate of imposex and the levels of butyltins were correlated with shipping density. In Sicily the omnipresent imposex could not be related to shipping density only, as several additional input sources of TBT appeared to be present in the coastal sampling areas, such as non–merchant marine shipping, smaller craft still using TBT, and runoff from harbors and ship repair facilities. The extent of such additional input sources of TBT was not inventoried during this project. Revisiting the North Sea, we found that a relationship with local hydrology was crucial for interpreting the dose-effect relationship in open sea. Deeper parts of the northern North Sea are often stratified in layers of different temperature or salinity, separated by a boundary layer (thermo- or halocline) that forms a barrier to a direct vertical transport of surface input to the local seabed (ten Hallers-Tjabbes et al. 2001).

The relationships between surface input and local effect was only clear in areas with a year-round mixed water column, because only then TBT can freely move from the water surface to the sea floor. In stratified waters TBT tends to remain longer in the upper layer, so the water mass may have moved away from source before the TBT reaches the sea bed. This has a major bearing on interpreting data of impact from surface-bound input sources and for monitoring efficacy of policy measures taken to reduce them.

The phenomenon also throws new light on the limitations of simulation models for predicting distribution of contaminants and effects. Our observations suggest there is a distinction between understanding deterministic processes and the potential to predict where such processes may end up. One must also appreciate the complexity of simulation models in order to evaluate what the models can and cannot do. Yet, time and again, such models have been fed into decision or management structures and models, without an awareness of what lies underneath them.

Communicating Findings

The communication part of our project aimed to encourage recognition of the problems associated with TBT antifouling in southern Europe and draw attention to continuing problems of TBT in the North Sea. A second goal was to develop a basis for assessment of the effectiveness of environmental policy measures, such as once the Convention on the Control of Harmful Antifouling Systems (IMO 2001) comes into effect.

We learned early on that awareness of the problems of TBT and of the regulations was indeed limited. Our Spanish partner learned this through informal inquiries among local fishers, other users, and authorities. He was inspired to develop a questionnaire about the impacts of TBT. A first mailing to two hundred target addresses met with a 40 percent return, of which only 20 percent indicated that they were aware of TBT, of the associated environmental problems, and of the existing and imminent regulations. Most users appeared not to be aware of the composition of the antifouling paints they used.

We scientists communicated our findings about the effects of TBT to users and regional policy makers and invited them to participate in informal discussions and in larger public meetings. After the project plans had been discussed in a wider environmental policy context, network members were fully participating in the project meetings and were explained the processes within IMO, related policy processes, and the concept of precaution in international marine policy and what mattered in successful implementation of regulations. They were speakers at the public meetings and at the special seminars, wrote background papers to clarify the processes in policy making, and facilitated the contacts between the scientist and national and EU policy makers. In between meetings they were frequently consulted by us and had an open ear for our partners, and some of them joined in research cruises. They also were crucial to inspire the joint submissions from our science directly into the process within IMO by our country governments.

The public was approached by organizing meetings in southern Europe, in Italy and in Spain, including a regional, cross-border Spanish-Portuguese meeting. Here, project results, background to the project, and policy consequences were presented to a broad regional audience. Speakers represented the scientific and the decision-making communities and government authorities. The meetings were attended by interested organizations—users of antifouling products, environmental and other nongovernmental organizations (NGOs), policy makers, people from shipping and harbors, paint merchants, environmental NGOs, scientists, and representatives of the press, many of them participating in the debate. The media outreach associated with each meeting was well received—for example, following a press conference in Spain, the meeting was widely covered by radio and television. Two stand-alone seminars for a broader audience were organized in the United Kingdom and at the World Maritime University in Sweden, in which project members, policy makers, and representatives from the shipping and paint communities presented their views.

The partners created regional project Web sites, linked to the general project Web site hosted by the project coordinator, which provides project reports and press information (www.nioz.nl/projects/tbt). A booklet on TBT from the Netherlands has been translated into Spanish and Italian and added to the Web sites.

What Scientists Learned and Achieved

We scientists learned, as we had through our research in Southeast Asia, how important regional policies and public processes could be. We also gained new skills in both science and communication. For example, those who were new to TBT research in the open sea gained practice in the multidisciplinary skills needed for complex marine investigations, including biological assessment as well as chemical expertise to detect burdens of TBT in marine biota and sediments. Even more challenging were the new skills needed in active communication with policy makers, users, and the public. As a result of the support we received for developing communication skills from the project network-steering group, our initial trepidations soon gave way to confidence. In particular, our Spanish partner developed new and original communication initiatives early on. He consulted with fishing communities about TBT and related problems, the findings of which motivated him to develop the questionnaire. Prior to the Spanish meeting he organized a press conference, resulting in a rewarding attendance of the press at the public meeting.

Finally, the independence of the participating scientists enabled them to explore an issue that had not emerged on the policy agenda, and in a way that was not steered directly by policy bodies, but yet had developed into science for informing policy. Government-related science institutes tend to confine their research to issues and developments that are already on the political agenda. In the case of TBT they would do research to monitor the impact of TBT in coastal snails, looking for signs of reduction in impact, since regulations banning TBT from small ships have been in place for several years. They would rarely explore the open sea or look for a host of other features of the marine environment. And while they have the advantage of a rather direct channel to the policy makers belonging to their government department, they are not stimulated to approach others. Moreover, such a direct link between a specific group of scientists and policy makers may obscure the issues from the public debate. We had the freedom to follow our scientific intuition as to what might matter and then seek for grants to support us in doing such research. Likewise, we had the freedom to search for policy makers who were important in the policy process relevant to our research and to establish links with them. And last but not least we had full freedom to enter our findings into the public debate.

Our independence meant that we had to invest more time and energy than those who work in government-linked institutes to establish our communication links and search for funding, but it opened the door to developing a new concept and learning from the process.

The Science-Policy Loop

During the EU project and in its forerunners, we employed a continuous feedback loop process: informing policy makers and receiving their responses, identifying further strategic research needs and finding material and political support for them, and reporting research findings directly back into the policy process through channels facilitated by decision makers in the international forum. This approach brings together the potential of different entities to identify problems and create change. It tends to create synergies that increase the effectiveness of scientists and policy makers alike. In this project, the whole truly became greater than its parts.

The TBT experience has shown that science does not only have to respond to policy needs, but can, with a proper understanding of the relevant policy processes, recognize what is called for in those processes and bring appropriate information to the attention of decision makers. This holds true in particular for our earlier investigations and their communication in Southeast Asia, which helped to create international support needed for a global convention.

Once the findings were brought forward beyond the scientific community, recognition of the problems caused by TBT increased. Regional policy makers acted on the emerging findings, resulting at the international level in policy contributions to IMO-MEPC, based on the research findings, by the Netherlands, Spain, Portugal, and Italy (1999: MEPC 44/INF.11; 2000: MEPC 46/INF.2) (IMO 1999, 2001). Government advisory groups on antifouling policies were established in Spain and Italy, and project scientists now participate in a senior advisory role. The strong communication process also created important press coverage in international, national, regional, and specialized journals, as well as on radio and television. TBT is now on the public agenda throughout the EU, not only in northern states.

Communication Challenges

To meet communication challenges it was crucial to create an atmosphere of trust and openness and of interest in another's "story." Establishing such trust and openness was enabled partly thanks to the people who joined—especially in the policy network (a matter of searching and inviting those who would be interested)—and also by starting with something informal and enjoyable (getting together over dinner the evening before the meeting). Setting an example ourselves also helped to prepare the soil for a fruitful exchange of ideas.

When scientists and decision makers cross paths, the perceptions they have about each other are frequently voiced in complaints, such as "policy makers ask for absolute answers and they want them yesterday," or "scientists forever discuss the results; they are never going to be certain." Such perceptions reflect different contextual frameworks and the different subsystem of society they belong to, all having their own internal dynamics (Glasbergen 1998). We were geared toward crossing barriers. Paramount to overcoming the tendency to infer assumptions on the others' way of dealing with the world was to have it on the table around which we all sat and to discuss its existence as something that is happening quite often in human communication. We asked the different participants to share how they perceive their world and how they perceive and maybe interpret the acts of others. As a background we gave examples and metaphors that could clarify what is actually happening when people assume they know from inference what and how others think.

As much as interpretation, coping with uncertainties benefits from being brought into the open, named, and recognized as something that we all have to deal with. This uncertainty has much to do with the complexity of marine systems (for us scientists) and of society (for policy makers) and can be also seen as intriguing; there is always more going on, and aren't we curious to learn more?

Paying attention to both voiced and unvoiced messages has played an important role in the debate on impact of TBT in open sea and its policy consequences. It is critical for scientists and policy makers alike to be attentive not only to what is being said in discussions about the environmental impacts and precautionary measures but also to what is *not* being said. The scientific tradition requires reporting on science as a successful set of hypothesis, method, and results, which are then discussed in the context of other scientific findings. Very little is said about the numerous and often not so successful trials to grasp the matter to study. We don't report normally how often we did not manage to find sufficient whelks and that this influences where we could do gradients along certain trajectories in the North Sea. Similarly, we rarely report what made us first believe we were on the right track and why we sensed we were, long before we found proof and did the required statistical testing. Likewise, policy makers work in a frame where politicians tend to speak in absolutes that then have to be filled in by the policy makers, without letting it out that they actually are not all that certain.

Both the marine environment and human society are dynamic systems, which we try to grasp in standing rules and descriptions. No wonder there is an intrinsic feeling of not being totally confident. In this project, as well as in the preceding scientific debate on impact of TBT from offshore vessels, daggers were often crossed over how much "absolute proof" we could bring forward. Opponents claimed that less-than-perfect evidence did not justify more stringent measures. Clearly, some scientists had been hired to disprove our results. (At least one of their experiments failed when the whelks they had brought into the laboratory died, as we learned in a desperate fax from a student researcher.) Wherever we spoke, we encountered some hostile questions, attacks on the integrity of our research, and certain repetitive questions and assertions. When meeting such opposition, we showed that we understood that there was an interest to lose, while bringing the issue itself back into the debate. This brought the debate automatically to a higher abstraction level, where more attendees could be comfortable and recognize the issue. By always being attentive only to communicate matters that were backed by our science, we maintained our integrity and met with a positive response from audiences.

The opponent body diminished through the years, which may have reflected some success in bringing alternative antifoulings to the fore. The policy process was never seriously delayed by such opposition, although some clever acts by dedicated policy makers have certainly helped to maintain the momentum.

Throughout the work of bringing the TBT problem to the public eye, the debate tended to become more emotional and charged once the question had been raised about consequences for humans, either as consumers of marine food sources or as workers with TBT-containing materials. Telling people that female snails developed male sexual characteristics got their attention, often making them giggle. The force of such underlying emotional links became evident in the emerging support for a global ban on TBT from Japan and Southeast Asia, which consume large amounts of fish and shellfish.

Conclusion

Our greatest achievement has been to move decision makers' attitudes to TBT from an assimilative capacity approach to a precautionary one. When TBT was first recognized as harmful in coastal, small-shipping areas, where clear proof was present of reproductive failure and decline of coastal snail populations, it was assumed that levels in the open sea were too low to cause harm and that local regulations would be sufficient to address the problem. In the early 1990s this attitude started to shift toward a precautionary approach, in which evidence of harmful effects, without full proof of a fatal impact, resulted in support for curbing the impact of TBT from the marine environment (Mee 1996). An International Convention on the Control of Harmful Anti-fouling Systems has now been established (5 October 2001; IMO 2001a).

The case supports an observation by Hey, who states, "It is now clear that timely certainty as to the limits of the assimilative capacity of the environment cannot be provided. The only certainty available is that, once detrimental effects are registered, the limits of the assimilative capacity have been surpassed" (Hey 1992).

A shift in perception of "content value" (what makes a message worthwhile, just numbers and facts or also underlying explanations) of the messages passed on by scientists is in agreement with the spirit of precaution (Tickner 1999). Such a shift in perception will assist in improving the mechanism of environmental protection (Santillo et al. 1998).

Precaution has been present in the political debate for many years. During the process we met with evidence that the perceptions long held by regulators and scientists played an important role in the effectiveness of interaction and communication, with respect to precautionary responses. Awareness of differences in how science, precaution, and policy are perceived is important, particularly when working in an international context, where a different cultural context may strongly influence differences in perception.

Those participating in the research on impact of TBT recognized the value of involving policy makers early in the process, of establishing communication channels that made science and policy transparent, and of creating sustainable participatory structures between scientists and policy makers. *Mutual confidence between scientists and policy makers can be a major asset in creating incentives for scientists to pass on early research findings that, according to their expertise and judgment, are cause for concern.* Instead of holding back information until full proof or statistical "certainty" (a contradiction in terms) has been established, scientists are able to issue early warnings and contribute to potential precautionary actions.

The EU project proved to be a successful experiment in what role scientists can play in supporting precaution. None of us knew beforehand whether it was going to work. We were able to inspire scientists to embark on self-initiated and maintained communication with policy (and the public) in the context of a specific policy process. And we could explore—in this case study where the science was not totally new and knowledge about the policy process existed—the potential to make it work and the mechanisms that make it work and then analyze the added value for other environmental cases. Crucial to the success was the understanding of the processes in science and in policy, of the interaction process of this study and the willingness to listen.

The concepts and tools developed in the project are potentially applicable to other projects and regions. During the project period, information about the project was transmitted to representatives of other governments, for example, Mexico (Report of the Second Progress Meeting, Spain 2000), Brazil (ten Hallers-Tjabbes 2000), and Australia, and to international policy makers outside our network. The model of identifying an environmental problem and investing in communication to policy makers and the public can be developed as a scheme for the role of science in precautionary environmental planning (ten Hallers-Tjabbes 20002b).

References Cited

Alzieu, C. 1991. "Environmental Problems Caused by TBT in France: Assessment, Regulations, Prospects." *Marine Environmental Research* 32: 7–17.

Bryan, G. W., P. E. Gibbs, G. R. Burt, and L.G. Hummerstone. 1987. "The Effects of Tributyltin (TBT) Accumulation on Adult Dog-Whelks, Nucella Lapillus: Long-Term Field and Laboratory Experiments." *Journal of the Marine Biological Association U. K.* 67: 525–44.

Cairns, J. 1977. "Aquatic Ecosystem Assimilative Capacity." *Fisheries* 2: 5.

CEFIC. 1996. "Use of Tributyltin Compounds in Anti-Fouling Paints: Effectiveness of Legislation and Risk Evaluation of Current Levels of Tributyltin Compounds in Coastal Waters." IMO, London, MEPC 38/INF.8.

Copenhagen Statement of Conclusions. 1993. Statement of Conclusions of the Intermediate International Conference on the Protection of the North Sea, Copenhagen.

Dundee, D. 1979. Personal communication. University of New Orleans.

Esbjerg Declaration. 1995. Ministerial Declaration of the Fourth International Conference on the Protection of the North Sea, Esbjerg, Denmark, 8–9 June 1995.

Evans, S. M., A. C. Birchenough, and M. S. Brancato. 2000. "The TBT Ban: Out of the Frying Pan into the Fire?" *Marine Pollution Bulletin* 40: 204–11.

Evans, S. M., P. M. Evans, and T. Leksono. 1996. "Widespread Recovery of Dogwhelks, *Nucella lapillus* (L.), from Tributyltin Contamination in the North Sea and Clyde Sea." *Marine Pollution Bulletin* 32: 263–69.

Evans, S. M., T. Leksono, and P. D. McKinnell. 1995. "Tributyltin Pollution: A Diminishing Problem Following Legislation Limiting the Use of TBT-based Anti-Fouling Paints." *Marine Pollution Bulletin* 30: 14–21.

Fent, K. 1996. "Ecotoxicology of Organotin Compounds." *Critical Reviews in Toxicology* 26: 1–117.

Freestone, D., and E. Hey. 1996. "Origins and Development of the Precautionary Principle." In *The Precautionary Principle in International Law,* edited by D. Freestone and E. Hey, 3–15. The Hague, Netherlands: Kluwer Law International.

Gibbs, P. E., and G. W. Bryan. 1986. "Reproductive Failure in Populations of Dog-Whelk, *Nucella lapillus,* Caused by Imposex Induced by Tributyltin from Antifouling Paints." *Journal of the Marine Biological Association U.K.* 66: 767–77.

Glasbergen, P. 1998. "The Question of Environmental Governance." In *Co-operative Environmemtal Governance: Public-Private Agreements as a Policy Strategy,* edited by P. Glasbergen, 1–18. Dordrecht: Kluwer Academic Publishers.

Gomez-Ariza, J. L., E. Morales, and I. Giraldez. 1999. "Uptake and Elimination of Tributyltin in Clams, *Venerupis decussata.*" *Marine Environmental Research* 47: 399–413.

Gray, J. S. 1996. "Integrating Precautionary Scientific Methods into Decision-Making." In *The Precautionary Principle in International Law,* edited by D. Freestone and E. Hey, 133–46. The Hague, Netherlands: Kluwer Law International.

Green, G. A., R. Cardwell, and M. S. Brancato. 1997. Comment on "Elevated Accumulation of Tributyltin and Its Breakdown Products in Bottlenose Dolphins *(Tursiops truncatus)* Found Stranded along the U.S. Atlantic and Gulf Coasts." *Environmental Science and Technology* 31: 3032–34.

Hague Declaration. 1990. Declaration of the Third International North Sea Conference on the Protection of the North Sea, Preamble, 8 March 1990. The Hague, Netherlands.

Heidrich, D., S. Steckelbroeck, F. Bidlingmaier, and D. Klingmueller. 1999. "Effect of Tributyltin (TBT) on Human Aromatase Activity." Paper presented at Congress of the Endocrine Society U.S.

Hey, E. 1992. "The Precautionary Concept in Environmental Policy and Law: Institutionalizing Caution." *Georgetown International Environmental Law Review* 6: 257–318.

IMO. 1990. Report of 30th Meeting Marine Environment Protection Committee. International Maritime Organization, London, MEPC 30.

———. 1994. Report of 35th Meeting Marine Environment Protection Committee. IMO, London.

———. 1996. "Harmful Effects of the Use of Antifouling Paints for Ships. Impact of Antifouling Paints in South-East Asian Seas." Submitted by the Netherlands. IMO, MEPC 38/INF.16. In Report of Meeting MEPC 38.

———. 1998a. Report of 41st Meeting Marine Environment Protection Committee. IMO, MEPC 41.

———. 1998b. Report of 42nd Meeting Marine Environment Protection Committee.

———. 1999. "Harmful Effects of the Use of Antifouling Paints for Ships. Information on TBT Levels and the Occurrence of Imposex in Certain Marine Species in the North Sea, the Mediterranean and the Coastal Waters of Portugal." Submitted by Italy, the Netherlands, Portugal and Spain. IMO, London, MEPC 44/INF.11.

———. 2001a. "Adoption of the Final Act of the Conference and Any Instrument, Recommendations and Resolutions. Resulting from the work of the Conference." International Conference on the Control of Harmful Antifouling Systems October, IMO, 5 October 2001. London.

———. 2001b. "Harmful Effects of the Use of Antifouling Paints for Ships. Information on TBT Levels and the Occurrence of Imposex in Certain Marine Species in the North Sea, the Mediterranean and the Coastal Waters of Portugal." Submitted by Italy, the Netherlands, Portugal, and Spain. IMO, London, MEPC 46/INF.2.

Jensen, A. 1989. "Co-operation of Policy and Science. Viewpoint of a Scientist." In *3rd North Sea Seminar 1989, Distress Signals, Signals from the Environment in Policy and Decision-Making*, edited by C. C. ten Hallers-Tjabbes and A. Bijlsma, 197–203. Amsterdam: Werkgroep Noordzee Foundation.

Kettle, B. 2000. Case Study. Current TBT issues in Australia. Presented at LC/SG23, Townsville, Australia, May 2000. LC Secretariat, LC/SG 23 INF, 1–17.

LC 23/INF.7. 2001. Relationship between the International Convention on the Control of Harmful Anti-fouling Systems, 5 October 2001 and LC 72; Proposal for the Jamaica Workshop. Submitted by IUCN, Agenda item 8.1 Technical Co-operation and Assistance.

London Convention. 1972. Convention on the Prevention of Marine Pollution by Dumping of Wastes and Other Matter and 1996 Protocol (1972), LC72.

London Declaration. 1987. Declaration of the Second International North Sea Conference on the Protection of the North Sea, London, 25 November 1987.

Mee, L. D. 1996. "Scientific Methods and Precautionary Principle." In *The Precautionary Principle in International Law*, edited by D. Freestone and E. Hey, 109–31. The Hague, Netherlands: Kluwer Law International.

Mensink, B. P. 1999. "Imposex in the Common Whelk, *Buccinum undatum*." Ph.D thesis, Wageningen University, Netherlands.

Mensink, B. P., J. M. Everaarts, J. Kralt, C. C. ten Hallers-Tjabbes, and J. P. Boon. 1996b. "Tributyltin Exposure in Early Life Stages Induces the Development of Male Sexual Characteristics in the Common Whelk, *Buccinum undatum*." *Marine Environmental Research* 42: 151–54.

Mensink, B. P., C. C. ten Hallers-Tjabbes, J. Kralt, I. L. Freriks, and J. P. Boon. 1996a. "Assessment of Imposex in the Common Whelk, *Buccinum undatum* (L.) from the Eastern Scheldt, the Netherlands." *Marine Environmental Research* 41: 315–25.

Mensink, B. P., C. C. ten Hallers-Tjabbes, A. G. M. Van Hattum, and J. P. Boon. 2002. "Imposex Induction in Laboratory Reared Juvenile *Buccinum undatum* by Tributyltin (TBT)." *Environmental Toxicology and Pharmacology* 11: 49–65.

OSPAR. 1992. *The Convention for the Protection of the Marine Environment of the North-East Atlantic*. OSPAR Convention, September 1992, Paris.

Pravdic, V. 1985. "Environmental Capacity: Is a New Environmental Concept Acceptable as a Strategy to Combat Marine Pollution?" *Marine Pollution Bulletin* 16: 295–96.

Resolution MEPC 46(30). 1990. Resolution on the Control of Harmful Antifouling.

Santillo, D., R. L. Stringer, P. A. Johnston, and J. Tickner. 1998. "The Precautionary Principle: Protecting Against Failures of Scientific Method and Risk Assessment." *Marine Pollution Bulletin* 36: 939–50.

Smedes, F. 1994. "Butylverbindingen in water, 4 jaar metingen (Butyltin compounds, four years of measurements)." Werkdocument RIKZ/IT 94.611x. Rijkswaterstaat, Netherlands.

Snoeij, N. J., A. H. Penninks, and W. Seinen. 1987. "Biological Activity of Organotin Compounds: An Overview." *Environmental Research* 44: 335–53.

Sperling, K. R. 1986. "Protection of the North Sea: Balance and Prospects." *Marine Pollution Bulletin* 17: 241–43.

―――. 1988. "The Dangers of Risk Assessment within the Framework of Marine Dumping Conventions." *Marine Pollution Bulletin* 19: 9–13.

Stebbing, A. R. D. 1992. "Environmental Capacity and the Precautionary Principle." *Marine Pollution Bulletin* 24: 287–95.

Stigliani, W. M., P. Doelman, R. Schuling, W. Salomons, G. R. B. Smidt, and S. E.

A. T. M. Van der Zee. 1991. "Chemical Time Bombs. Predicting the Unpredictable." *Environment* 33: 5–30.

Swennen, C., N. Ruttanadakul, S. Ardseungern, H. R. Singh, B. P. Mensink, and C. C. ten Hallers-Tjabbes. 1997. "Imposex in Sublittoral and Littoral Gastropods from the Gulf of Thailand and Strait of Malacca in Relation to Shipping." *Environmental Technology* 18: 1245–54.

ten Hallers-Tjabbes, C. C. 1979. "The Shell of *Buccinum undatum* L. Shape Analysis and Sex Discrimination." Ph.D. Monograph. University of Groningen, Netherlands.

———. 1997. "Tributyltin and Policies for Antifouling." *Environmental Technology* 18: 1265–68.

———. 2000. "Environmental Effect of TBT Antifouling from Commercial Shipping and Associated Sources, Such as Harbour Dredge Material." TBT Seminars, Ministry of the Environment, Brazil, 28–30 August 2000, 1–10. Rio de Janeiro and Brasilia.

ten Hallers-Tjabbes, C. C., and J. P. Boon. 1995. "Whelks (*Buccinum undatum* L.) or Dogwhelks (*Nucella lapillus* L.) and the Partial Ban on TBT: A Cause for Confusion." *Marine Pollution Bulletin* 30: 675–76. Also LC/SG 19/INF.

ten Hallers-Tjabbes, C. C., J. P. Boon, J. L. Gomez Ariza, and J. F. Kemp. Forthcoming. "Communicating the Harmful Impact of TBT: What Can Scientists Contribute to EU Environmental Policy Planning in a Global Context?" In *Ocean Yearbook Volume 17.*

ten Hallers-Tjabbes, C. C., J. M. Everaarts, B. P. Mensink, and J. P. Boon. 1996. "The Decline of the North Sea Whelk (*Buccinum undatum L.)* between 1970 and 1990: A Natural or Human Induced Event?" *Marine Ecology Pubblicazioni della Stazione Zoologica di Napoli* I 17: 333–43.

ten Hallers-Tjabbes, C. C., J. F. Kemp, and J. P. Boon. 1994. "Imposex in Whelks *(Buccinum undatum)* from the Open North Sea: Relation to Shipping Traffic Intensities." *Marine Pollution Bulletin* 28: 311–13.

ten Hallers-Tjabbes, C. C., J. F. Kemp, B. Van Hattum, and J. P. Boon. 1995b. "Report of TBT Concentrations in Whelks in the North Sea." (DGSM 3110 [70/4/145]), in MEPC 42/INF.6.

ten Hallers-Tjabbes, C. C., B. Van Hattum, and J. P. Boon. 1995a. "Imposex and TBT in *Buccinum undatum* and in Sediments in the North Sea." Report regarding STOF*CHEMIE, RIKZ-RWS. LC/SG 19/INF.

ten Hallers-Tjabbes, C. C., J.-W. Wegener, A. G. M. Van Hattum, J. F. Kemp, E. ten Hallers, T. J. Reitsema, and J. P. Boon. In Press. "Imposex and Organotin Concentrations in *Buccinum undatum* and *Neptunea antiqua* from the North Sea: Relationship to Shipping Density and Hydrographical Conditions." *Marine Environmental Research.*

Tickner, J. 1999. "A Map Toward Precautionary Decision-Making." In *Protecting*

354   VI. Science to Support Precautionary Decision Making

Public Health and the Environment: Implementing the Precautionary Principle, edited by C. Raffensperger and J. Tickner. Washington, D.C.: Island Press.

UNCED. 1992. Rio Declaration at the United Nations Conference on Environment and Development. Rio de Janeiro, Brazil.

Ward, J. 1988. "Antifouling Paints Threaten Fisheries Resources." In *Naga, The ICLARM Quarterly (Journal of the International Center for Living Aquatic Resources Management),* 15. Philippines: International Center for Living Aquatic Resources Management.

Whalen, M. M., B. G. D. Logathan, and K. Kannan. 1999. "Immunotoxicity of Environmentally Relevant Concentrations of Butyltins on Human Natural Killer Cells in Vitro." *Environmental Research* 81, no. 2: 108–16. http://www.nioz.nl/projects/tbt.

CHAPTER 23

Whose Scientific Method?
Scientific Methods for a Complex World

Richard Levins

Opponents of the precautionary principle frequently express two criticisms that have to do with science. They argue that the precautionary principle itself is not "scientific" because it calls for action without full proof that the substance or activity in question is really harmful. These critics call, instead, for action based on objective evidence obtained through the "scientific method." And then they explain this failing of the precautionary principle by accusing supporters of the principle of being "antiscience" or "antitechnology."

In this chapter I mainly address the first reproach. But I would like to address, briefly, the charge that advocates of the principle are antiscience, conjuring visions of Luddite machine-smashers who hearken back to an imaginary golden age—a nostalgia for the man with the hoe on the part of people who have never tilled the soil. Proponents of the status quo imagine that there is only one kind of progress, their own version of it, and that the only alternative is stagnation. They contrast their modern pesticides to doing nothing about pests, their modern antibiotics to newts' eyes and holly, their plastic cities to living in caves. Yet progress does not proceed along a single line from backward to advanced. Rather, progress is a branching pathway with choices along the way. The conflict is about those choices.

The progress of science is not the smooth penetration of light into dark corners. "Technology" in contemporary polemics usually stands for "high technology," that is, methods based on electronics, biochemistry, and informatics. But technology is merely the ability to do something, whether it is controlling mosquitoes by spraying with synthetic molecules or by putting a tight-fitting cover on the well. *The conflict is not between science and antiscience but between different pathways for science and technology; between a commodified science-for-profit and a gentle science for humane goals; between the sciences of the smallest parts and the sciences of dynamic wholes.*

This chapter discusses the social construction of scientific production and the pattern of strength and weakness to which this leads. I offer proposals for a more holistic, integral approach to understanding and addressing environmental issues.

What Is the Scientific Method?

Working scientists know that there is no such thing as *the* scientific method. There are many scientific methods, depending on the nature of the problem and the stage of the investigation.

Certain scientific traditions have developed to avoid at least the most obvious kinds of errors:

- We know that if B follows A, it does not necessarily mean that A caused B. Therefore, we carry out controlled experiments. We might treat two groups of patients identically except that one group receives a new drug and the other receives the standard treatment or a placebo. Then we can claim with some confidence that the new drug does or does not cure the disease. (But only with some confidence: if the two groups of patients were both badly undernourished it may be that neither treatment helps or that joining a study of any kind improves the patients' condition. The things that are held constant can also affect the results.)
- Scientists learned early that the expectations of participants can influence the outcomes of experiments, so that it is important for patients not to know whether they are receiving the new drug. Then they learned that the expectations of researchers can also affect the outcome, and the double blind was invented in which the researchers in contact with the patients do not know which ones belong to which group.
- All sorts of factors might affect outcomes, so we have learned to match study populations for age, sex, income, and other characteristics of possible relevance.
- Two groups of similar animals might differ just by chance. Therefore, we make use of replication and statistical analysis, mathematical models that ask "what if" questions, and formal procedures of hypothesis testing.
- Social criteria have also been developed: scientific evidence must be public, repeatable, and subject to criticism. Ideas must be examined independently of their source. Investigators should know and reveal their biases and conflicts of interest.

Formal discussion of "the scientific method" usually focuses on the last stages of an investigation: hypothesis testing. But investigative methods vary considerably. In some sciences, a single critical experiment, after it is repeated in other labs, can be decisive. In others, knowledge is gained through the accumulation of many different kinds of evidence.

Thus the case for global climate change is not only the direct observation of a temperature rise but also the retreat of glaciers, the melting of icecaps, the upward and northward expansion of plant and animal species, the decline of coral reefs, fossil climates revealed in ice cores, tree rings, geological deposits, pollen succession, and so on. All of this is reinforced by a theoretical basis for climate change in the accumulation of greenhouse gases. Each kind of evidence may be questioned for specific technical reasons, but the massive accumulation of evidence and the diversity of the arguments are overwhelming.

Similarly, the evidence for the environmental causation of our cancer epidemic is based on its historical association with the oil industry and use of pesticides, the geography of cancers related to specific industries or molecules, the history of the rise in cancer in relation to specific carcinogenic exposures (such as the increase in smoking in the twentieth century), differences among generations of immigrants, population studies, and laboratory research.

The various sciences have their own methods and conventions. Some sciences, such as physics and chemistry, are experimental in the narrow sense that allows replication. In these fields, relatively few kinds of objects occur in vast numbers and can be manipulated many times. Astronomy and geology, in contrast, do not allow for direct experiments, but the large number of stars permits controlled comparisons, and geological processes can be duplicated in the laboratory even if geohistory cannot. Anthropology has no precise replication, and the number of distinct peoples in the world is small compared to numbers of molecules or mosquitoes. Further, the anthropological observer is very much a part of the system studied. (Quantum theory and relativity both introduce the observer as part of the system studied. It is now recognized as relevant to all sciences.) Human physiology and psychology often make use of animals, which are studied as proxies for humans. They differ from people, of course, but it is hoped that they are similar enough to permit transfer of conclusions to humans.

Failures of the Scientific Method

The development of the scientific method is aimed at avoiding the kinds of errors we have learned to worry about. And indeed, it has been successful in catching ordinary sloppiness, dirty glassware, division by zero, wishful thinking, and the individual biases of scientists or their economic stake in their findings. It has been less successful at recognizing the shared biases of a whole scientific community, the beliefs that are so much a part of the common sense of the community that they are not even recognized as biases. Therefore, it is instructive to note the history of some major collective scientific errors and failures.

The doctrine of the epidemiological transition proposed that infectious diseases would decline in importance over time (Omran 1971; Levins et al. 1994). Yet since the mid-1980s, the percent contribution of infectious diseases to total mortality has been increasing. This is true even in developed countries and even if AIDS is excluded. Advocates of this doctrine failed to take into account the broad epidemiological pattern of waves of surging and ebbing human diseases. They also failed to examine thoroughly the disease profiles of other species; the ecology of disease, including the lesson that any major change in land use, vegetation, climate, human settlement, economics, or technology may also cause major changes in our relations with vectors and pathogens; the rapid evolution of resistance to antibiotics and pesticides; and the vulnerability of a socially and economically stratified population. They assumed that the toxicological fact that a chemical could kill insects in a bottle implied the ecological fact that widespread use of pesticides would control populations of the vectors that spread disease.

In agricultural science, the theory that mechanization, monoculture, and widespread use of chemicals would allow increased production to eliminate hunger failed to consider how pesticides exacerbate pest problems by disrupting the communities of species and provoking resistance, that monoculture increases vulnerability to invasion by pests and to unexpected climatic events, and that agribusiness disrupts rural life and displaces populations. In all these cases, enthusiasm for a new, promising tool mitigated against critical examination of the tool's limitations or harmful consequences.

The commodification of science produces a growing sophistication on the small scale and in the laboratory but an increasing irrationality in the scientific enterprise as a whole. Economic concerns encourage the narrowness that runs through the major debacles of modern applied science. The quest for profits exaggerates the benefits of innovations, belittles the dangers, and claims more knowledge and control than is possible. Furthermore, as science becomes an increasingly commodified knowledge industry, the criteria of quality shift from validation by open and critical peer testing to those applied to any commodity: Does it sell? How profitable is it? Crucial knowledge is withheld in the name of proprietary information. The most egregious example is perhaps the tobacco industry's concealment of smoking's long-known toxic effects.

Any serious scientific method must correct not only individual error but also the prevailing collective biases. I would propose several guidelines.

Preparing for Surprises

Science must recognize Hegel's principle that the truth is the whole. A problem must be posed broadly enough to accommodate a solution. Even then, we must acknowledge that there will be surprises. The best that science can do is study the unknown by pretending provisionally that it is like the known. This is the case often enough to make science possible. But the unknown is also unlike the known, deeply enough to make science necessary. Especially in times of rapid change, there will be surprises. It is therefore useful to ask, how do other species cope with uncertainty? We must be prepared to be surprised, and we must develop science equipped to deal with the unknown, using the following approaches.

Detection of a Problem and Response

In order to be effective, the response must be rapid enough to prevent vast damage and to operate before things have changed again. Foraging ants return to the nest when the day gets too hot. Plants wilt when water loss becomes excessive. Insulin is released when blood sugar levels rise. Prey species flee their predators. In all these cases the response comes after the changed condition has been detected. In order for this to be successful, detection must be accurate and rapid enough in relation to the duration of the threat. Otherwise, organisms would always be responding to the previous situation.

In preparing for new disease problems, the public health system is concerned with efficient surveillance and reporting programs. It is important that these programs operate rapidly enough to stay ahead of the spread of disease. Similarly, long delays in recognizing problems with toxic chemicals lead to a dynamic state in which new pollutants are introduced into the environment as fast as or faster than old ones are detected and removed.

Prediction

Many insects with short generations such as aphids prepare for winter by becoming dormant. However, the signal that starts dormancy is not winter itself, that is, low temperature or lack of food. Such signals would not be reliable, since cold spells could occur in midsummer. Rather the insects respond to the shortening days of autumn as a predictor of winter. But the silkworm responds differently. It has only two generations per year. The first emerges in early spring when the short days indicate that it is the first generation and can have time for a second one, whereas the long summer days inform it that there will not be time for another generation. It is the information content of the environmental signal that matters. There is no necessary relation between the signal and the condition it predicts (Levins 1968). For human well-being we need short-term prediction of the usual sort, for example: if a few cases of West Nile virus appear, there may be more. We also need longer-range predictions: if there are more mosquitoes, expect mosquito-borne disease. More rain will mean more mosquitoes and maybe more disease. If we build dams, dig irrigation ditches, and kill mosquito predators, we may have more mosquitoes and mosquito-borne disease. If we allow a depletion of biodiversity, exotic invaders are more likely to be successful.

There is also need for long-range, evolutionary prediction based on comparative epidemiology. We should identify those groups of insects that are potential disease vectors or major agricultural pests, that have the flexibility to extend their host ranges or modes of transmission or to adapt to new habitats. Most vectored human diseases are transmitted by flies, including mosquitoes, or by ticks. Among the plant viruses, the majority are transmitted by homopterans (aphids, scale insects, mealybugs). Like mosquitoes, they are sucking insects that remove liquid from their hosts and return liquid to offset the vacuum their sucking creates. We should know which mammals are good reservoirs of viruses that could also infect people, which groups share more diseases with humans, and which kinds of diseases are likely to be shared. For instance, the human gut and the gut of a cow are quite different, but our lungs are similar, so we might be alert to the spillover of respiratory diseases.

Broad Tolerance

Plant breeders distinguish between "vertical resistance" and "horizontal resistance." Vertical resistance, provided by a single gene, gives complete protection but only to one pathogen type. It lasts only until the pathogens overcome that resistance. Horizontal resistance usually depends on many genes. It protects against a broad range of enemies, and it persists because pathogens or insects have to do many things, more than they are capable of doing, to overcome it. In general, biodiversity provides horizontal resistance. It reduces the vulnerability of communities and is an important element of ecosystem health.

We need to design horizontally resistant systems that will defend against most diseases and poisons, even if we do not know what these might be or when they might appear. The vulnerability of our eco-social systems thus becomes a major target for research. It is especially important to take into account Schamlhausen's Law (Lewontin and Levins 2000): systems in extreme or unusual conditions or at the boundary of their tolerance are more sensitive to all environmental factors.

Poor and oppressed populations are more vulnerable to slight changes. This shows up in increased variability of health status. For instance, for any age group, blood pressure is more variable among African Americans than among whites, and life expectancy varies more across cities. In evaluating the potential effects of environmental or socioeconomic changes, we must look at the most vulnerable groups.

Prevention

Organisms can change their environments in ways that make potential threats less likely to arise. For instance, trees create the forest environment, which modulates the flow of water and slows down strong winds. Ants nesting in the soil create environments where temperature extremes are less frequent.

The precautionary principle encourages prevention as far upstream as we can reach. Instead of looking only at how to regulate an industrial toxin, we adopt a strategy that starts by asking, "Is this product necessary?" Then if it is, in fact, necessary, we consider the best way to make and use it, with the least damage to the source of materials, the workers who produce it, the consumers, and the general environment.

Prevention includes doing the right kind of research so that practical and reasonable alternatives are available. The organization of research, the setting of priorities, and the allocation of resources should be a matter of public debate. We must not continue to follow the spontaneous and haphazard course of investigations that lead to marketable commodities but ignore the complex relationships that might thwart the beneficial effects of invention, bring unexpected "side effects," and divert scientific attention from less intrusive, gentler technical possibilities.

Mixed Strategy

The best strategy combines all these approaches. It prepares us to cope with the possibility that, even after we have used our best judgment and proposed the best solution we can find, we may still be wrong. It cautions us not to be caught up in technological fashions but to reserve the capacity to prepare for something different.

Rejecting Reductionism

The demand for a science of wholeness is a reaction against the very powerful but limited reductionism that sees the truth residing in the smallest pieces of reality. A science of wholeness does not reject *reduction*—taking something apart to see what it is made of—as a research tactic, but it does reject *reductionism*—the illusion that once that has been done, the rest is an exercise for the reader.

The conflict between reductionism and a complex science of wholes has emerged from the domain of abstract philosophical contention to become a practical political issue. It expresses the division between those who see the world as separate objects to be domesticated and processed for market and those who see the world as themselves and the places they live; between those who are reluctant to take responsibility for their actions beyond the most narrow, immediate effects of their products and those who see themselves responsible for their actions as they percolate through the world to the furthest detectable consequence.

Recognizing Connections

Things are more connected than we realize, even across disciplinary boundaries. Human biology is a socialized biology that varies according to our positions in the world. Real physiological differences exist among social classes in such things as the cortisol (stress hormone) cycle, immune system, and balance of excitatory and calming neural activity (Goodman et al. 2000). The common dichotomies of social/biological, genetic/environmental, physiological/psychological, lifestyle/social circumstance are false and misleading. The task of science in studying the complex ecosocial phenomena of concern is not to assign relative weights to various factors but to understand how the wholes fit together. These wholes include physical, biological, and social processes.

Recognizing Bias and Partisanship

We are part of the system. The agendas and methods of science have their history, their strengths, and their biases. Science has a dual nature as part of the generic progress of human understanding and the product of a knowledge industry that helps its owners maintain their prosperity, power, and self-justification. Therefore, we need to look at our own biases as well as those of the fields we work in and not accept as given the rules that represent conventional wisdom.

It is important to acknowledge that science is often a battleground in which interests are defended in the guise of seeking truth. While striving for objectivity is necessary in science, neutrality is not. Just as industrial science serves industry, critical, dissenting scientists should acknowledge a frank and joyful partisanship: we do science in order to understand the rich complexity of the world; to have a deeper appreciation of the beauty and excitement of a marvelous, intricate, and spontaneous biosphere and to cherish and protect it; and to benefit humanity, but especially the people most excluded from the benefits of our society.

Partisanship also implies a departure from the Earth Day injunction not to point fingers. The history of tobacco, asbestos, PCBs, pesticides, auto tires, and many other industries shows that corporations lie to protect profits. They mobilize staffs of public relations experts, lawyers, and politicians to prevent interference with the quest for profit. While it is sometimes possible to demonstrate to an industry that improved environmental protection is also profitable, this is not always so, and profit is still the goal of business.

So we are in a conflictive relationship with private, corporate industry. While ecology demands the limiting of growth to what is needed, corporations require growth. While ecology seeks to minimize inputs of energy and materials, industry seeks new products for turning resources into commodities. While ecology recognizes the unique qualitative properties of different species, habitats, resources, and people, a commodified economy measures them all on the single scale of economic value and treats them as interchangeable. While corporations must seek to limit their responsibility to the narrowest possible effects of their actions, an ecological approach traces the consequences of industrial activity through the complex networks of the biosphere.

The corporate economy sees people as labor power or consumers, and spends millions to make us insecure enough to need its products, discover new uses for the raw materials corporations own, and invent new needs regardless of impact on health and well-being. Therefore, any suggestion of potential harm is regarded as "bad for the economy." But the ecological perspective sees resource use justified only insofar as it improves human life. Thus, we do have to point fingers and hone our science for battle.

Democratization

Democratizing science would help to mobilize the collective intelligence of our species for the solution of shared problems. The discovery that every place is different, every forest and beach unique, every species special, means that there will never be enough scientists to analyze and study every phenomenon. We need everybody's imagination, experience, and knowledge. Each time a previously excluded group—people of color, women, working people—has been able to enter science, new insights have emerged that previous positions of privilege obscured.

Democratization has three elements: a democracy of recruitment into science; the popularization of science in a way that respects people's capacity as well as acknowledges their past limited access; and encouragement of non-professional participation in setting agendas, gathering and examining information, and analyzing results. Many examples illustrate the value of popular participation in science. Here are several:

- In environmental struggles such as those at Love Canal in New York State or Woburn, Massachusetts, neighbors recognized a problem long before epidemiologists began to think about it (Reich 1991). The environmental justice movement has identified patterns of discrimination and inequality. Groups such as the River Network monitor local pollution (http://www.rivernetwork.org).
- Ornithology has a long history of amateur participation. In just one Ornithological Laboratory survey of backyard birds, more than 16,000 people submitted over 53,000 checklists reporting 4.5 million birds belonging to 442 species (*Birdscope* 2001). Despite uneven sampling and possible misidentifications, this mass effort gave a good sense of the status and changes in U.S. bird populations.
- In Cuba, epidemiologists organized teams to survey more than 5 million potential mosquito breeding sites around Havana, both to evaluate the effectiveness of control programs and to judge current problems. In 72 days, a recent outbreak of *Aedes aegypti* was stopped by the combined efforts of some 11,000 people who identified and eliminated breeding sites (Granma 2002).

Address Biases in Experimentation and Hypothesis Testing

Currently accepted methods of experimentation and hypothesis testing create certain biases that are so standard that they are seldom examined but rather are taken for granted.

Limited Hypotheses

A hypothesis is tested and shown to be likely only in comparison to certain other hypotheses. Therefore, the choice of which hypotheses to compare is crucial. When the chemical industry claims that new genetically engineered varieties are needed to save the world from hunger, they compare the expected yields of their products to doing nothing to improve yields. Or the industry may compare applying pesticides to letting the bugs run rampant. Such choices ignore such alternatives as diversifying plantings, encouraging predators and parasites of the pests, and strengthening the physiological resistance of crops through nutritional management of soil. Similarly, pharmaceuticals are tested against placebos or standard treatments but not usually against holistic approaches. The design of hypothesis testing requires a bold imagination to create real alternatives worth testing.

Healthy Populations

Statistical design is usually improved by reducing sources of extraneous variation. Thus, laboratory animals are of a uniform genetic makeup, are in good health, and are raised under constant conditions. Human populations are selected with careful exclusion of people who may have extraneous conditions. They are matched for such things as age, sex, smoking habits, number of children, body weight, dietary habits, or anything else the investigators think might confound the results, provided the data are available. Often convenience, such as likelihood of being able to monitor people for a long time, determines the class or occupation of the participants. In occupational work, researchers want to examine people who have been at the same job for a long time. Thus, we miss the effects of a pollutant against a background of many pollutants, unstable employment, and varying economic conditions.

In order to avoid this type of error, studies must be conducted with the most vulnerable populations, often minority and poor communities. The variance of data exhibited in such studies is not simply noise interfering with detection of "main effects," but also an object of interest in its own right, an indicator of stressful conditions leading to greater vulnerability. For instance, my colleagues and I found that when we grouped human settlements by size (central metropolitan area, smaller metropolitan area, smaller cities, and rural), across the populations within each category the variance of mortality was greater for African Americans than for whites. African Americans are more vulnerable to economic differences and to differences in the structure of racism that has had much less effect on whites (Levins 2002).

Multiple Risks

In a population subject to multiple risks, statistical tests are less sensitive. For instance, suppose that in an exposed population a toxin causes 20 deaths per hundred and, in a control population, only 10. The relative risk is 2. But if, in addition to deaths caused by the toxin under study, there are 100 deaths due to other causes, the total deaths are 120 and 110. The relative risk has fallen to 1.1. Dioxin has not become less toxic for the exposed population; rather, the environment has become worse. Each environmental toxin helps to mask the effects of the others.

Sample Size

Statistical tests all depend on sample size. Studies often do not show sufficient numbers of positive results to achieve statistical significance. A lack of statistical proof is often misinterpreted as evidence that there is no problem. Instead, the appropriate description of negative findings is that *with this particular sample* a suspected effect has not been detected. A more robust, interdisciplinary approach might detect a problem that a single statistical test might miss.

Confidence Levels

Estimates of the outcome take the form of a most likely effect, a comparison of difference (where zero means no difference, no effect), or a risk ratio (where a value of 1 means equal risk with and without the suspected toxin). This is presented with a confidence interval. If the confidence interval includes zero (for differences) or 1 (for risk ratios), we conclude that the suspected effect is not demonstrated. The press and industry then interprets this as proving there is no effect. But suppose that the risk analysis comes out 1.3 ± 0.7 (with 1.3 being the relative risk and 0.7 the confidence level). That means the result is compatible with a risk level of 1 (no effect) or as much as a doubling of risk. It is even compatible with the conclusion that the substance is beneficial. However, in such cases of high uncertainty, a *possible* doubling of risk may be sufficient grounds for action.

Separate Variables

All statistical tests examine the relationship between some possible cause and an outcome of concern. They presume some model that relates the suspected cause to its effect. The model usually identifies a dependent variable or outcome, such as cancer, and independent variables that might influence it, such as pesticide use. The statistical analysis attempts to show whether or not there is an effect and how great it is. Then the model assumes that the prevalence of cancer is equal to the impact of a unit of variable 1 multiplied by the degree of exposure to that variable, plus the impact of a unit of variable 2 multiplied by exposure to this variable, and so on. If there is no impact, that variable contributes nothing to the prevalence of the cancer.

Adversary Statistics

Adversary statistics is the use of statistics by opposing parties, each selecting new sets of data until they are satisfied with the results. They are fundamentally different from studies in which a design procedure is set in advance, followed, and conclusions drawn. If a contending party does not like the results, a new study is done, and the process is continued until the results satisfy the sponsors. These are then reported as newer research that refutes "generally held beliefs."

Long after Gregor Mendel's classic experiments with pink and white peas that established the foundation of Mendelian genetics, it was observed that the data were suspiciously close to the 3:1 ratio that confirmed Mendel's hypothesis. Further research suggested that Mendel was so sure of his result that he continued counting peas until they came as close to 3:1 as he could reasonably expect, and then stopped. This did not constitute cheating, since Mendel was not guided by statistical theory, but it was erroneous.

This revelation opened up interest in optional stopping and the field of "sequential analysis." The criteria for accepting hypotheses when we are free to carry out as many studies as we want are different from fixed-sample statistics in which it is decided before the study how much data to examine. This process is more complex than the now familiar optional stopping of sequential analysis, in which there is only one decision maker and no conflict of goals. There is at present no general statistical theory to address adversary statistics in which contending parties carry out additional studies until they like the outcomes.

These considerations refer to duplications of previous research. A whole new domain opens up when continued research uncovers new harmful effects of chemicals, and the terms of the dispute evolve along with the information. As one question is settled, new ones arise.

Besides recognizing and addressing these standard biases, I would propose two further improvements in scientific method. First, incorporate nonprofessional data. A democratized science with nonprofessional data collection has special statistical properties. On the one hand, technical errors are likely to increase the standard error, expressed as the confidence interval within which the "true" number that we are trying to estimate will lie. On the other hand, the increased number of replications reduces the error. Thus, if the error is increased 10-fold and the sample size k-fold, the new confidence interval will be multiplied by $10/\sqrt{k}$. And if k is greater than 10 the estimation is improved. We gain more than we lose by mobilizing community data collection. Moreover, if we want to determine trends, the heterogeneity of the data (times, local site conditions, etc.) increases the variability of the data and reduces resolving power, but also increases the numbers of possible observations. Further, the heterogeneity itself is a virtue if the conditions under which samples are collected are recorded.

Second, use historical analysis as a useful supplement to traditional testing. The history of most molecules under scrutiny is that, over time, the number of deleterious effects attributable to them has increased as we have looked more closely, tested for more effects, and understood more of the subtleties of interaction. Further, molecules in the same families often differ in the magnitude of their effects but tend to be qualitatively similar. For example, molecules in which the carbon atoms form a branching structure instead of a straight chain (phthalates, for example, or hydrocarbons combined with chlorine such as dioxins) do not occur naturally in animals. We have not evolved enzyme systems for dealing with them, but they are chemically active and can enter our metabolism. Therefore, they should be suspect even without any direct evidence. Guilt by association is legitimate for molecules.

The argument of this chapter grows out of an ecological approach that sees understanding dynamic complexity as the central scientific problem of our time. It looks at science itself as an object of study, a historically developed way of producing knowledge that creates a rich mix of insights and confusions. This argument is frankly partisan, rejecting the notion that feeling is the enemy of reason or that a commitment to human well-being is an enemy of objectivity. The suggestions outlined above would get us closer to a good, combative, perceptive scientific method that is more reflective of the complex, dynamic world in which we live and more supportive of precautionary decisions.

References Cited

Birdscope. 2001. Spring 15, no. 2. Available at http://birds.cornell.edu/publications/birdscope/Spring2001/gbbc.html.

Goodman, E., B. C. Amick, M. O. Rezendes, S. Levine, J. Kagan, W. H. Rogers, and A. R. Tarlov. 2000. "Adolescents' Understanding of Social Class: A Comparison of White Upper Middle Class and Working Class Youth." *Journal of Adolescent Health* 27, no. 2: 80–83.

Granma (the official daily newspaper of the Communist Party in Cuba), 28 March 2002.

Levins, R. 1968. *Evolution in Changing Environments.* Princeton: Princeton University Press.

Levins, R., T. Awerbach, and U. Brinkmann. 1994. "The Emergence of New Diseases." *American Science* 82: 52–60.

Lewontin R., and R. Levins. 2000. "Schmalhausen's Law." *Capitalism, Nature, Socialism* 11, no. 4: 103–8.

Omran, A. R. 1971. "The Epidemiological Transition: A Theory of the Epidemiology of Population Change." *Milbank Memorial Fund Quarterly* 49, no. 4: 509–38.

Reich, M. R. 1991. *Toxic Politics: Responding to Chemical Disasters.* Ithaca, N.Y.: Cornell University Press.

PART VII

Conclusions and Afterword

Precaution, Environmental Science, and Preventive Public Policy

Joel A. Tickner

The chapters in this volume reflect hundreds of person-years of experience by leading scientists, policy analysts, and legal scholars working to protect health and ecosystems in the face of highly uncertain, complex, and often controversial risks. Although they have struggled with similar issues and barriers in conducting science for policy, these authors, representing a variety of disciplines and regions of the world, had rarely communicated with one another. As a result they have lacked the benefits and strength of accumulated understanding and experience. The desire to remedy this situation was the initiative that eventually led to this book.

On 20 to 22 September 2001, the Lowell Center for Sustainable Production (www.uml.edu/centers/lcsp/precaution) hosted the International Summit on Science and the Precautionary Principle. The summit brought together scientists, philosophers, legal scholars, and other environmental and health professionals to explore the relationship between science and the precautionary principle as well as to develop a vision for scientific methods, tools, and policies that would more effectively support precautionary decision making, particularly in the face of complex, highly uncertain human and ecosystem health risks. Eighty-five representatives from government agencies, academic institutions, research consultancies, professional societies, and nongovernmental organizations attended. They came from seventeen countries and represented the fields of medicine, public health, epidemiology, toxicology, ecology, molecular biology, chemistry, botany, law, philosophy, physics, sociology, psychology, economics, geography, conservation biology, evolutionary biology, wildlife biology, virology, marine science, agronomy, and political science.

The summit was the first international opportunity for leading scholars to discuss the role of science in implementing the precautionary principle. The primary goal of the summit, and ultimately of this book, was to build understanding and support for the role of science in implementing the precautionary principle. The summit established a new community of scientists and other health and environmental professionals, linking individuals working on evaluating risks with individuals working on developing preventive solutions.

Summit discussions—also evident in the chapters included in this volume—reflected both the complexity of the precautionary principle and the need to understand its role in science and policy. What does the precautionary principle mean in practice? How can it be implemented in everyday practice when there are numerous countervailing strains on resources? The summit discussions did not challenge the importance of precaution in environmental and health decision making under uncertainty, although participants believed that its application at this most basic level needed further elaboration.

A major focus of the summit discussions was the critical need for more funding to be channeled to interdisciplinary research, examining broader hypotheses. Unfortunately, resources and incentives are generally unavailable for this type of research that would help scientists and decision makers more comprehensively understand complex risks as well as more effectively support precautionary decisions. Yet funding is critical to defining the questions asked in scientific research. For example, a large percentage of government funding goes into research that is mechanistic, narrow in scope, and limited in time. If funding is not available for research that is interdisciplinary, that is systems based, or that asks broader questions about problems, prevention opportunities, and long-term goals for human and ecosystem health, such research is not likely to occur.

Participants identified several research and science policy needs that would support more precautionary approaches to policy under uncertainty and complexity:

1. Increased use of integrative assessment frameworks
2. More use of science to begin with the goals of desired health and environmental outcomes and "backcast"—take steps to reach those outcomes—rather than just forecast damage in the future
3. Development of processes to allow for speculation in environmental research, as well as additional funding for identification of early warnings and situations in which damage could occur
4. Development of new language to express conclusions and discuss uncertainty and limits in science for policy

5. Establishment of better methods to assess alternative technologies and activities
6. Establishment of means to address the problem of a fragmented knowledge base, that is, to overcome the tendency to view problems narrowly and within separate disciplines
7. Ways to educate students in interdisciplinary problem solving, including bioethics

Nonetheless, participants believed that it is currently dangerous for students and researchers to be multidisciplinary or to engage in public interest research. The way we govern and fund science, including the incentive system in research and academia, must be changed to address this problem.

Recommendations

Some of the summit recommendations to overcome barriers and build momentum toward a vision for science and policy that better reflects uncertainty and complexity of natural systems include:

- *The need for a new community that links scientists working on evaluating risks with scientists working on preventive solutions.* The most important outcome of the summit was the creation of a new community of scientists, dedicated to exploring ways to improve the use of science in decision making, where scientists can support one another's (often controversial) work, as well as candidly discuss ideas and policies. It was noted that science has an important role in both studying the impacts of human development and generating the solutions that would lead us toward more sustainable modes of production and consumption.
- *The need for transdisciplinary approaches to science and policy.* Participants found that interdisciplinary collaborations are the most effective and robust way to conduct science for policy in the face of uncertainty. However, these types of collaborations are the exception rather than the rule and are often frowned upon by funding agencies, government authorities, and professional societies. Participants concluded that "we need to make it safe for transdisciplinarity."
- *The need for critical self-reflection and discussion.* Participants noted that there are insufficient opportunities for scientists to think about their methods, tools, and the implications of the research they do. It is necessary to identify opportunities for scientists to step back from everyday practice and think about whether their work could more effectively support precautionary policies. It is also important to find language to reach out to different scientific communities, to engage them in thinking about new ways of conduct-

ing science for policy. Case studies and examples can provide one such vehi-
cle for communication.

- *The need to develop specific case examples of ways in which science can better support precautionary, preventive policies.* Such cases would examine aspects of scientific methods that could change, including broader hypothesis development, more effective communication of uncertainty, integration of qualitative and quantitative data, and a more effective integration of science in policy. In addition, there is a need to examine problems created by the way science has traditionally been conducted and applied in policy. These case studies can serve as useful educational tools.
- *The need to identify opportunities in research and regulatory structures for developing and promoting a new vision for science and policy that supports precaution.* For example, there may be ways to advocate for changes in government research funding structures so that more interdisciplinary, innovative methods could be undertaken.

Participants agreed that the most effective way to proceed with this critical self-reflection and vision is through education, outreach, and debate within
professional societies, educational institutions, and government agencies.
Ongoing discussion, dialogue, and critical analysis within this new community
of scientists are also essential so that knowledge and understanding of these
challenging issues can be shared and updated. This discussion should be broad-
ened to include more disciplines (particularly in the social sciences) and more
regions of the world, where substantial cultural differences may exist in the
application of environmental science in policy.

A Vision

The purpose of the International Summit on Science and the Precautionary
Principle and of this volume was to engage scientists, policy makers, advocates,
and students in a broad public discussion about whether the tools and methods of environmental science and its integration in policy are adequate to
address complex, highly uncertain environmental and health risks. We believe
implementing the precautionary principle requires rethinking environmental
science for policy. Through the summit and case studies and overviews presented in this book, participants were able to articulate a consensus on problems with the current paradigm of science, precaution, and policy making, as
well as some clear ideas for what a new paradigm could look like—a vision
for environmental science and precaution.

This vision is expressed in the Lowell Statement on Science and the Precautionary Principle (appendix). It reflects the diverse dialogue and input of a
vast range of scientists, scholars, experience, and expertise. As one signatory

stated, "On its surface the statement is quite rational and logical, but underneath it far-reaching implications for science and policy."

We believe that the Lowell Statement is a critical document in advancing the discussion on precaution. It outlines areas for action and further research and, most important, notes that precaution is entirely consistent with good science and good policy. In this respect, it serves to dissolve the ideological argument that precaution is based on emotion or bad science and refocuses the debate about precaution into one of scientific methods, tools, and policies, as well as their use in decision making.

Scientists and the Real World

Lee Ketelsen

Scientists, by definition, seek the truth. Truth is exactly what we need most in the United States, particularly on the question of how human activities can and do impact our health and environment. Instead, what we generally have, at least from "official sources," is very far from a truthful depiction of the impacts of industrial activities on health and ecosystems.

Public Perception of Science, Regulation, and Safety

One big obstacle to achieving precaution in the U.S. is that most Americans believe they already have it. They think that they have policies that *watch* for potential hazards and make the safest choices. Americans want precaution, they support it—but most mistakenly think they already have it.

I believe that if the American people can learn the truth about their lack of precaution, about the harm they are are doing to themselves, they can organize the power to change the system of health and environmental policies and the science that informs them.

First, we have to change the picture most Americans have of environmental policy. Many Americans, while they recognize that big business has big power in the U.S and do not trust big business to police itself, think that the government is doing a good job doing the policing. They see a big EPA— thousands of regulators running everywhere, with limits on everything, a scrubber on every stack, and at least lots of forms to fill out that must tell us everything we need to know. They see a huge Food and Drug Administration, a Consumer Product Safety Commission. Americans believe that pollution is controlled to safe limits. Above all, they believe that their food and their products are safe, that someone has tested the chemicals that go into them and has thought about real-world conditions of use and exposures.

Just as many Americans were very surprised, after the September 11 terrorist attack, to hear the head of the FAA (Federal Aeronautics Administration) say the agency had never "contemplated" suicide terrorist attacks, so most citizens are surprised when we tell them that tens of thousands of chemicals that are in use in their environment and their products were never tested for safety.

Americans also think that scientists are working to protect their health—thousands of scientists studying every organism, every behavior, and every chemical. They see stories in the paper about scientific discoveries exploring the very minutiae of life. They logically assume that if scientists understand the human genome, certainly they know what different doses of arsenic will do to us.

And then citizens see the big environmental groups starring in a press story every week, demanding this, winning that, so they assume that things must be pretty much in control by now.

This is the picture most people have in their heads when they think about health and the environment. They see a picture of big government and big environmental groups and big academia—all employing lots of highly trained scientists—watching over big business. And they believe that this has given us a healthy balance—a fair balance between economic progress and a healthy environment—a healthy balance of risk and benefit.

It is important to note that this misconception is not rooted in apathy. It is not a result of Americans not valuing their health and their future, or not paying attention. Americans are being aggressively misinformed. Corporations, of course, work the hardest to proclaim the safety of all that they do. But perhaps the more damaging problem is that their government works hard to assure us that there is no danger, that it has everything under control—or at least *if* there is some risk of harm, it has diligently made sure that it is in our best interest and analyzed all the costs and benefits to do what was best for the public as a whole. Government officials spend a lot of time defending this perception, because they need to defend their job performance.

When Americans try to make sense of all this, they put their faith in science, the great arbitrator, the guardian of the truth in this battle of interests. Surely scientists can give us the answers. And unfortunately, many scientists assert that they do give us the answers.

What galls advocates the most about the current use of quantitative risk assessment as a scientific tool is that the risk assessors, and those who employ them, claim at the end of the process that they know they have accurately predicted what the impact of a chemical sent out in the world will be. And these scientists also defend themselves, their judgments, and their products.

But this misinformation has not convinced everyone. We have hope of change in the U.S. because there are many Americans who have seen the truth of their situation and have had a reason to learn that they are not protected. Some learned because of where they live; a toxic threat hit home to them because they live near a toxic facility or dump, or fought one proposed for their town. Or they learned because of where they work—they learned that OSHA has not protected their health. Or they learned because they got cancer and then learned how many cancer-causing chemicals are allowed in their environment. This has given birth to a large grassroots citizens environmental movement that has campaigned on every toxics issue for greater protections.

We can therefore start with this base of those who know and work to educate others. We here are starting a call for precautionary action and precautionary guiding policies—a campaign to implement the precautionary principle in policy. In Massachusetts we have a coalition forming to do just that.

What Scientists Can and Should Do

Because of the widespread belief that we are already protected, we know we must start by educating the rest of the public on the flaws that we see in our system of protections. This is a job for activists. But as we go out to inform the public, we need scientists to remember that we need you by our side. Citizens believe that scientists speak the truth. They put great trust in science to solve our problems. But don't be intimidated by my use of the word "truth"—I do not mean that you have all the answers. In fact, I mean the opposite; we need you to explicitly say *what you don't know and cannot know.*

If I had to pick out one piece of misinformation that is doing the most damage, it would be this statement: "This is safe; this is a safe level." Practically the whole U.S. regulatory system is centered on finding that elusive "safe level." What makes that statement false is that the complexity of the real world, the complexity of the real human body, is not "all known" by science.

We need you to communicate that in the way you design your studies and present your results. We need science that works to improve the methods for studying complexity, interactions, synergy, cumulative effects, and more. We need multidisciplinary groups that work with affected communities to collect both the quantitative evidence as well as the knowledge, observation, and experience that are not so easily captured by "experts," yet critical in more thoroughly understanding a complex reality.

We will join with you to proclaim that we need this better science, more science, truthful and transparent science, information put in full context. Can we call it the "Science of Real Life"? Some scientists already practice this Science of Real Life and are working to improve it, designing a vision for a better science. This vision will be important in a campaign for change. But it will not be enough. It is only the easy part for you.

My plea to scientists is to do the hard part also, the part that will take the most courage. There are two courageous, necessary tasks that scientists need to do—courageous because they will *not* be popular with fellow scientists (or the biggest sources of research funds).

First, scientists must point out publicly what others are saying that misleads us. When they say it is a safe level, scientists must point out that the science does not back it up. Scientists must be the ones to point out what others are *not* telling us and that what they are not telling us has already hurt us and can hurt us again.

The second unpopular job is to point out that science cannot rule alone. We must decide our course together through democracy. Scientists must admit that they cannot provide all the answers that will show us the right way to go. Yes, more science is needed, yes, better science is needed, but the decisions about our health, our values and goals, our lives belong to us all.

This won't be popular with all scientists, but it will be applauded by all of us who see clearly our current level of vulnerability and want to make better choices for the future. We are ready to stand with scientists to call for an enhanced, expanded Real World Science and for the opportunity to make truly *precautionary* decisions *together.*

Lowell Statement on Science
and the Precautionary Principle

December 17, 2001

Growing awareness of the potentially vast scale of human impacts on planetary health has led to recognition of the need to change the ways in which
environmental protection decisions are made and the ways that scientific
knowledge informs those decisions. As scientists and other professionals committed to improving global health, we therefore call for the recognition of the
precautionary principle as a key component of environmental and health
policy decision making, particularly when complex and uncertain threats must
be addressed.

We reaffirm the 1998 Wingspread Statement on the Precautionary Principle and believe that effective implementation of this principle requires the following elements:

- Upholding the basic right of each individual (and future generations) to a
  healthy, life-sustaining environment as called for in the United Nations Declaration on Human Rights
- Action on early warnings, when there is credible evidence that harm is
  occurring or likely to occur, even if the exact nature and magnitude of the
  harm are not fully understood
- Identification, evaluation, and implementation of the safest feasible
  approaches to meeting social needs
- Placing responsibility on originators of potentially dangerous activities to
  thoroughly study and minimize risks and to evaluate and choose the safest
  alternatives to meet a particular need, with independent review
- Application of transparent and inclusive decision-making processes that

increase the participation of all stakeholders and communities, particularly those potentially affected by a policy choice

We believe that effective application of the precautionary principle requires interdisciplinary scientific research as well as explicitness about the uncertainties involved in this research and its findings. Precautionary decision making is consistent with "sound science" because of the large areas of uncertainty and even ignorance that persist in our understanding of complex biological systems, in the interconnectedness of organisms, and in the potential for interactive and cumulative impacts of multiple hazards. Because of these uncertainties, science will sometimes be incapable of providing clear and certain answers to important questions about potential environmental hazards. In these instances, policy decisions must be made on the basis of sound judgment, open discussion, and other public values, in addition to whatever scientific information is available. We believe that waiting for incontrovertible scientific evidence of harm before preventive action is taken can increase the risk of costly mistakes that can cause serious and irreversible harm not only to ecosystem and human health and well-being, but also to the economy.

Some of the ways that scientific information is currently applied in formulating policy can work against the ability to take precautionary action, for example by misrepresenting limitations in the state of scientific knowledge. Decision makers frequently look for high levels of proof of causal links between a technology and a risk before acting, so that their decisions will be protected from accusations of being arbitrary. But often, high levels of proof cannot be achieved and are not likely to be forthcoming in the foreseeable future. A more complete and open presentation from scientists on the current limitations in understanding of environmental risks will encourage the acceptance on the part of government decision makers and the public of the idea that precautionary action is a prudent and effective strategy when potential risks are large and uncertainties are large as well.

It is not only the communication between scientists and policy makers, however, which needs improvement. We believe that there are ways in which the current methods of scientific inquiry may also retard precautionary action. For example, research frequently focuses on narrow, quantifiable aspects of problems, thus inadvertently excluding from consideration potential interactions among different components of the complex biologic systems of which humans are a part. The compartmentalization of scientific knowledge further impedes the ability of science to detect and investigate early warnings and develop options for preventing harm when far-reaching health and environmental risks are involved. Unfortunately, government decision makers, scientists, and proponents of hazardous activities often misinterpret limitations in scientific tools and in the ability to quantify causal relationships as evidence of safety. However, not knowing whether an action is harmful is not the same thing as knowing that it is safe.

We contend that effective implementation of the precautionary principle demands improved scientific methods and a new interface between science and policy that stresses the continuous updating of knowledge as well as improved communication of risk, certainty, and uncertainty. With these objectives in mind, we call for a reevaluation of scientific research agendas, funding priorities, science education, and science policy. The ultimate goals of this effort would include:

- A more effective linkage between research on hazards and expanded research on primary prevention, safer technological options, and restoration
- Increased use of interdisciplinary approaches to science and policy, including better integration of qualitative and quantitative data
- Innovative research methods for analyzing the cumulative and interactive effects of various hazards to which ecosystems and people are exposed, for examining impacts on populations and systems, and for analyzing the impacts of hazards on vulnerable subpopulations and disproportionately affected communities
- Systems for continuous monitoring and surveillance to avoid unintended consequences of actions and to identify early warnings of risks
- More comprehensive techniques for analyzing and communicating potential hazards and uncertainties (what is known, not known, and can be known)

We understand that human activities cannot be risk free. However, we contend that society has not realized the full potential of science and policy to prevent damage to ecosystems and health while ensuring progress toward a healthier and economically sustainable future. The goal of precaution is to prevent harm, not to prevent progress. We believe that applying precautionary policies can foster innovation in better materials, safer products, and alternative production processes.

We urge governments to adopt the precautionary principle in environmental and health decision making under uncertainty when there are potential risks as well as to take timely preventive and restorative actions in cases where damage has been demonstrated. The elements of decision-making processes incorporating the precautionary principle, as outlined here, represent necessary aspects of sound, rational processes for preventing negative impacts of human activities on human and ecosystem health. This approach shares the core values and preventive traditions of medicine and public health.

For a list of signatories see www.uml.edu/centers/lcsp/precaution.

ABOUT THE AUTHORS

JUAN ALMENDARES, a physician, is former Dean of the Medical School of Honduras and former President of the National University of Honduras. He is an environmental advocate and the author and coauthor of several articles and books on ecology and related issues. He is currently a Professor of the Medical School of Honduras.

KATHERINE BARRETT is a Research Associate with the POLIS Project on Ecological Governance at the University of Victoria. She was Project Director with the Science and Environmental Health Network, a consortium of nongovernmental organizations. Her Ph.D. dissertation in botany and applied ethics at the University of British Columbia examined the Canadian regulatory system for genetically modified organisms. She has published several research papers, book chapters, and general interest articles on the precautionary principle in environmental and agricultural policy.

KAMALJIT BAWA is a Distinguished Professor at the University of Massachusetts at Boston. He has worked in conservation and the environment for the past thirty years in Central America and South Asia, and he has authored or coauthored many papers in professional journals and edited several monographs and books. He has been a Bullard Fellow at Harvard University, a Guggenheim Fellow, and a Pew Scholar in Conservation and the Environment. He serves on many advisory and editorial boards and is the President of the Ashoka Trust for Research in Ecology and the Environment.

FINN BRO-RASMUSSEN holds an M.Sc. in Chemical Engineering and a Ph.D. in Nutrition and Human Physiology. He has worked as Professor and Chair in Environmental Science and Ecology at the Technical University of Denmark (emeritus from 1998) and has served as a member and chair of the WHO/Joint Meeting on Pesticide Residues and the Scientific Advisory Committee to the EU Commission on Toxicity and Ecotoxicity of Chemicals.

DONALD A. BROWN is director of the Pennsylvania Consortium for Interdisciplinary Environmental Policy. Before holding this position, he was Program Director for United Nations Organizations at the U.S. Environmental Protection Agency's Office of International Environmental Policy. He has written extensively about the need to integrate environmental science, economics, and law with ethics.

THEOFANIS CHRISTOFOROU is a Legal Advisor with the European Commission in Brussels in the area of European Comission external trade and economic relations, including the WTO, where he is responsible for legal issues concerning the implementation of the Uruguay Round agreements, in which he participated. He previously worked in the Legal Service in the areas of health and environmental protection, consumer protection, transport, and agriculture. He has acted as counsel in more than fifty cases before the European Court of Justice and the Court of First Instance and has handled as lead counsel several dispute settlement cases before GATT 47 and WHO panels and the Appellate Body.

TERRY COLLINS is the Thomas Lord Professor of Chemistry and the Director of the Institute for Green Oxidation Chemistry at Carnegie Mellon University as well as Honorary Professor of the University of Auckland, New Zealand. He specializes in inorganic and green chemistry and has written widely on the need for the chemical community to adapt itself to the technological problems of sustainability. In 1999 the Collins group received the Presidential Green Chemistry Challenge Award (USA) and the award of the Society of Pure and Applied Coordination Chemistry (Japan) for its work on environmentally friendly oxidation systems. Professor Collins is a Dreyfus Teacher-Scholar and a Fellow of the Alfred P. Sloan Foundation.

BARRY COMMONER is a prominent environmental scientist, writer, and lecturer on the relation between environmental and energy problems and economic and political issues. Dr. Commoner has published a number of books, including *Making Peace with the Planet* (1990), *The Politics of Energy* (1979), *The Poverty of Power* (1976), and *The Closing Circle* (1971). Dr. Commoner's current research activities include the origin of dioxin in incinerators, the long-range air transport of toxic chemicals such as dioxin, and inherent uncertainties in genetic modification.

CARL F. CRANOR is Professor of Philosophy at the University of California, Riverside, specializing in legal and moral issues arising in the legal and scientific adjudication of risks of toxic substances, including theoretical issues in risk assessment, as well as the philosophy of science in regulatory and tort law. He has published numerous articles in these fields and has authored *Regulating Toxic Substances: A Philosophy of Science and the Law* (1993), edited *Are Genes Us? The Social Consequences of the New Genetics* (1994), and coauthored the U.S. Congress' Office of Technology Assessment report, *Identifying and Regulating Carcinogens* (1987).

DAVID GEE is Information Needs Analysis and Scientific Liaison at the European Environment Agency, a European Commission information-providing body, in Copenhagen. He previously worked for more than two decades on occupational and environmental risk reduction with United Kingdom trade unions and Friends of the Earth. He is currently working on indicators for chemicals; environment and health issues; particularly related to children; eco-efficiency and associated indicators, integration of environment into economic sectors; and the practical application of the precautionary principle.

ELIZABETH A. GUILLETTE is Assistant Research Scientist at the Department of Anthropology, University of Florida, and Visiting Scholar at the Tulane/Xavier Center for Bioenvironmental Research. She is known for her anthropological approaches, particularly rapid assessment techniques, to the interplay between contaminants and human health. Her manual, *Performing a Community Health Assessment,* written for those concerned with local contamination issues, has been requested by groups throughout the world. She has been an invited speaker at national and international conferences on environmental health.

SHEILA JASANOFF is Professor of Science and Public Policy at Harvard University's J. F. Kennedy School of Government and is affiliated with the School of Public Health and the Department of the History of Science. Dr. Jasanoff's research interests center on the interactions of law, science, and politics in democratic societies, focusing on environmental governance. She has lectured widely, published many articles and book chapters, and authored or edited several books, including *Science at the Bar* (1995), which received the Don K. Price award of the American Political Science Association. She has served on the Board of the American Association for the Advancement of Science and as President of the Society for Social Studies of Science.

MATTHIAS KAISER is Professor for Philosophy of Science and Director of the National Committee for Research Ethics in Science and Technology in Norway. He is also Acting Director of the Norwegian Board of Technology, Adjunct Professor for the Doctoral Education Program at the Oslo College for Architecture, and Chairman of the Standing Committee on Responsibility and Ethics in Science. His research is mainly on philosophy of science, ethics of science, and technology assessment. He has published two books, coauthored a third, edited several publications, and published more than fifty articles, mostly in international journals and anthologies.

LEE KETELSEN is New England Director of Clean Water Fund. She has twenty-four years of citizen-organizing experience, beginning with neighborhood organizing in an African American community in Chicago and including six years at Massachusetts Fair Share, first as a regional organizing director and then as statewide toxics campaign director. Since 1987, she has been working for Clean Water Fund in Boston, directing community organizing and statewide campaigns. Ms. Ketelsen is also codirector of the Massachusetts Precautionary Principle Project.

STUART LEE is an independent scholar who specializes in research on the societal influences and applications of science. Within this broad arena, he writes about the impact of sciences' cultural assumptions on natural resource management and environmental aspects of genetically modified organisms. He has published in *Social Studies of Science* and other academic journals and is looking forward to writing for a broader audience in the future.

RICHARD LEVINS is a tropical farmer turned ecologist, working in evolutionary and agricultural public health ecology. His central interest are understanding complexity in ecosocial systems and learning how to intervene gently in system processes. Dr. Levins teaches human ecology at Harvard School of Public Health and is active in development of Cuba's commitment to ecological development.

MARY O'BRIEN is the Ecosystem Projects Director of the Science and Environmental Health Network (SEHN). She has served as staff scientist with grassroots organizations throughout the past twenty years: Northwest Coalition for Alternatives to Pesticides, Environmental Law Alliance Worldwide, Environmental Research Foundation, and Hells Canyon Preservation Council. She is the author of *Making Better Environmental Decisions: An Alternative to Risk Assessment* (2000) and is currently coediting, with Nancy Myers of SEHN, a book on ecosystem decision making and the precautionary principle.

ROMEO F. QUIJANO, a medical doctor and toxicologist, is a Professor at the College of Medicine, University of the Philippines, Manila. He is also a Lecturer with the National Poison Control and Information Service, a Consultant to the Fertilizer and Pesticide Authority, and a Member of the Forum Standing Committee, Intergovernmental Forum Chemical Safety.

REINMAR SEIDLER is a graduate student in the Environmental Biology Department at the University of Massachusetts, Boston. His research interests incorporate neotropical butterfly migration, tropical forest management, and land-use change. He has published articles on forest management in *Conservation Biology* (1998) and the Academic Press *Encyclopedia of Biodiversity* (1999) and is coeditor with Kamaljit Bawa of *Dimensions of Sustainable Development,* a volume in the forthcoming UNESCO publication *Encyclopedia of Life Support Systems.*

ANDREW STIRLING is a Senior Lecturer at the Science Policy Research Unit (SPRU) at Sussex University in Brighton, with a background in the natural sciences, archaeology, social anthropology, and science and technology policy. His earlier work was on nuclear disarmament and energy issues, and he was a board member of Greenpeace International. His work at SPRU focuses on precaution and wider public engagement in technology and risk policy. He collaborates on this with a variety of academic, government, industry, and nongovernmental organization bodies and has served on a number of policy advisory committees.

CATO TEN HALLERS-TJABBES, marine biologist and former science policy maker, is at CaTO Marine Ecosystems, associated with the Royal Netherlands Institute for Sea Research, doing research on the decline of predatory snails and the impact of tributyltin and of fisheries. She has worked on policy process in the International Maritime Organization to prevent harmful environmental effects of antifouling paints on seagoing ships and coordinated activities with scientists from southern Europe to raise public and policy recognition. She continues to represent the IUCN (World Conservation Union) at London Convention Meetings.

BOYCE THORNE-MILLER is a consultant and marine biologist who has worked as an environmentalist since 1988 for organizations including the Oceanic Society, Friends of the Earth, and SeaWeb on the precautionary principle, coastal pollution, and fisheries and aquaculture. She has served on delegations at the London Dumping Convention, the Conference on the Protection of the Marine Environment from Land-Based Activities, the International POPs Convention, and the North Atlantic Salmon Conservation Organization. She is the author of two books on marine biodiversity, *The Living Ocean: Understanding and Protecting Marine Biodiversity* (1991; 2nd ed. 1999) and *Ocean* (1993).

JOE THORNTON is an Assistant Professor in the Department of Biology at the University of Oregon. He is the author of *Pandora's Poison: Chlorine, Health, and a New Environmental Strategy* (2000), an analysis of the health impacts of global chemical pollution, its technological and political causes, and the policy reforms necessary to address it. Dr. Thornton's laboratory research addresses gene family evolution, particularly the molecular evolution of hormone receptors. His continuing work in environmental policy focuses on precautionary management of organohalogens, endocrine disrupting chemicals, and other major chemical hazards.

JOEL A. TICKNER is a Research Assistant Professor at the Lowell Center for Sustainable Production, University of Massachusetts, Lowell where he also lectures in the Department of Work Environment. He has served as an advisor and researcher for several government agencies, nonprofit environmental groups and trade unions both in the United States and abroad during the past nine years. He was co-coordinator of the Wingspread Conference on the Precautionary Principle and coeditor of the book *Protecting Public Health and the Environment: Implementing the Precautionary Principle.*

REGINALD VICTOR is Director of the Centre for Environmental Studies and Research and Professor of Biology at Sultan Qaboos University, where he was responsible for the establishment of the graduate program in environmental science. Dr. Victor is a freshwater biologist who has made significant contributions to the study of microcrustacea in Southeast Asia, floodplain fisheries management, and riverine ecology, with reference to macroinvertebrates, fish communities, and pollution issues in Africa. In Oman, Dr. Victor has studied aquatic ecology and water quality issues from management and conservation perspectives. He was a member of the drafting committee of the National Biodiversity Strategy and Action Plan for Oman.

ALISTAIR WOODWARD is Professor of Public Health at the Wellington School of Medicine and Health Sciences. He is an epidemiologist whose research interests include environmental health, particularly air pollution, radiation, and climate change, and more wide-ranging questions of how science and policy relate to each other. He contributed to the Second and Third Assessment Reports of the Intergovernmental Panel on Climate Change. Dr. Woodward is a member of New Zealand's National Health Committee.

INDEX